Cryosphere and Earth Science

Cryosphere and Earth Science

Edited by **Cortez Ford**

SYRAWOOD
PUBLISHING HOUSE
New York

Published by Syrawood Publishing House,
750 Third Avenue, 9th Floor,
New York, NY 10017, USA
www.syrawoodpublishinghouse.com

Cryosphere and Earth Science
Edited by Cortez Ford

© 2016 Syrawood Publishing House

International Standard Book Number: 978-1-68286-020-5 (Hardback)

Contents

Preface

Cryosphere is a specific area of study under the umbrella of earth science. It is concerned with those sections of earth where water is available in solid form. This book focuses on some of the diverse geographic regions that fall under cryosphere like permafrosts, ice cover in sea and oceans, river and lake ice, etc. It strives to provide a fair idea about this field and to provide information on the current technologies and instruments used for measurements and observations. The core areas along with researches of this field presented herein will provide in-depth knowledge to the readers.

The information shared in this book is based on empirical researches made by veterans in this field of study. The elaborative information provided in this book will help the readers further their scope of knowledge leading to advancements in this field.

Finally, I would like to thank my fellow researchers who gave constructive feedback and my family members who supported me at every step of my research.

Editor

Observing Muostakh disappear: permafrost thaw subsidence and erosion of a ground-ice-rich island in response to arctic summer warming and sea ice reduction

F. Günther[1], P. P. Overduin[1], I. A. Yakshina[2], T. Opel[1], A. V. Baranskaya[3], and M. N. Grigoriev[4]

[1]Alfred Wegener Institute Helmholtz Centre for Polar and Marine Research, Potsdam, Germany
[2]Ust-Lensky State Nature Reserve, Tiksi, Yakutia, Russia
[3]Lab. Geoecology of the North, Faculty of Geography, Lomonosov Moscow State University, Moscow, Russia
[4]Melnikov Permafrost Institute, Russian Academy of Sciences, Siberian Branch, Yakutsk, Russia

Correspondence to: F. Günther (frank.guenther@awi.de)

Abstract. Observations of coastline retreat using contemporary very high resolution satellite and historical aerial imagery were compared to measurements of open water fraction, summer air temperature, and wind. We analysed seasonal and interannual variations of thawing-induced cliff top retreat (thermo-denudation) and marine abrasion (thermo-abrasion) on Muostakh Island in the southern central Laptev Sea. Geomorphometric analysis revealed that total ground ice content on Muostakh is made up of equal amounts of intrasedimentary and macro ground ice and sums up to 87 %, rendering the island particularly susceptible to erosion along the coast, resulting in land loss. Based on topographic reference measurements during field campaigns, we generated digital elevation models using stereophotogrammetry, in order to block-adjust and orthorectify aerial photographs from 1951 and GeoEye, QuickBird, WorldView-1, and WorldView-2 imagery from 2010 to 2013 for change detection. Using sea ice concentration data from the Special Sensor Microwave Imager (SSM/I) and air temperature time series from nearby Tiksi, we calculated the seasonal duration available for thermo-abrasion, expressed as open water days, and for thermo-denudation, based on the number of days with positive mean daily temperatures. Seasonal dynamics of cliff top retreat revealed rapid thermo-denudation rates of $-10.2 \pm 4.5 \, \mathrm{m\,a^{-1}}$ in mid-summer and thermo-abrasion rates along the coastline of $-3.4 \pm 2.7 \, \mathrm{m\,a^{-1}}$ on average during the 2010–2013 observation period, currently almost twice as rapid as the mean rate of $-1.8 \pm 1.3 \, \mathrm{m\,a^{-1}}$ since 1951. Our results showed a close relationship between mean summer air temperature and coastal thermo-erosion rates, in agreement with observations made for various permafrost coastlines different to the East Siberian Ice Complex coasts elsewhere in the Arctic. Seasonality of coastline retreat and interannual variations of environmental factors suggest that an increasing length of thermo-denudation and thermo-abrasion process simultaneity favours greater coastal erosion. Coastal thermo-erosion has reduced the island's area by 0.9 km^2 (24 %) over the past 62 years but shrank its volume by $28 \times 10^6 \, \mathrm{m^3}$ (40 %), not least because of permafrost thaw subsidence, with the most pronounced with rates of $\geq -11 \, \mathrm{cm\,a^{-1}}$ on yedoma uplands near the island's rapidly eroding northern cape. Recent acceleration in both will halve Muostakh Island's lifetime to less than a century.

1 Introduction

Muostakh Island in the southern Laptev Sea is a prominent example (Are, 1988a, b; Romanovskii et al., 2000; Grigoriev et al., 2009) of thousands of kilometres of unstable unlithified coastline along arctic shelf seas (Lantuit et al., 2011a; Overduin et al., 2014). Along this coast, cliffs border marshy coastal tundra lowlands and islands that are underlain by continuous permafrost and composed of continental late Pleistocene ice-rich permafrost sequences called Ice Complex deposits (Schirrmeister et al., 2013). During summer, this coast is no longer protected by sea ice and retreats at erosion rates of -2 to $-6\,\mathrm{m\,a^{-1}}$ (Grigoriev et al., 2006). Large areas of up to $-3400\,\mathrm{m^2\,km^{-1}}$ of coastline are lost annually and currently this rate is more than twice as rapid as historical erosion (Günther et al., 2013a). The distinguishing feature of polar coasts is the presence of a variety of ice types on and ground ice below the earth surface (Forbes and Hansom, 2011). Ogorodov (2011) emphasises the influence of hydrometeorological conditions on the development of coastal thermo-erosion, in particular of thermal and wave energies, both of which are linked to sea ice extent and duration. For the Laptev Sea, Markus et al. (2009) report that the duration of the sea-ice-free season increased on average by 10 days over the last decade, exceeding the average increase of 2 days around the Arctic Ocean. Especially when considering the warming trend of cold continuous permafrost (Romanovsky et al., 2010) and the vulnerability of deep organic carbon to mobilisation (Grigoriev et al., 2004; Grosse et al., 2011), it is important to assess the impact of current climate warming in the northern high latitudes not only as an external disturbance force on ice-bonded permafrost coasts, but also on permafrost-thaw-related land surface lowering of presumably undisturbed adjacent territories. However, such information is practically non-existent for Siberia.

As a consequence of coastal erosion, clastic material enters the near-shore zone (Are, 1998; Jorgenson and Brown, 2005), where it is deposited, reworked and transported (Overduin et al., 2007; Winterfeld et al., 2011). Because ground ice occupies a large proportion of the land's volume above and below sea level, a much smaller amount of material is removed by wave action after thaw than along ice-free coastlines and high rates of coastline retreat are the result (Zhigarev, 1998). Are et al. (2008) conclude that it is mostly thawed material that is being eroded, rather than permafrost.

Coastal thermo-erosion includes two related processes that work temporally and quantitatively differently together. Thermo-denudation (TD) is comprised of the thawing of exposed permafrost, the upslope or inland propagation of a retreating headwall and the transport of material downward to the bottom, all under the influence of insolation and heat flux on the slope (Mudrov, 2007). Thermo-abrasion (TA), on the other hand, is defined as the combined action of mechanical and thermal energy of sea water at water level (Are, 1988a). Despite temporal variations in their intensity, both processes are interconnected, since TD sooner or later becomes inactive after TA comes to a standstill.

Multitemporal applications of remote sensing data are of particular interest for assessing permafrost-related natural hazards such as erosion of frozen sea coasts and thaw subsidence (Kääb, 2008). Numerous recent change detection studies exist and aim to identify coastline variations in different permafrost settings (Lantuit et al., 2013). In concert with time-lapse photography, Jones et al. (2009a) analyse the coastal erosion development around Cape Halkett using high-resolution remote sensing data of the northern Alaska sea coast. They find that, after increasing slightly over the last 5 decades, annual erosion accelerated abruptly and almost doubled, reaching $-13.8\,\mathrm{m\,a^{-1}}$ from 2007 to 2009. They attribute this increase to more frequent block failure as a consequence of higher sea surface temperatures and longer fetch, which potentially create more erosionally effective storm events (Jones et al., 2009b). Lantuit et al. (2011b) study storm climatology and use a set of aerial photographs and satellite images to investigate erosion rates around the entire Bykovsky Peninsula near Tiksi in the Laptev Sea over six consecutive time periods. They show a clear dependency of coastal erosion on backshore thermokarst geomorphology, but do not find either a pronounced temporal trend in the mean annual coastal retreat rate over 55 years ($-0.59\,\mathrm{m\,a^{-1}}$), nor a relation to storm activity. For the western coast of the Yamal Peninsula, where retreat rates range from -0.8 to $-2\,\mathrm{m\,a^{-1}}$, Vasiliev et al. (2006) rely on long-term observational data of the polar station Marre Sale, where the length of the warm period is 102–137 days long, while the open water season lasts for 70 days, on average, generating different preconditions for TD and TA. Although the Kara Sea region experiences frequent storms of long duration (Atkinson, 2005), Vasiliev (2003) finds that only in occasional cases up to 20 % of coastal retreat can be attributed to storms. Arp et al. (2010) report on recent erosion for the Alaskan Beaufort Sea coast, where they observe even more rapid rates of up to $-17.1\,\mathrm{m\,a^{-1}}$, but find little correlation to sea surface and soil temperatures and, in particular, no consistency with storm events. Although potential local controls on erosion such as ground ice content have been identified (e.g. Dallimore et al., 1996; Vasiliev, 2003), it is difficult to establish a relationship of erosion of permafrost coasts to one or another external factor. Moreover, since current environmental changes are expected to intensify coastal erosion, there is a sustained need for information on coastline recession rates in conjunction with seasonal observations in order to better understand the mechanisms driving thermo-erosion and subsequent land loss along permafrost-affected coasts. In addition, however, the influence of coastal thermo-erosion on permafrost degradation processes in backshore areas has received scant investigation.

The main objective of this paper is to systematically analyse seasonal thermo-erosion dynamics and backshore degradation for a ground-ice-rich permafrost coast in the Laptev Sea. We use a set of contemporary very high resolution satellite imagery, repeated geodetic surveys in the field and historical aerial photographs to provide current (2010–2013) and historical (since 1951) quantification of planimetric land loss, volumetric coastal erosion, and land surface lowering due to thaw subsidence in backshore areas. In conjunction with digital elevation models (DEMs), we use a geomorphometric method for assessing macro ground ice content of Ice Complex deposits, in order to consider this factor for the estimation of the mass of material that must be reworked by coastal thermo-erosion following thaw and the resulting sed-

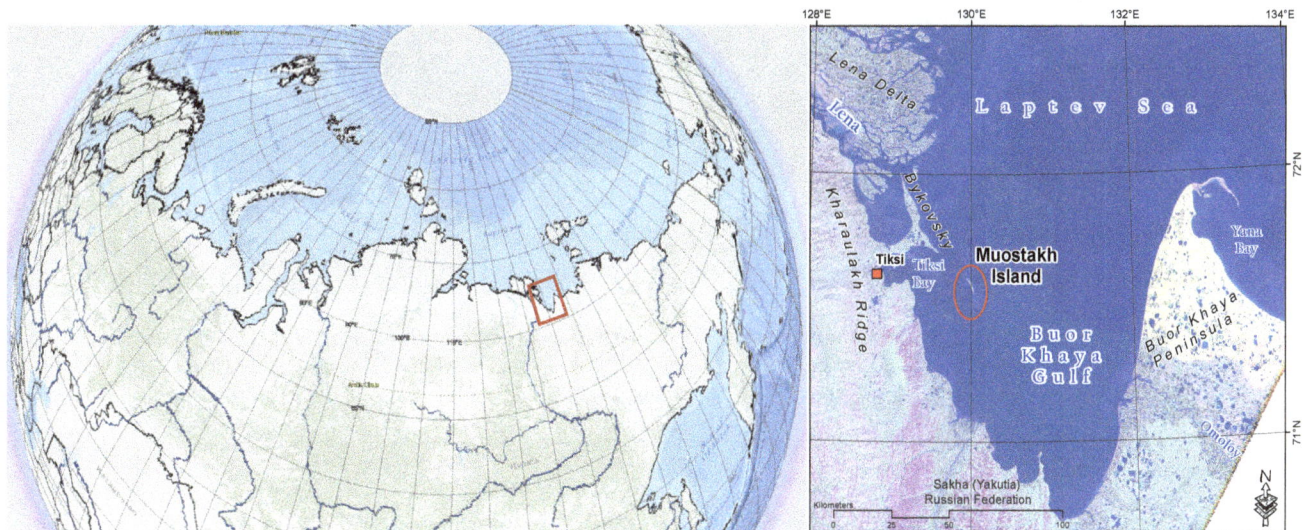

Figure 1. Left: situation of the right-hand map in East Siberia (Russia; source: ESRI). Right: location of Muostakh Island within Buor Khaya Gulf, central Laptev Sea (September 2010 Landsat-5 imagery as background)

iment supply to the nearshore zone. Using time series of local sea ice concentration and air temperatures, we apply normalisation to coastal retreat observations over seasonal and inter-annual periods to identify their seasonal intensity and to discuss environmental controls on processes involved in coastal thermo-erosion development.

2 Study site

Muostakh is a small island (70°35′ N, 130°0′ E), in the Buor Khaya Gulf of the southern central Laptev Sea (Fig. 1), located 40 km east of the harbour town Tiksi in northern Yakutia (Russian Federation). Though situated on the ocean, the severe subpolar climate with mean annual air temperatures in Tiksi of −12.9 °C (1933–2013), where the warmest month does not exceed 10 °C, is continental due to prolonged sea ice cover. Muostakh lies within the northern tundra zone. The vegetation cover is characterised by moss-grass, lichens and dwarf shrub tundra. Cryogenic micro relief features are widespread and include mud boils, frost cracks, peat mounds, thermo-erosional gullies, high-centred polygons on inclined surfaces and thermokarst mounds (baydzharakhs) on coastal bluffs. The island has an elongated narrow form oriented SSE–NNW and is approximately 7.5 km in length with a maximum width of ≤ 500 m at sea level. At the southern margin, next to the former polar station Muostakh, Ostrov, a lighthouse marks the navigable channel into the sheltered Tiksi Bay. As a continuation of the island, an interrupted sand spit chain extends another 5.2 km southwards.

Grigoriev (1993) supposed that Muostakh Island was formerly connected with the Bykovsky Peninsula further in the north (Grosse et al., 2007), but nowadays they are separated by a distance of 15.8 km. Both Bykovsky and Muostakh

consist of Ice Complex deposits and their sedimentological and cryolithological structures suggest simultaneous formation (Slagoda, 2004). According to the Mamontovy Khayata section on Bykovsky, the Ice Complex in this area formed from 58.4 to about 12.2 ka BP (before present) (Schirrmeister et al., 2002), the clastic material is of local origin (Siegert et al., 2000), and accumulated during the subaerial exposure of the East Siberian shelf. Subsequent Ice Complex degradation through thermokarst resulted in alternating relief of depressions (alas) and uplands (yedoma) (Morgenstern et al., 2011, and references therein). Peat and wood on the base of the Holocene cover on Muostakh showed ages in the range of 2–7 ka BP. Muostakh represents a remnant of the late Pleistocene accumulation plain that remained after the sea level drew to near the current level 8 ka ago (Gavrilov et al., 2006), and the highstand of the Holocene transgression was reached 5 cal. ka BP (Bauch et al., 2001). It serves as a witness for the widespread occurrence of Ice Complex islands on the shelf that have been completely destroyed by coastal thermo-erosion (Gavrilov et al., 2003). Ice-poor sands of Pliocene–early Pleistocene age underlay Ice Complex deposits (Slagoda, 2004). Ice complex thickness on Muostakh is 31 m, 10 m of which extend below sea level (Kunitsky, 1989), providing very favourable conditions for TA.

According to Kunitsky (1989) the permafrost temperature of the non-degraded yedoma on Muostakh at the depth of zero amplitude is −10.4 °C, which is cold permafrost and a typical value for yedoma uplands at this latitude in northern Yakutia (Romanovsky et al., 2010). Since Ice Complex deposition took place under permafrost temperatures of −25 to −28 °C (Konishchev, 2002), it therefore has already undergone considerable thermal degradation. Along 600 m on the west coast of Muostakh, a semicircular fragment of an

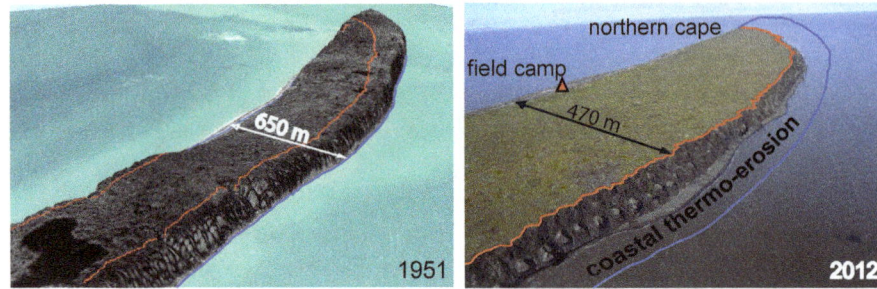

Figure 2. Coastal thermo-erosion over time in the northern part of Muostakh Island. Historical cliff bottom (1951, blue) and current cliff top line (2012, red) border the subaerial coastal thermo-erosion zone. Left: orthophoto of historical aerial imagery draped over 1951 DEM. Note: the lake in the lower left has been drained during the observation period. Right: oblique photograph taken from helicopter in August 2012. Location of 2011 and 2012 field camp, maximum island width for scale.

Figure 3. Appearance of the 21 m high east coast close to the northern cape at the same season in 2 consecutive years. Top photograph: mud flows covering coastal bluff in 2011 indicates thermo-denudation (TD) surpassing thermo-abrasion (TA). Bottom photograph: nearly vertical ice-wall undercut by thermo-erosional niches in 2012 indicates TA surpassing TD

alas depression is preserved, where Kunitsky (1989) reports a temperature increase from −9.4 to −6 °C at 20 m depth along a transect from the yedoma top down to the alas bottom close to the coast. However, according to Romanovsky et al. (2010), alas temperatures at this latitude are around −9 °C, suggesting that permafrost in the coastal zone has undergone additional thermal degradation. Also in the south of the island, around the former polar station, Slagoda (2004) identified fragments of an alas.

Regular stationary monitoring of coastal erosion in the Laptev Sea is conducted only in two places: Mamontovy Khayata on the Bykovsky Peninsula and on Muostakh Island (Grigoriev, 2008). Based on these time series, Muostakh is famous for very high erosion rates, where the northern end of the island for example retreated by about 25 m in 2005 and the nearby east-facing coast by 11 m (Fig. 2). Along with rapid erosion rates, the morphology of the coastal cliff may substantially change its appearance (Fig. 3).

3   Data and methods

3.1   Field work

During a joint Russian–German expedition to Muostakh in August 2011, a network of well-distributed geodetic anchor points was established as a precondition for consistent repeat topographic surveys and their transformation to an absolute coordinate system (Günther et al., 2013b). During a subsequent expedition in August 2012 (Opel, 2015), a repeat survey was conducted. We used a ZEISS ELTA C30 tacheometer for distance and height measurements. Concentrated mainly along the coast, 2392 points were measured that cover about two-thirds of the island's 15 km coastline perimeter. In local project coordinates the point cloud was highly self-consistent, while the absolute geocoding accuracy had a root mean squared error (RMSE) of 1.36 m. Dur-

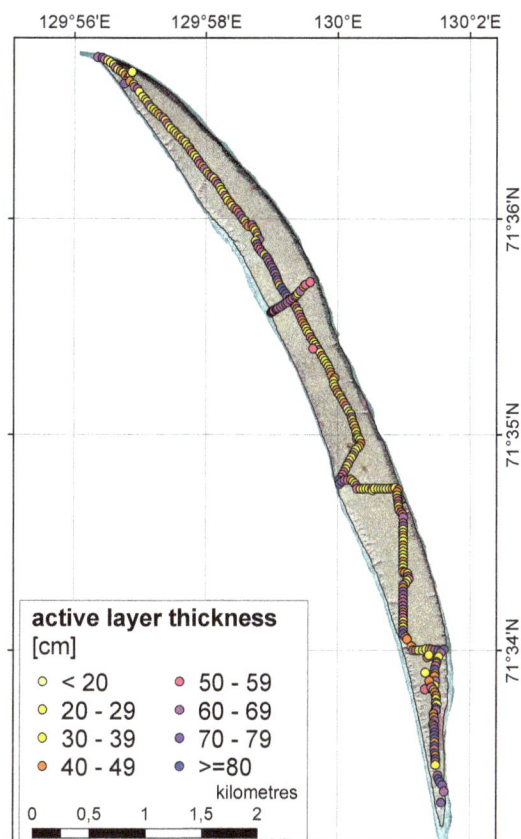

Figure 4. Map showing locations of active layer thickness measurements on Muostakh made during field work in August 2012. Sampling points were classified and colour coded in steps of 10 cm. 2012 GeoEye image as background.

ing a 1-day visit in 2013, selected points were resurveyed within the long-term monitoring.

Measurements of active-layer thickness (ALT) were conducted during the 2012 expedition for the period from 15 to 23 August. In order to capture spatial variability of ALT, altogether 323 ALT measurements were made by mechanical probing. Mapping of ALT across the island was done along transects of 9 km length with an equidistancy of ≤30 m and comprised all soil and vegetation associations (Fig. 4).

3.2 Data fusion and change detection

Remote sensing data were acquired on different dates in order to create a time series of images that was integrated into a geographic information system (GIS), to detect and measure land loss resulting from coastline position changes. In this study, GIS serves as a basis for combining field survey data, historical aerial imagery, contemporary satellite images, and products generated from these data such as DEMs, orthoimages, and digitisation records.

The use of satellite images from different sensors with varying spatial, spectral, and radiometric properties repre-

sents challenges for change detection. This is especially true when using very high resolution image data with a ground resolution of < 1 m (Dowman et al., 2012), not only because pixel-based approaches have been designed for low- to medium-resolution imagery (Hussain et al., 2013) but, specifically, because of the different acquisition geometries that must be considered, requiring careful geometric rectification and topographic correction. Although our study area is flat tundra lowland, our main object of interest, the upper coastline, is always located on the sharp edge of coastal cliffs where abrupt elevation changes occur. In addition to temporally very dynamic elevation changes, steep thermo-abrasional cliffs and baydzharakhs on thermo-denudational coastal bluffs cause large dynamic shadow effects due to different illumination angles of the low solar elevation in high latitudes. Varying conditions of sea-ice-covered and sea-ice-free coastal waters, as well as the presence or absence of banks of snow at the cliff bottom, lead to large reflectance variabilities between acquisitions, making radiometric calibration almost impossible. These conditions lead to problems with automated change detection techniques, examples of which are given in Kääb et al. (2005).

Satellite images must be georeferenced for spatial calibration of multitemporal and multisensor data for change detection. While georeferencing corrects for most distortions connected with the acquisition system, orthorectification corrects for relief-induced displacement effects and creates calibrated satellite image products with the geometry of a map, allowing for distance and area measurements.

GIS-related work was done using ESRI ArcGIS 10.1. Cliff top line positions were manually mapped in orthorectified imagery at different points in time. Unlike other studies on coastal thermo-erosion using the transect method (Günther et al., 2013a), we derived vector data of areal land loss and subsequently calculated seasonal variations of TD. This approach is also used by Aguirre et al. (2008) and Tweedie et al. (2012) for monitoring changes along an arctic coastline but on the basis of DGPS (differential GPS) measurements. Since we also aim to analyse very short time periods of a few days, for which erosion is expected to be rather chaotic, this approach ensures that every event is captured. We divided the studied coastline into 118 segments of 50 m width, with 29 located along the western coast and 89 on the eastern coast. Normalisation of eroded area by baseline length of each segment provided absolute linearised coastal retreat in metres and normalisation over time rates in metres per year. Cliff bottom line position changes are regarded as baselines for historical and subdecadal TA dynamics. We use TD and TA to refer to the rates of coastline position change per year.

3.2.1 Aerotriangulation of historical air photo strip

Aerotriangulation, or block adjustment of a bundle of rays from object to image coordinates, is a standard method in photogrammetry (Konecny and Lehmann, 1984). Aerial photogrammetry is well-suited to quantify historical decadal-scale temporal change (Kääb, 2008). Although coastal erosion studies in the East Siberian Arctic using early aerial photography are common (Grigoriev, 1993; Are, 1999; Lantuit et al., 2011b; Pizhankova, 2011), few use stereophotogrammetry, generally because image parameters for old aerial photos are unknown. Therefore, valuable elevation information available from these data sets still remains untapped.

Five airborne images covering Muostakh Island taken on 9 September 1951 along one flight strip were utilised in this study. Hard copies of 180×180 mm edge length were scanned using a photogrammetric scanner at 14μm scan resolution, corresponding to ≈ 0.4 m on the ground. No information on focal length, principal point offset and radial lens distortion was available. However, the latter two can be compensated to some extent with exterior orientation (spatial location of the projection centre and camera's view direction), for which camera focal length is required (Jacobsen, 2001). Whether or not the correct focal length of the air survey camera is used, calculation of the flight altitude is necessary (Knizhnikov et al., 2004), which can be done by determining the scale of the frame photography. The scale number S_a of the frame photography was roughly estimated following

$$S_a = S_m \cdot \frac{d_m}{d_a}, \tag{1}$$

by measuring the same distance between two objects within the original photograph (d_a, cm) and on-screen (d_m, cm) within a contemporary orthoimage with a map scale (S_m) set to 1 : 10 000. This approximation resulted in a large scale of 1 : 28 000. Using the Aerial Photography model in PCI Geomatica's 2013 module OrthoEngine, we collected fiducial marks in each image to visually define the principal point, collected a set of 25 stereo GCPs (ground control points) and automatically computed over 850 tie points (TPs) using cross-correlation for strip stabilisation. Bundle block adjustment was performed iteratively with the focal lengths of air survey cameras that existed at that time according to Shcherbakov (1979). The best overall solution was achieved with a focal length of 100 mm (likely Liar-6, 104° wide angle lens, used for topographic medium-scale mapping), yielding a RMSE for GCP locations of 2.4 m and TPs of 0.3 m. For the exterior orientation parameters, this corresponds to a flight altitude of around 2600 m, which is consistent with the theoretical flight height of 2800 m calculated from the approximate photo scale following

$$h = S_a \cdot c_k, \tag{2}$$

where c_k is the focal length and h flight height, both in metres. The air survey strip constellation provides along-track stereo and triple overlap situations, with base-to-height-ratios of 0.7 and 1.4, respectively, allowing for height parallax measurements and DEM extraction of steeper slopes as well as over flat terrain. Vertical accuracy of the DEM was estimated to be 2.3 m, based on the difference of input and calculated elevations at stereo GCP locations. Rectangular corners of footprints verify nadir viewing geometry and a robust model, which did not result in overfitting outside the GCP cloud, an important precondition for reconstructing the former shape of the island and subsequent change detection. Based on the DEM of 1951, the aerial photographs were orthorectified and stitched together in a seamless orthomosaic.

3.2.2 Multisensor block adjustment

Multisensor data fusion, in this study of GeoEye (GE), QuickBird (QB), WorldView-1 (WV-1) and WorldView-2 (WV-2), offers the opportunity to merge images collected from different satellites and different orbits in one triangulation process. According to Toutin (2004), the simultaneous solution of an entire image block offers several advantages, for example the number of GCPs can be reduced, better relative accuracy between images can be obtained and finally more homogeneous orthoimages over large areas can be produced. Satellite sensor models described by rational polynomial coefficients (RPC) provide a high potential of simple and accurate geopositioning (Fraser et al., 2006), are ideally suited for block adjustment of narrow field of view sensors (Grodecki and Dial, 2003), but require some bias correction (Fraser and Ravanbakhsh, 2009), and generally serve only as an approximation of physical sensor models when orbital information is not provided in the metadata (Poli and Toutin, 2012). We performed block adjustment and subsequent orthorectification using our own ground control within the Rational Functions model (RPC-based) in OrthoEngine.

Prior to further geometric processing, we applied single-sensor and single-date image fusion to QB, GE, and WV-2 imagery using the enhanced pan-sharpening method of Zhang (2004). Nine very high resolution images were acquired as standard/orthoready products (Table 1), with panchromatic (PAN) imagery resampled with sinusoidal kernels, for better representation of sharp features (Toutin, 2011). Due to varying moisture and illumination conditions between acquisitions, we found that not all GCPs collected in the field could be identified unambiguously in each image, resulting on average in about four GCPs per image. The RPC model is a viable alternative for rigorous sensor models (Cheng et al., 2003), and several studies show that the effect of the number of GCPs on 3-D RPC block adjustment is limited, yielding almost no further improvement, if configurations of more than four GCPs are used (Fraser and Ravanbakhsh, 2009; Aguilar et al., 2012). Based on our topographic reference measurements, 37 elevation TPs were additionally incorporated into the block, to achieve higher redundancy in RPC bias correction and mainly to better align images to each other. The zero-order polynomial turned out to be the most stable and best possible solution of the block, yielding a submetre accuracy within the entire block of 0.81 m RMSE (Table 1).

Table 1. List of very high resolution satellite imagery used for change detection and summary of multisensor bundle block adjustment.

Sensor	Date	Incidence angle (°)	Resolution (m)	Number of GCPs	Number of TPs	GCP RMSE (m)	TP RMSE (m)	RMSE (m)
QuickBird-2	23 May 2010	12	0.6	2	16	0.7	1.06	1.03
QuickBird-2	15 Jun 2010	15.3	0.6	3	22	0.88	0.71	0.73
WorldView-2	29 Jun 2010	18.8	0.5	3	18	1.05	0.71	0.77
GeoEye	13 Jul 2010	18.9	0.5	4	21	1.27	0.59	0.87
WorldView-1	8 Aug 2010	15.6	0.5	4	34	1.14	0.61	0.68
WorldView-1	28 Jun 2011	17	0.5	4	18	1.11	0.69	0.78
GeoEye	7 Sep 2012	14	0.5	6	18	0.78	0.54	0.61
GeoEye	17 Jul 2013	28	0.5	2	17	0.99	1.03	1.03
GeoEye	17 Jul 2013	19	0.5	2	16	0.59	0.72	0.71
Entire block				7	37	1.07	0.74	0.81

According to Günther et al. (2013a), the mutual RMSE of each data set (Table 1) is then considered to be the relative georeferencing uncertainty in the determination of the cumulative uncertainty in coastline position, which results out of the combination of this error, the ground resolution, and the additional 2-D positional error introduced by the DEM used for orthorectification, depending on the incidence angle. Uncertainties in change rate calculation over six periods between 15 June 2010 and 17 July 2013 are also applied following Günther et al. (2013a) and were in the range of 0.47–0.65 m.

3.2.3 True orthorectification

The 2013 GE images were acquired as a stereo pair and used for DEM extraction using the stereo model of the entire image block. We applied a DEM editing procedure in order to remove noise, interpolate areas of unsuccessful matching, and for low-pass filtering. The final DEM features high detail and its spatial resolution is 1 m. Using 1158 survey points from 2011 that were located within the 2013 cliff top line extent, the mean elevation difference was 0.06 m, which reflects the good match of both data sets in absolute reference height. Accordingly, the vertical accuracy of the 2013 DEM was evaluated as $\sigma = 0.64$ m. However, the 2013 DEM could not be used for orthorectification of the 2010–2012 satellite images, because of its mismatch in cliff top line position compared to earlier dates.

Due to this, radiometric similarities for multidate single-sensor (e.g. QB–QB) and multidate multisensor (e.g. WV–GE) constellations were evaluated with regard to stereoscopic interpretation at an earlier date. Through pansharpening of GE's near-infrared band (0.45–0.8 μm) with the PAN band (0.78–0.92 μm wavelength), the spectral range was adjusted to WV-1 PAN imagery (0.38–0.88 μm) to achieve enhanced image matching. Epipolar image matching of GE acquired on 13 July 2010 and WV-1 acquired 26 days later resulted in visually good topography and showed the

least standard deviation between input and calculated elevations for 3-D TPs.

Final DEM generation was performed on a hybrid vector–raster data basis. Contour lines from the stereoscopic DEM, cliff bottom and top lines (digitised in ellipsoid-based orthoimages considering geoid height offset), and all point data of the 2011 and 2012 topographic field surveys (that are necessarily within the 2010 cliff top line extent), were incorporated into a terrain interpolation procedure according to Hutchinson and Gallant (2000). The final DEM represents the island's state in the early summer of 2010. The vertical accuracy of the DEM at survey point locations was $\sigma = 0.3$ m, therefore introducing ≤0.1 m 2-D positional uncertainty along the cliff top line through subsequent orthorectification, instead of ≤7.2 m random terrain-induced displacement for each of the very high resolution images when using the initial RPC reference height of the image products.

3.2.4 Elevation difference uncertainty assessment

Differencing of multitemporal DEMs was done using the 2013 and 1951 data sets. The relative vertical uncertainty between both DEMs in the island's interior was -0.4 ± 2.2 m. However, according to the strategies of Nuth and Kääb (2011) and based on our reference data from topographical surveys, a height-dependent bias in DEM difference of 0.47 mm m^{-1} could be identified. Günther et al. (2012) also report systematic underestimations of height measurements in historical aerial photography stereo pairs. DEM errors may result from inexact matching and a lack of contrast within and similarity between stereo images (Nuth and Kääb, 2011). The systematic bias was corrected using an empirical equation of second polynomial order (Fig. 5).

The vertical accuracy of the 2013 and 1951 DEMs was determined to be 0.64 and 2.3 m, respectively. Regarding the 1951 DEM, this is not particularly meaningful, because reference heights were derived from the survey data, collected 60 years later. Restricting the search matrix size for cross-correlation of the 1951 aerial photographs to a modern elevation maximum of 21 m a.s.l. resulted in an almost complete failure of height parallax measurements north of the alas. For this reason, we did not restrict the search matrix size and obtained a consistent 1951 DEM with maximum elevations of 24.9 m a.s.l., which is in accordance with a topographic map from 1953. The observed maximum elevation change is −25.5 m, which illustrates that bias correction is robust also in former island areas not covered by contemporary reference data. Since almost the whole terrain of Muostakh may potentially be susceptible to permafrost thaw subsidence or other causes of elevation change, no long-term reference height points were available, except for two elevation indications of 20.9 and 25 m a.s.l. in the 1953 map. The positional inaccuracy of 75 m related to the topographic map was evaluated using the centroids of four lakes that were visible in the 1951 imagery and marked in the topographic map, but gradually drained during the 1960s and 1970s. With respect to the flat relief on yedoma uplands, elevation mismatch was evaluated within buffers of 150 m diameter around both points and revealed an absolute DEM difference uncertainty of −1.56 ± 0.78 m. Within the already degraded alas depression at 3–4 m a.s.l., where no substantial changes are expected, elevation differences were −1.01 ± 0.55 m on average. Together, both indicators suggest a small residual height dependent bias of −0.028 m m^{-1} and revealed a mean error of 1.31 m that possibly corresponds to a slight overestimation of negative terrain height changes.

3.3 Environmental parameters

We use environmental observations to relate coastal dynamics to its potential drivers of atmospheric warming and sea ice reduction. As a proxy for marine abrasion along the cliff bottom line, we use the inverse of daily sea ice concentration that is open water extent per day. To relate the rate of thaw along the cliff top line, we use air temperature (T_{air}) data as positive mean daily temperatures. These data are then used for correction of coastal erosion rates from different image acquisition periods. To identify single wind events and general wind patterns during the period of open water that potentially influenced coastal retreat we use wind data.

3.3.1 Sea ice concentration data

Daily percent sea ice concentrations from 1992 to 2013, based on Special Sensor Microwave Imager (SSM/I), were used to calculate the open water fraction in percentage per day. Derivation of total sea ice concentration from SSM/I data uses dual polarisation measurements with the 19 and 37 GHz channels of 25 km spatial resolution since 1978. Higher resolution of 12.5 km is available using the 85.5 GHz high frequency channel that did not work before 1992. Although the 85.5 GHz product covers a shorter period of time and might be affected by larger uncertainties over lower sea ice concentrations and open water (Lomax et al., 1995),

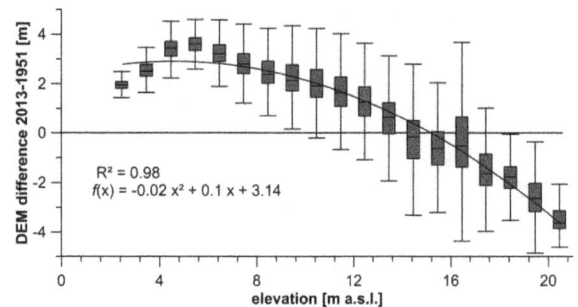

Figure 5. Elevation dependent bias in 2013–1951 raw DEM difference and correction using an empirical function of second order polynomial.

25 km resolution is too coarse to study the coastal zone around Muostakh. We worked with the 12.5 km data product of the ARTIST (Arctic Radiation and Transport Interaction STudy) sea ice algorithm (Kaleschke et al., 2001) distributed by Ifremer/CERSAT (2000), which is based on a hybrid model and provides reliable results (Ezraty et al., 2007). In very few cases, the data can also accept negative and positive values outside the 0–100 % range (Andersen et al., 2007), requiring correction. Data was masked by land mass and limited to a 100 km radius around Muostakh Island, spatially corresponding to the Buor Khaya Gulf (Fig. 6). Daily sea ice coverage was smoothed using a 7-day running mean and converted to data of open water fraction in order to determine and count open water days (OWDs).

3.3.2 Air temperature data

The spring to fall seasonal cycle in the Lena Delta region features rising T_{air} during spring until the end of snowmelt, T_{air} well above the freezing point during summer, and fall is characterised by the beginning of refreeze (Langer et al., 2011). In order to evaluate the response of TD to T_{air} over time, we use positive mean daily T_{air} and positive degree-day (PDD) sums (Braithwaite, 1995), obtained from the temperature curve integral above 0 °C in Kelvin days (Kd) (Jonsell et al., 2013). The hydrometeorological observatory Polyarka near Tiksi (WMO # 21824) measured T_{air} 3 times a day from 1932 until 1936, every 6 h until 1970, and has measured it since then every 3 h. Data were downloaded from the electronic archive of observations at Tiksi (Ivanov et al., 2009a, b). For different erosion observation time ranges within the period from 1951 to 2013, PDD were calculated using mean daily T_{air} data as annual sums and as PDD sums over certain seasonal and interannual periods.

3.3.3 Wind data

Wind blowing over open water generates waves breaking on the shore face. Wave height, energy and erosion potential of TA is proportional to wind speed. Wind data of Tiksi from

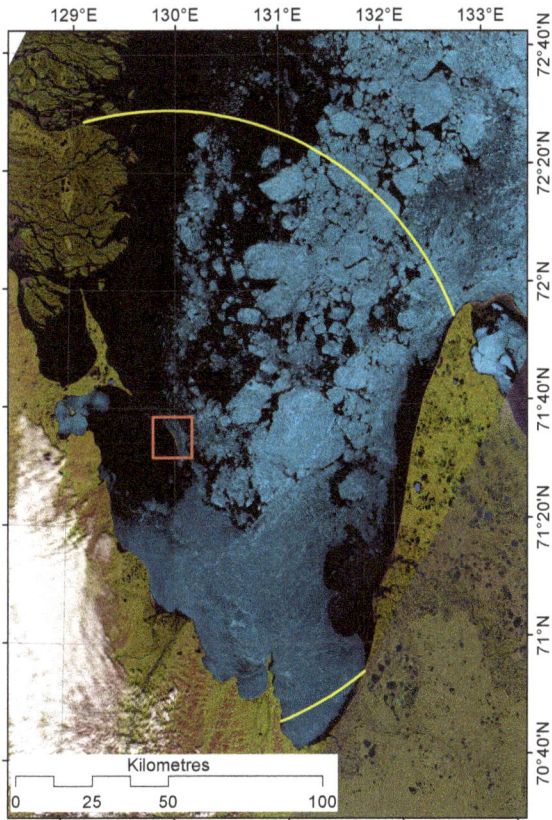

Figure 6. Detail of a Landsat-8 image $(5, 4, 2\,\text{CIR})$ showing sea ice break-up around Muostakh Island (red square, size corresponds to one 12.5 km SSM/I Pixel) on 18 July 2013. The open water fraction on this day, 100 km around Muostakh (yellow circle), was 20 %. The sea-ice-free season started 19 days later on 6 August 2013.

1992 until 2012 were examined for direction, speed, and single strong wind events. We focus on this period to relate wind data to daily sea ice conditions. Based on measurements every 6 h, mean wind speeds were classified into eight wind directions and analysed for observations during sea ice break-up and the sea-ice-free period of a particular year.

3.4　Local parameters

Permafrost is prone to thawing because its core element is the occurrence of ground ice. Ice wedges constitute a large fraction of the subsurface volume. They extend in different generations and stratigraphic units from the top of the permafrost to below sea level, and their size directly determines how much clastic material must be thawed and subsequently removed by coastal thermo-erosion. On Muostakh, they are syngenetic ice wedges. Visually estimating macro ground ice content based on the fractional ice-wedge volume (for example, from photographs of the coastal cliff) is often complicated by slope and perspective, by debris material that obscures undisturbed in situ material, as well as by the fact that the collapse of thermo-abrasional cliffs occurs along ice-

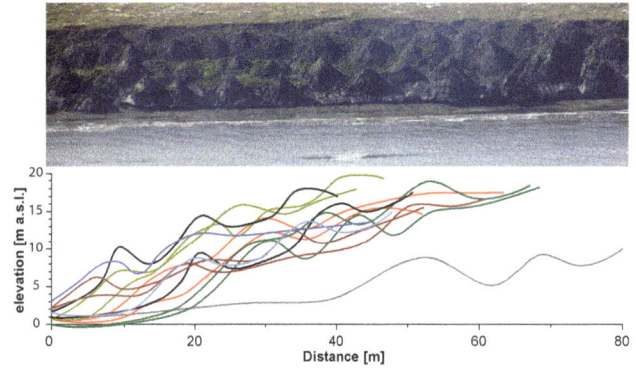

Figure 7. Top: photograph of a baydzharakh field on the east coast of Muostakh Island. Bottom: examples of slope profiles across baydzharakhs, showing differences in baydzharakh spacing.

wedge axes (Are, 1988b), implying that an exposed wall of ice may not serve as a representative random test for the geological subsurface.

3.4.1　Macro ground ice

Baydzharakhs are a characteristic ephemeral cone-shaped thermokarst landform and represent the remnant frozen sediment core and geometric centres of thawed ice-wedge polygons (Mudrov, 2007). During field work in 2011, twenty coastal slope profiles were surveyed between and across baydzharakhs at different locations, where we observed that baydzharakh spacing varied and might do so depending on their fractional volume of the subsurface (Fig. 7).

Using terrain-corretced satellite imagery, we extended baydzharakh mapping to erosional coast segments (Fig. 8). Taking baydzharakh centres as seeds, we subdivided the surface into cells of a Voronoi diagram (Reem, 2010), which we use as an estimate of polygon morphology prior to thaw. The largest possible circle within a cell was calculated using the maximum Euclidean distance of each cell as radius. Based on the assumption that the sediment centre of each polygon has a cylindric form, macro ground ice content as fractional volume (V_{wm}) is calculated from

$$V_{\text{wm}} = 1 - \frac{A_{\text{circle}}}{A_{\text{polygon}}}, \tag{3}$$

where A_{circle} is the surface of the sediment centre and A_{polygon} the total polygon area in square metres.

3.4.2　Subsidence potential

According to Katasonov (2009), the porosity of Ice Complex is very large, due to excess ice and fine particle size. However, natural sediment deposition forms cavities, and this fraction of the total porosity must be disregarded in terms of subsidence. We assume a porosity of 0.4, following

$$\phi = 1 - \frac{\rho}{\rho_O}, \tag{4}$$

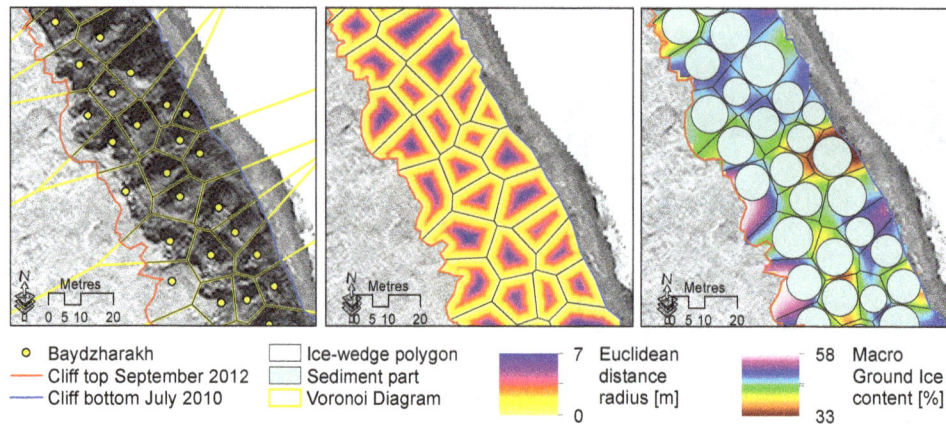

Figure 8. Left: points mark mapped baydzharakh centre locations, used for derivation of the Voronoi diagram. Middle: determination of the Euclidean distance within each polygon. Right: construction of largest possible circles within polygons using maximum Euclidean distance as radius, representing the sediment component of the subsurface. Calculation and interpolation of macro ground ice content between circles based on the ratio of area occupied by circles and total polygon area (8 August 2010 WorldView-1 imagery as background).

where ρ is mean bulk density of the Ice Complex on Muostakh of $1.6 \pm 0.25 \times 10^3 \, \text{kg m}^{-3}$, according to Solomatin (1965), and ρ_O the particle density of non-porous clastic material ($2.65 \times 10^3 \, \text{kg m}^{-3}$, Strauss, personal communication, 2012). Assuming that all pores are filled with ice (Strauss et al., 2012), the pore ice fraction of intrasedimentary ice does not contribute to subsidence. Therefore, the relative subsidence potential of thawing Ice Complex deposits was calculated following

$$\delta z = V_{\text{wm}} + (V_{\text{s}} \cdot W_{\text{is}}) - (V_{\text{s}} \cdot W_{\text{is}} \cdot \phi), \qquad (5)$$

modified after Mackay (1966) and Are (2012), where V_{wm} is volumetric macro ground ice, V_{s} volume of the sediment part, W_{is} intrasedimentary ground ice and ϕ porosity.

4 Results

4.1 Historical erosion development

The development of thermo-erosion on Muostakh and its shaping of the island was analysed over a period of time of more than half a century. Starting with the first aerial photographs from 1951 and ending with the most recent GeoEye image of 2013, erosion was quantified. We analysed areal land loss and the associated volumetric land losses over 62 years for the entire island. Based on this, we concentrated on the eroding portion of the coastline, where 118 coastline segments were studied in more detail. Each segment corresponds to a 50 m coastline length at beach level. Squares used for symbolising erosion in Fig. 9 (middle) show the spatial distribution of all studied coastline segments.

4.1.1 Mass movements

The volume of Muostakh Island has decreased by 40 % between 1951 and 2013, based on multitemporal DEMs calculated for those years. According to the overall volume change (Table 2), the calculated rate of mean annually eroded volume on Muostakh is $0.45 \times 10^6 \, \text{m}^3 \, \text{a}^{-1}$, corresponding to $0.36 \times 10^6 \, \text{t a}^{-1}$ of annual ground ice thaw and $0.16 \times 10^6 \, \text{t a}^{-1}$ of sediment displacement. About two-thirds of this results from erosion on all sides of the island and about one third from an overall degradation through surface lowering. The cumulative eroded volume within the coastal segments studied in detail was $16.3 \times 10^6 \, \text{m}^3$, which means that 90 % of coastal erosion was recorded by our coastline subsample (Fig. 9, middle). Within each of the the 50 m coastal segments, annually eroded volumes per segment varied broadly and were on average $-2240 \, \text{m}^3 \, \text{a}^{-1}$. Very high values of up to $-28\,300 \, \text{m}^3 \, \text{a}^{-1}$ were detected at the north cape. Generally, eroded volumes were close to the median of $-1280 \, \text{m}^3 \, \text{a}^{-1}$, because most of the coastline is eroding more slowly and has a lower backshore height than the northern cape.

4.1.2 Land subsidence and active layer thickness

In addition to large elevation decreases along the eroded coastline, we observed land subsidence across the entire island (Fig. 9, right). The mean elevation of Muostakh in 2013 was 14.4 m a.s.l. The 1951–2013 DEM difference raster was clipped to the interior of the cliff top area of 2013, in order to exclude the influence of coastal cliffs for further analyses of this phenomenon. The mean elevation change was -3.56 ± 1.8 m. Based on these data, except for a very limited area around the former polar station, the island experienced land subsidence at a mean rate of $-5.8 \pm 2.9 \, \text{cm a}^{-1}$. In particular, in the northern part elevation decreases were large

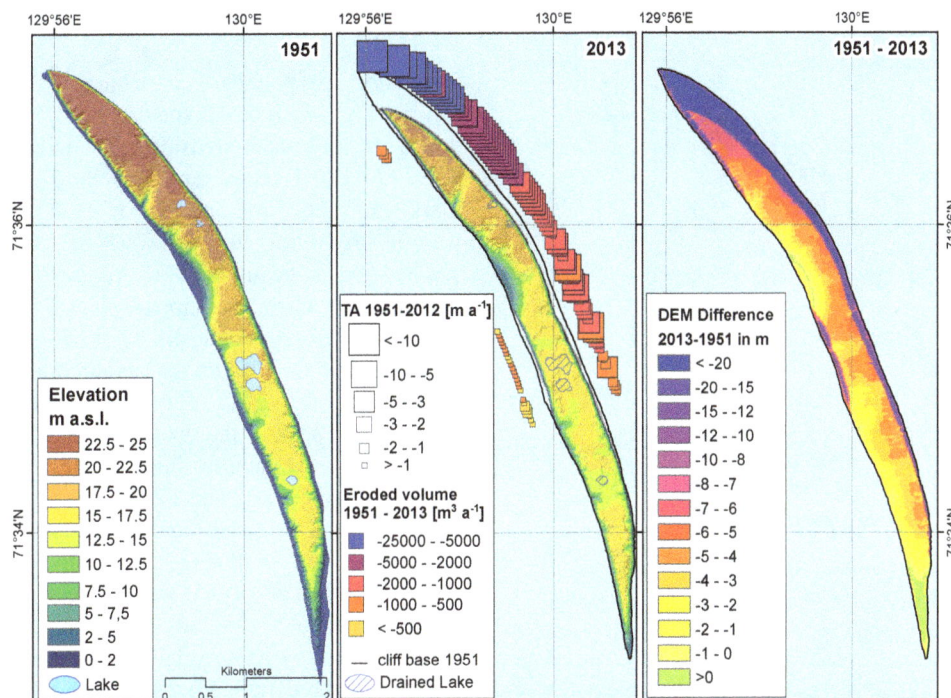

Figure 9. Left: DEM from 1951 stereoscopic aerial photography. Middle: DEM from 2013 GeoEye stereo pair. Symbol size is the classified planimetric coastal erosion rate. Colour-code displays volumetric erosion from 1951 until 2010 for 118 coastline segments. Right: difference raster from multitemporal DEMs representing elevation changes over 62 years.

Table 2. Volumetric losses and associated mass displacement on Muostakh Island, based on DEMs of 1951 and 2013 for different compartments of the subsurface, assuming fractional volumes of 44 % macro ground ice and 43 % intrasedimentary ground ice.

	1951	2013	Total loss
Total volume ($m^3 \times 10^6$)	69.6	41.6	28.0 ±4.9
Surface layer ($m^3 \times 10^6$)	2.4	2.0	0.4
Macro ground ice ($m^3 \times 10^6$)	29.6 ±3.1	17.4 ±1.8	12.2 ±1.7
Intrasedimentary ice ($m^3 \times 10^6$)	28.6 ±2.4	16.9 ±1.3	11.7 ±1.6
Clastic material ($m^3 \times 10^6$)	9.0 ±0.8	5.3 ±0.4	3.7 ±0.6
Total ground ice ($t \times 10^6$)	53.4 ±2.5	31.4 ±1.4	21.9 ±2.3
Clastic material ($t \times 10^6$)	23.9 ±2.1	14.1 ±1.1	9.8 ±1.6

and land subsided by around $-10.9 \pm 0.6 \, cm \, a^{-1}$ over the last 62 years. Quite unexpectedly, the spatial pattern shows land subsidence is more active close to erosive parts of the coastline and intensifies the more rapidly coastal thermo-erosion is proceeding (Fig. 9, right). This becomes particularly evident not only in the north where rapid rates of coastal erosion coincide with strong subsidence, but also in the middle part of the island, where erosion from both sides leads to stronger subsidence compared to neighbouring non-erosive coastal segments and the adjacent hinterland.

Based on 326 equally spaced transect measurements, ALT on Muostakh in 2012 was on average $47 \pm 19 \, cm$. Generally, large ALTs were clustered on well-drained slopes close to the coast, on slopes of the alas, and on the southern tip of the island (Fig. 4), the only place not affected by subsidence. Land subsidence observations over the historical time period were then linked to current measurements of ALT (Fig 10). In contrast to the predominant image of larger ALT causing permafrost thaw and subsequent subsidence, we found, for example, that a shallow ALT of $\leq 20 \, cm$ is associated with intensive subsidence of around $-7.5 \pm 0.4 \, cm \, a^{-1}$, while at locations of deep ALT $\geq 80 \, cm$ mean subsidence was only $-3.4 \pm 1.9 \, cm \, a^{-1}$.

4.1.3 Coastline changes

TA was analysed over 61 years, where the start and end points of the observation period are in early September, meaning there is no shift with respect to season.

Günther et al. (2013a) found that TD and TA along Ice Complex coasts are interconnected: TA is the limiting component for coastal thermo-erosion intensity on the long-term scale. TA also better reflects the overall land loss of the base area of the island. The base area extent of Muostakh in 1951 was $3.8 \, km^2$. By 2012 it had shrunk by -23.7% to $2.9 \, km^2$, which corresponds to a mean land loss of $-14\,700 \, m^2 \, a^{-1}$. Given Muostakh's 2012 cliff bottom line

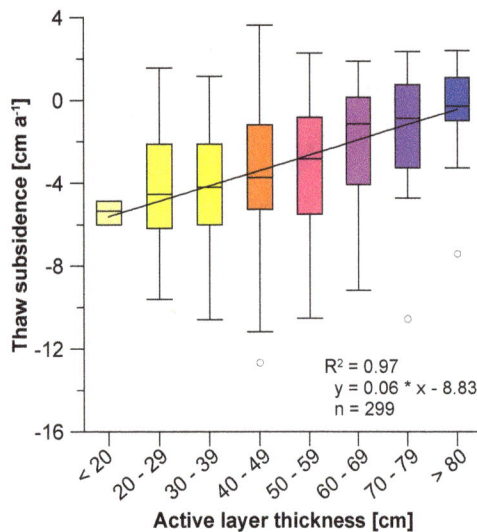

Figure 10. Classified observations of active layer thickness (ALT) in August 2012 in relation to mean annual permafrost thaw subsidence from 1951 to 2013 show intensified subsidence in places of shallow ALT. The outliers are indicated by open circles. The colour code is adjusted to ALT classification in Fig. 4. Inverse relation of ALT and subsidence indicates water drainage on the permafrost–active layer interface and consequently irreversible ground ice thaw.

perimeter of 15.5 km, this corresponds to a mean coastline retreat rate due to TA of $-0.95 \, \mathrm{m \, a^{-1}}$ when examined for the entire island including non-erosive coastline sections.

For erosive sections, areal land loss was mapped within 118 segments. Transformation of 2-D areal data to mean distance measurements was done individually via the baseline length of a particular segment. The uncertainty of cliff bottom position change is $\pm 1.7 \, \mathrm{m}$, for TA $\pm 0.04 \, \mathrm{m \, a^{-1}}$. Along the 5.8 km coastline covered by our segmentation (corresponding to the former 6.4 km in 1951), absolute TA had a mean of $-109.7 \pm 80.6 \, \mathrm{m}$, whereas annual rates were in the range from -0.2 to $-7.2 \, \mathrm{m \, a^{-1}}$ with a $-1.8 \pm 1.3 \, \mathrm{m \, a^{-1}}$ mean. According to Fig. 11 (left), 71 % of TA rates over the historical period were clustered towards slower rates of $\geq -2 \, \mathrm{m \, a^{-1}}$; for comparison, only 59 % of modern TA rates were slower than the mean value, suggesting coastal erosion over the historical period progressed relatively uniformly, or that temporal averaging of several erosion events occurred. Although the northern cape is eroding at a different angle than the rest of the north-eastern coastline, it has traditionally been of interest as it reflects the increasing distance between Cape Muostakh on the adjacent Bykovsky Peninsula and Muostakh Island. Tracing the position change of the exposed northern cape of the island between 1951 and 2012, maximum absolute cliff bottom line recession was $-585 \, \mathrm{m}$, which is equivalent to $-9.6 \, \mathrm{m \, a^{-1}}$.

4.2 Interannual and seasonal erosion development

Open water and positive T_{air} are unequal in duration and time of year (Fig. 12). In 2010, the first two images were acquired during ice melt and increasing mean daily T_{air} (15, 29 June). The 13 July 2010 image represents the T_{air} summer peak and marks the start of the open water season. The fourth image was acquired during the open water season, when mean daily T_{air} had already begun to fall (8 August 2010). Together with the previous image, it completely spans a period when both TD and TA are active. The 2010 fall season was bracketed by the fourth and the fifth image, acquired in early 2011 during ice melt and rising T_{air}. The 7 September 2012 image, acquired at the peak of the open water season and falling T_{air}, together with the previous image, captured almost two complete seasonal cycles. The 17 July 2013 image was acquired prior to late sea ice break-up in 2013, and completes not only fall 2012, but captured also spring 2013.

Sums of PDD and OWD were correlated (Fig. 13) and are probably generally correlated, since sea ice melt is driven to a great degree by heat exchange with the atmosphere. This means that we expect TD and TA to be correlated insofar as they are driven by PDD and OWD, respectively. Due to the strong seasonal constraints on the development of coastal thermo-erosion, the discrepancy between the start and end points of the observation periods and the duration of the season when TD and TA are able to proceed may result in an over- or underestimation of rates. In cases of mismatch between two acquisition dates, instead of direct change rate calculation only, we corrected calculated rates over time using a season factor. Season factors were derived from the ratio of either the number of days of open water or of positive mean T_{air} during the specific observation period to a perennial reference period. Season factors are used to calculate the actual coastal erosion velocity over a particular period of time and to compare velocities between periods.

4.2.1 Current thermo-abrasion

We examined current dynamics of TA using GE images of 13 July 2010 and 17 July 2013 as the data set spanning the longest period of the recent past, for which the cliff bottom is free of snow. The base area reduction of the entire island was $-22\,300 \, \mathrm{m^2 \, a^{-1}}$ during the last 3 years (compared to $-14\,700 \, \mathrm{m^2 \, a^{-1}}$ on average from 1951 to 2012; Table 3). Of this erosion, 89 % occurred within our 118 coastline segments for detailed study. During the last 3 years, mean TA was $-3.4 \pm 2.7 \, \mathrm{m \, a^{-1}}$ and therefore currently 1.9 times faster than over the historical time period (Fig. 11, left). Of all segments, 19 % experienced slight deceleration, while only at a few segments TA rates remained almost unchanged. However, of note is the fact that this almost doubling of coastal erosion is not due to outliers, but derives from a broad acceleration at segments previously eroding in a narrow range from -0.5 to $-2.5 \, \mathrm{m \, a^{-1}}$ to currently -1.5 to $-8 \, \mathrm{m \, a^{-1}}$.

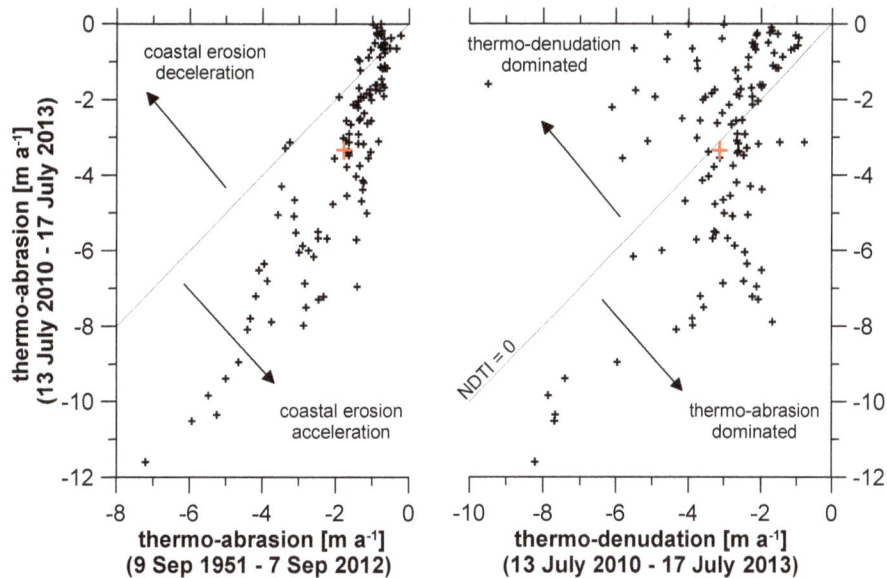

Figure 11. Point by point coastal thermo-erosion over time, red crosses indicate mean values. Left: historical thermo-abrasion (TA) from 1951 to 2012 in relation to current changes from 2010 to 2013, showing TA acceleration at 95 out of 118 coastal segments. Line of equal rates. Right: current thermo-denudation (TD) in relation to TA, showing TD- and TA-dominated coastal erosion regimes were around the same frequency. Virtual normalised difference thermo-erosion index (NDTI) zero-line for differentiation.

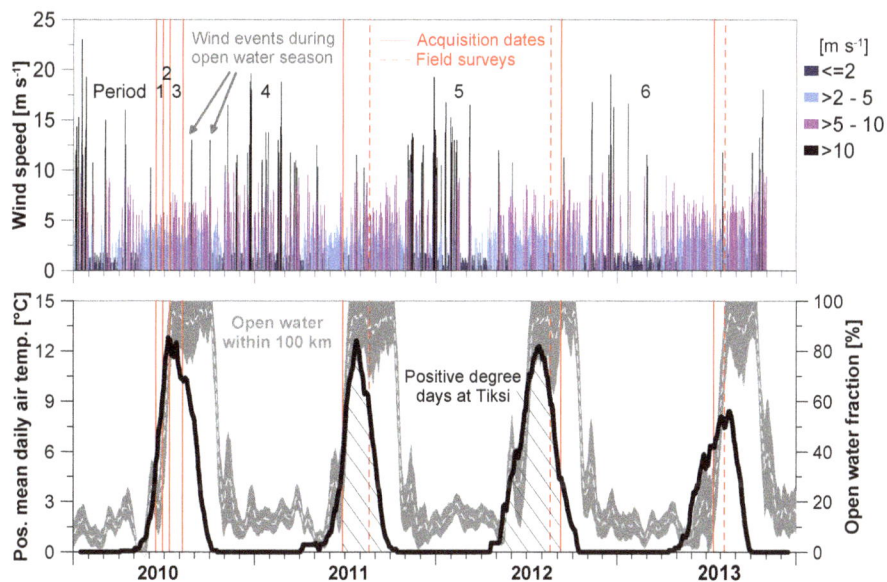

Figure 12. Hydrometeorological data over time; straight-through lines mark satellite image acquisitions, dashed lines topographic surveys during on-site visits, all with uneven distribution. Top: bars of average 6 h wind speeds measured in the nearby town of Tiksi. Bottom: seasonal fluctuations of positive mean daily T_{air} at Tiksi (black curve) and open water fraction within a 100 km radius around Muostakh Island (white dashed curve, grey standard deviation range), both plotted as a 15-day running mean. Integration under the Tair curve was used as positive degree-day sum for normalisation of observation period specific thermo-denudation rates (shaded area as example for period 5). Note the time shift of positive T_{air} and open water period.

Between July 2010 and 2013 the northern cape retreated at −51 m, which corresponds to a TA rate of −17 m a^{-1} (compared to −9.6 m a^{-1} over the historical period). Erosion at the cape determines the dynamics of the sand spit forma-

tion next to it. At this location, additional topographic survey data of the expeditions in 2011 and 2012 were available and covered down to the cliff bottom (Fig. 14). Interannual variations in coastline position change were large. Between

Table 3. List of time periods used for coastal thermo-erosion change detection, bracketed by data acquisitions. Periods referred to as letters correspond to observations covering more than or exactly 1 year. PDD sums were considered as indicators for possible ground ice melt. TD was season-corrected using the number of days with positive mean daily T_{air} occurrence, while TA observations along the cliff bottom did not require correction through open water fraction, because of identical observation period start and end dates.

Numbering (type)	Time period	Land loss (m²)	Erosion (m)	PDD sum (Kd)	Open water (% d⁻¹)	Season factor
A (TA)	9 Sep 1951–7 Sep 2012	735000	109	40260	–	1
B (TA)	13 July 2010–17 July 2013	58700	10.1	2739	36.6	1
C (TD)	29 June 2010–28 June 2011	27300	4.8	937	37.7	1
1 (TD)	15 June–29 June 2010	2000	0.5	85	19.7	0.34
2 (TD)	29 June–13 July 2010	4300	1.0	149	61.4	0.34
3 (TD)	13 July–8 Aug 2010	12400	2.1	330	95.6	0.34
4 (TD)	8 Aug 2010–28 June 2011	10700	1.9	445	32	0.67
5 (TD)	28 June 2011–7 Sep 2012	30000	5.2	1636	46.6	0.79
6 (TD)	7 Sep 2012–17 July 2013	1300	0.3	305	23	0.4

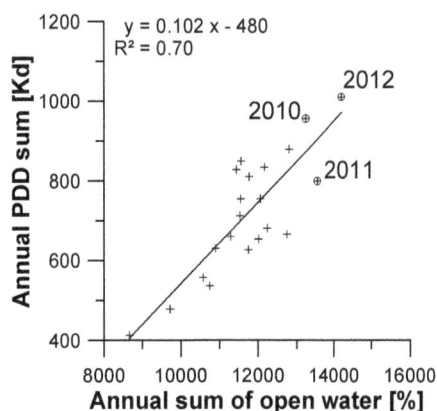

Figure 13. Open water fraction and positive degree-days (PDDs) as annual sums for the reference period of both records (1992–2013).

2010 and 2011, the northernmost point of the island retreated by only −4.3 m, while it was −39.4 m during the following period between the two subsequent expedition surveys. However, further away from the cape the opposite picture emerged. Little erosion occurred from 2011 until 2012 when compared to the previous year. Due to these effects, we considered the surrounding 50 m coastline segment as the northern cape area where mean TA over the whole 2010–2013 observation period was −11.6 m a⁻¹, being still the most rapid erosion along the entire coast of Muostakh Island (Fig. 11, left). Also of note is erosion that occurred between the survey in August 2012 and the GE image, acquired 3 weeks later on 7 September, when the coast next to the northern cape was eroded by up to −7.9 m, because of block failure and collapse due to the deep thermo-niche that existed before in this area (Fig. 14, right), leaving ground ice debris on the sand spit (Fig. 14, left). Despite strong east winds shortly following over 11–13 September 2012 (Fig. 12), TA until July 2013 was only −1.9 m at the cape.

4.2.2 Current thermo-denudation

Using a chronologically consecutive approach of frequent closely spaced TD measurements, we identified seasonal variations for 2010 and interannual variations during the 2010–2013 period. Using intervals of 14–437 days, we dealt with the problem of incomparability of land loss measurements along the cliff top line. The first two consecutive periods 1 and 2 were both 14 days long, while mean absolute TD had doubled from −0.47 ±0.51 in period 1 to −0.98 ±0.52 m in period 2. Period 3 was longer (26 days) and consequently absolute TD was −2.13 ± 0.95 m. Period 4, as the last period which extended over less than a year, absolute TD decreased to −1.84 ± 1.22 m, despite its longer lasting duration of 324 days, 83 days of which have to be considered as the TD active season (50 days in 2010, 33 days in 2011). Period 5 showed mean absolute TD of −5.18 ± 2.88 m over 437 days, 201 days of which nevertheless fell into the TD active season. Period 6 was characterised by very slow TD of −0.3 ± 0.47 m, which is a mixed signal of the remaining 22 days of TD activity in 2012 and 44 days in early 2013. In other words, almost 50 % of the observed absolute TD during the last 3 years happened in 2010.

In order to compare TD intensity across the purely seasonal periods (1–4 and 6), rate calculation of erosion over time through simple annualisation failed, due to large overestimations. For example, simple annualisation would result in mean TD of −29.9 ±13.3 m a⁻¹ during period 3 (summer), while an incredible rapid single value of −79.7 m a⁻¹ would have been measured in period 1 (starting TD in spring) at the northern cape. As a purely descriptive metric, normalisation of TD rates based on specific sums of PDD and OWD for a certain period showed different results for both environmental parameters. OWD-sum-normalised mean TD rates for all periods ranged from −0.2 to −6.6 m a⁻¹ and were on average −2.6 ± 1.4 m a⁻¹ (Fig. 15, right). OWD sum normalisation of TD worked for periods that cover at least one

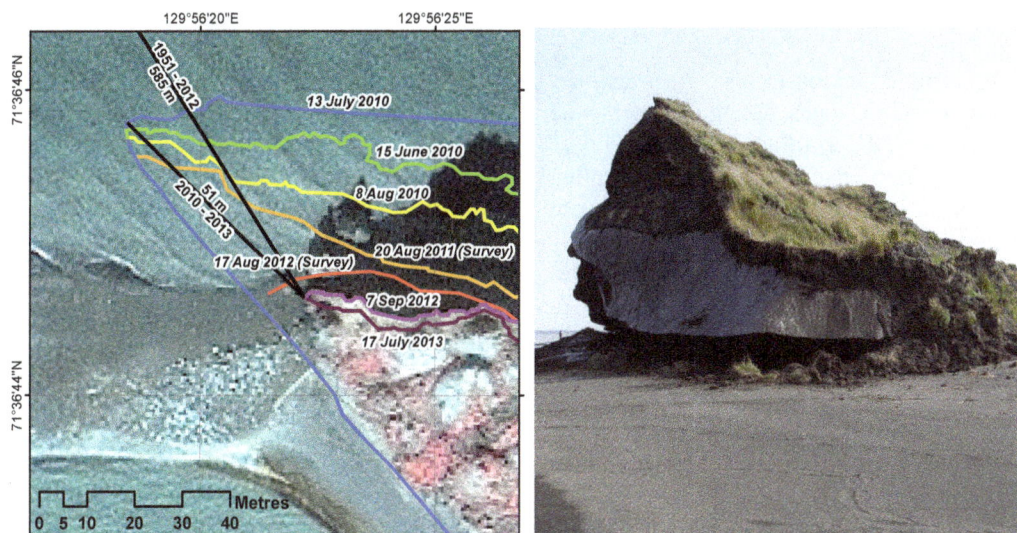

Figure 14. Close up view of coastal erosion at the northern cape of Muostakh Island. Left: selected cliff top position lines of the 2010–2013 period. Blue line outlines cliff bottom position in early 2010. Ends of cliff top lines are at sea level, marking the northernmost point of the island. Note stranded ground ice debris on the sand spit. Right: photograph of the northern cape (8 August 2012).

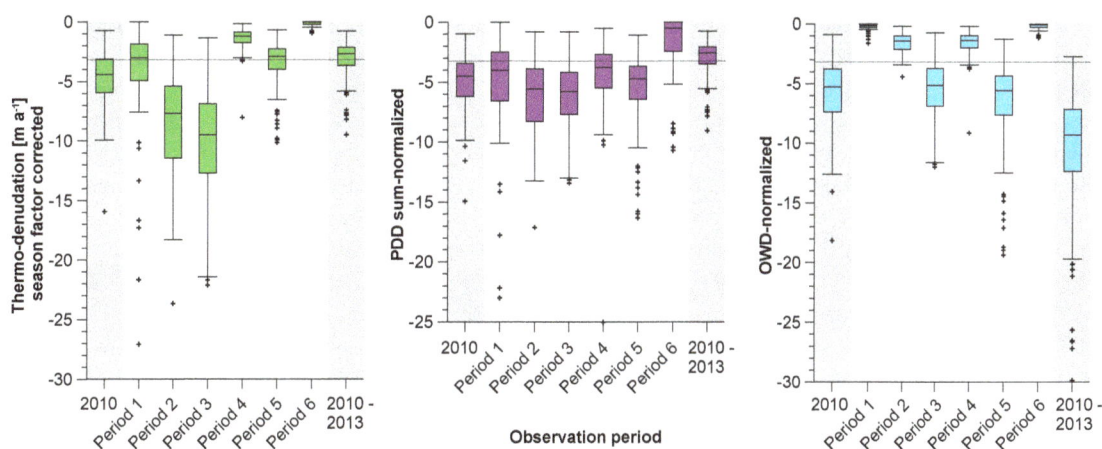

Figure 15. TD rates for eight different time periods: 2010, 1 (spring 2010), 2 (summer 2010), 3 (summer 2010), 4 (fall 2010), 5 (2011–2012), 6 (fall 2012–spring 2013), and the entire observation range 2010–2013 (with a mean rate of $-3.1 \pm 1.6\,\mathrm{m\,a^{-1}}$ included as horizontal reference baseline for current TD). Left series: correction to the length of the TD-active season (number of days with mean $T_{air} > 0\,°C$) during the observation period yield actual TD velocity. Middle: TD rates normalised using PDD sums for each period, which produces roughly comparable rates. Right series: similar normalisation using OWDs.

sea ice cycle and for period 3, as the only period when TD and TA proceeded unrestricted simultaneously. PDD-sum-normalised rates ranged from -1.7 to $-6.3\,\mathrm{m\,a^{-1}}$ and were on average $-3.5 \pm 1.6\,\mathrm{m\,a^{-1}}$ (Fig. 15, middle). The very low standard deviation of $\sigma = 1.6\,\mathrm{m\,a^{-1}}$ compared to initially $\sigma = 11.5\,\mathrm{m\,a^{-1}}$ for simply annualised rates, indicates a high degree of levelling between periods and demonstrated TD's dependency on T_{air}.

Based on this finding, we carefully applied a season correction factor into TD rate calculation that accounts for the fraction of number of days with positive mean daily tempera-ture occurrence during the observation period compared with the total annual number (for periods shorter than 1 year) or the annual mean total number (for periods longer than 1 year) of these days (Table 3). Accordingly, the season factor must be ≤ 1. The resulting season-factor-corrected rates were considered to be the actual TD velocity that had taken place. This approach was validated by the fact that the reference annual erosion cycle 2010 (29 June 2010–28 June 2011) turned out to be not affected by the correction approach, since simply annualised and season-factor-corrected TD rates over 2010 were equal, because of a season factor of 1 (Table 3). The spatio-temporal comparison of TD rates for all periods is shown on maps in Fig. 16.

Based on comparable season-factor-corrected rates (Fig. 15, left), we found TD rates during summer were always more rapid (from -8.7 ± 4.6 to $-10.2 \pm 4.5\,\mathrm{m\,a^{-1}}$ for periods 2 and 3, respectively) than during spring with $-4.1 \pm 4.5\,\mathrm{m\,a^{-1}}$ (period 1) or during fall with $-1.4 \pm 0.9\,\mathrm{m\,a^{-1}}$ (period 4) or even $-0.2 \pm 0.2\,\mathrm{m\,a^{-1}}$ (period 6) (Fig. 16). Analyses of interannual variations revealed more rapid TD rates over 2010 of $-4.8 \pm 2.3\,\mathrm{m\,a^{-1}}$, compared to $-3.4 \pm 1.9\,\mathrm{m\,a^{-1}}$ during 2011–2012. The 3 years of TD observations, from 2010 to 2013, revealed a mean rate of $-3.1 \pm 1.6\,\mathrm{m\,a^{-1}}$ (included in Fig. 15 as reference baseline for current TD), which is a little slower than TA over the same period (Fig. 11, right). Although rapid, it is evident that even on a short-term scale the longer the observation period, the more consistent TD rates will be.

The quality of current thermo-erosion was assessed using the NDTI according to Günther et al. (2012). It turned out that 55 % of the segments from 2010 to 2013 eroded under prevailing TD with an average NDTI of 0.39, while the largest positive TD-indicating NDTI values were observed in the middle of the island, parallel along both the western and eastern coasts with NDTI > 0.5. Along TA-dominated segments, NDTI was -0.34, while on the north-eastern coast, where erosion is most rapid, NDTI was closer to zero, meaning constant erosion of large volumes (Fig. 11, right).

We related positive mean daily T_{air} to TD and TA rates of all observation periods. As a result, we found that a continuous increase of mean daily T_{air} by 1 °C throughout the TD active period is responsible for an acceleration of coastal erosion by $-1.2 \pm 0.55\,\mathrm{m\,a^{-1}}$ (Fig. 17).

4.3   Environmental parameters

4.3.1   Open water days

Open water fraction ($\%\,\mathrm{d^{-1}}$) was calculated based on SSM/I sea ice concentration data for the past 22 years from 1992 to 2013 in order to identify OWDs and to understand basic characteristics and interannual variability of sea ice in the Buor Khaya Gulf. Background noise present in the data was quantified as $11 \pm 5.5\,\%\,\mathrm{d^{-1}}$, by evaluating the open water fraction from December until April, when the land-fast ice zone can be assumed to be completely covered with sea ice, and the sea ice concentration of August, when the coastal waters can be assumed to be free of sea ice. The influence of the perennial Laptev Sea polynya (Reimnitz et al., 1994; Dmitrenko et al., 2005) on the open water data was excluded by taking a 100 km reference zone (Fig. 6). Meier and Stroeve (2008) found SSM/I sea ice concentration data of 15 % matches with the sea ice edge location for the Laptev and East Siberian seas. Adding $11\,\%\,\mathrm{d^{-1}}$ uncertainty due to underestimation, we assumed days with $< 26\,\%$ within the 100 km to be OWD.

As first-order approximation, we assumed the mean annual open water fraction as indicator for seasonal duration available for TA. Over the past 2 decades, mean open water fraction was $32.1\,\%\,\mathrm{d^{-1}}$, corresponding to 117 OWD $\mathrm{a^{-1}}$ in the Buor Khaya Gulf, including break-up and freeze-up transition periods. Based on the $< 26\,\%\,\mathrm{d^{-1}}$ threshold, the core open water season is on average $81 \pm 15\,\mathrm{d\,a^{-1}}$ long and lasts from 21 July until 8 October (Fig. 18). The difference between 117 and 81 OWD probably reflects 36 days of sea ice drift. The open water season for the current investigation period 2010–2012 was 96 ± 2.5 days long and lasted from 11 July to 14 October. For example Fig. 6 shows Muostakh and the surrounding coastal waters during the very late break-up in 2013. Although the open water season was short in 2013 (73 days), it featured the latest freeze-up (Fig. 18) and was another unusual year, also in terms of mismatch between the TD and TA active seasons (Fig. 12).

4.3.2   Positive degree days

Time series of T_{air} in Tiksi were used to calculate annual PDD sums in Kelvin days (Kd) over the historical time period from 1951 to 2013. The mean seasonal duration based on the first and last occurrence of positive mean daily T_{air} is 133 days, where the season accordingly starts around 17 May and ends on 27 September. However, during this period, days with negative mean daily T_{air} still occur. Mean annual PDD sum over the last 62 a was 660 Kd, the number of days with positive mean T_{air} occurrence was 112 and, accordingly, positive mean daily T_{air} was 5.9 °C. Accounting for the difference of 21 days and assuming a spring–fall partition coefficient of 1 : 3 according to Fig. 19, the season available for TD starts roughly on 1 June and ends on 21 September. Current seasonal and interannual variations of PDD sums from 2010 to 2012 showed that the mean PDD sum was 922 Kd, with a mean daily T_{air} of 7.3 °C, corresponding to seasonal duration of 126 days (Fig. 19). In 2012, the annual PDD sum reached 1010 Kd and exceeded 1000 Kd for the first time in the period of record in Tiksi. A total of 134 days with positive mean daily T_{air} also lengthened the overlap period between the TD and TA active seasons (Fig. 12).

4.3.3   Wind

Wind speed data was cross-checked with data on sea ice concentration and T_{air} for the 2010–2012 study period and for the time series overlap since 1992. During our current study period strong breezes with wind speeds $\geq 10\,\mathrm{m\,s^{-1}}$ were observed twice during the open water period of 2010 and once in 2011, 2012, and 2013 each (Fig. 12). Severe storm events with wind speeds $\geq 24.5\,\mathrm{m\,s^{-1}}$, measuring 10 or higher on the Beaufort wind force scale, usually occurred only during the winter. Over the previous decades (since 1951), during a generalised open water season from 15 July to 15 October, severe storm events in the Buor Khaya Gulf could be

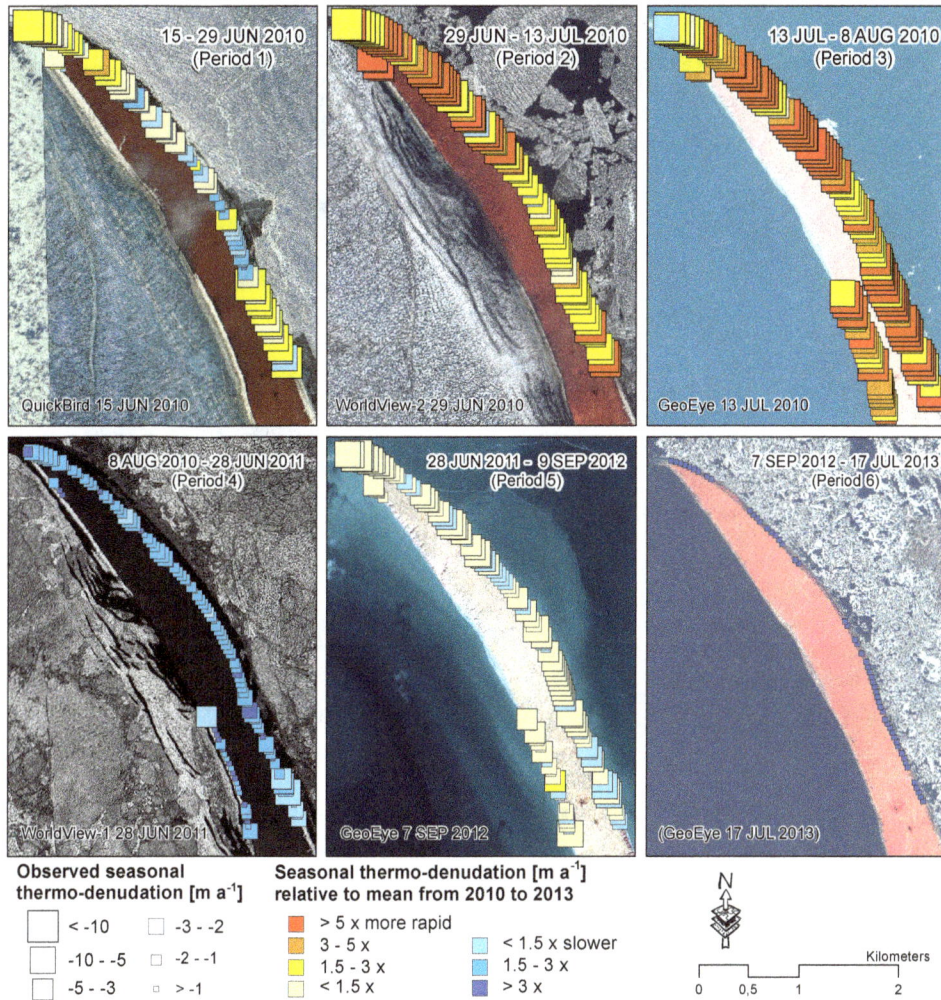

Figure 16. TD during current observation periods from 2010 to 2013 using very high resolution remote sensing data. Symbol size is equivalent to season-factor-corrected TD rates in metres per year (periods 1–6), colour coding expresses proportion between observed seasonal TD rates and mean TD rates over the July 2010–July 2013 period. Seasonal observations show differences of varying TD intensities during short summer (period 3) vs. fall (periods 4 and 6), while interannual observations revealed homogeneous erosion (period 5).

expected to recur at most every 5 years. They had almost exclusively SW direction (Fig. 20), while only < 10 % had northern direction, causing the water level to rise in the Buor Khaya Gulf. Our current study period falls into this storm gap. Generally 6 h wind directions during the TD-active season were diverse, but with prevailing winds from north and north-east and an intensification of strong east winds during the current observation period, which probably enhanced wave action and turbulent heat flux on Muostakh's erosive east facing coast (Fig. 20). Also of note is the increase of T_{air} during the 2010–2012 period that is associated with southerly winds. Despite dominant winds from the north during the sea ice break-up period, usually larger open water fractions and sea ice export out of the Buor Khaya Gulf were associated with winds from the south. In contrast, during the current observation period winds from the north-east and east seem to have favoured sea ice break-up (Fig. 20).

4.4 Ground ice and sediment budget

Average macro ground ice content in the subsurface was calculated as $44 \pm 4.6\%$ ($n = 1264$) by volume. The sediment occupied 56 % of the ground volume. According to Schirrmeister et al. (2003, 2011), the gravimetric ground ice content of Ice Complex sediments on Muostakh between 0 and 10 m a.s.l. is 108 wt %. Taking this value for the entire Ice Complex series and assuming different densities for ice and solid material according to Strauss et al. (2012), 76 % of the sediment volume is occupied by intrasedimentary ground ice. Combining these 43 % intrasedimentary and 44 % macro ground ice, altogether, 87 % by volume of the geological subsurface is composed of ground ice.

We applied the concept of critical ice content of Are (2012), using the newly available information on topography and ground ice contents. The relative subsidence potential

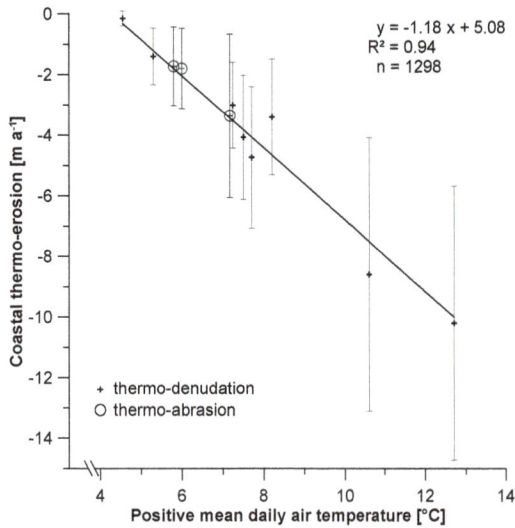

Figure 17. Mean annual coastal thermo-erosion (eight data points for TD and three for TA) vs. positive mean daily T_{air}. Each erosion data point is a mean of all 118 coastal sections; error bars indicate standard deviation of rates. T_{air} was calculated for specific observation periods using only days with mean T_{air} >0 °C.

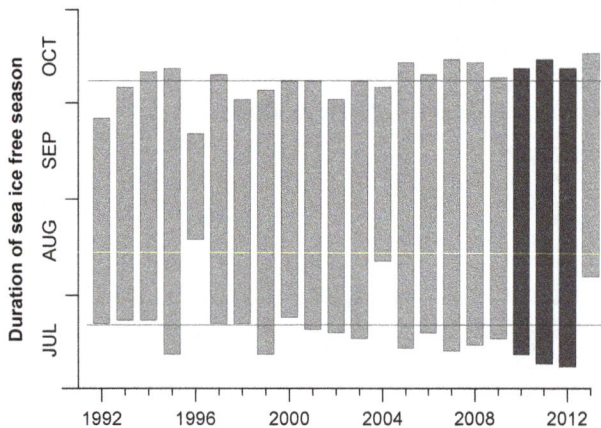

Figure 18. Seasonal duration of the sea-ice-free period 100 km around Muostakh Island in the Buor Khaya Gulf. Top and bottom horizontal lines indicate mean start (21 July) and end (8 October) days according to the reference time period 1992–2013 with mean seasonal duration of 81 days.

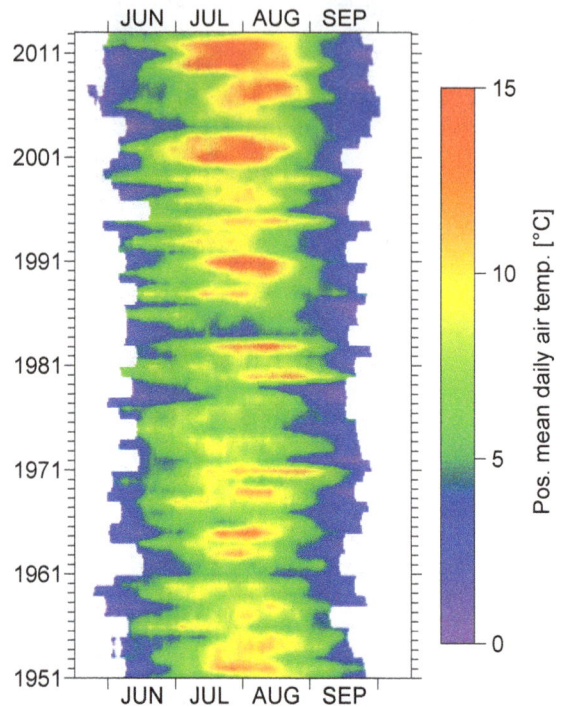

Figure 19. Time series of positive mean daily T_{air} over the period from 1951 to 2013 show lengthening and intensification. Seasonal duration available for thermo-denudation lengthened from 111 days for the entire period to 127 days on average for complete years of the current observation period (2010–2012). Simultaneously, positive mean daily T_{air} was currently 1.4 °C higher and increased from 5.9 to 7.3 °C.

was calculated as $\delta z = 0.69$. According to field observations, the ice-poor surface layer was on average 0.7 m thick, the upper two-thirds of which is the active layer. The surface layer was assumed to be mainly composed of organic soil material and therefore has been deducted from the depth of the Ice Complex section. For the northern part of the island, where the base of Ice Complex deposits was detected at 10 m below sea level (Kunitsky, 1989; Grigoriev, 1993) and cliff height is 21 m a.s.l., potential subsidence of the entire 30.3 m Ice Complex thickness equals 20.9 m. If the ground ice would completely melt out, the top of the remaining thawed mate-

rial would be situated exactly at sea level. As a possible scenario for the southern part of the island, where no information on the lower Ice Complex boundary is available, we applied the subsidence factor ($\delta z = 0.69$) to the mean elevation of the island of 14.4 m, deducted the surface layer, and assumed the Ice Complex base at sea level. Accordingly, potential subsidence is 9.5 m, resulting in a top of thawed material 4.3 m a.s.l. Both scenarios demonstrate that thawing results in a much reduced volume of sediment to be removed from the cliff bottom by waves. In the first scenario, where subsidence would extend almost down to sea level, Are (2012) emphasises that this would mean coastal thermo-erosion may proceed almost exclusively based on its thermal component.

In addition to subsidence, ground ice can result in thermo-erosional niches at the water level that undercut the ice-rich cliff. Their theoretical maximum depth is determined by the ice-wedge polygon size (Hoque and Pollard, 2009). Further investigation of the Voronoi diagram was done with a sub-sample of ice-wedge polygons that are entirely surrounded by other polygons. As a result, each polygon has 5.6 neighbours, on average, which corresponds to the common hexag-

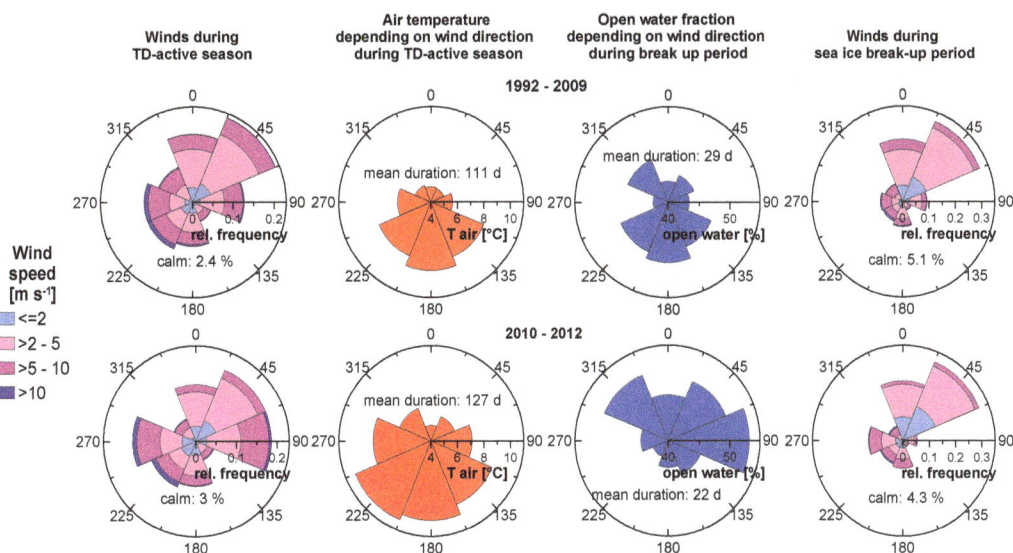

Figure 20. Wind charts and rose diagrams for the long-term reference time period (top row) and the current observation periods (bottom row). Left: wind speed and direction during days with mean daily positive air temperature (T_{air}), defined as season available for TD to proceed. Right: wind speed and direction during sea ice break-up, defined as time period when open water fraction increases from 30 to 90 %. Middle left and right: mean T_{air} and mean open water fraction depending on prevailing wind direction during TD-active season and sea ice break-up, respectively. Note, although directed away from the erosive coastline, winds from the south also provide favourable conditions for coastal thermo-erosion through higher T_{air} and enhanced sea ice export.

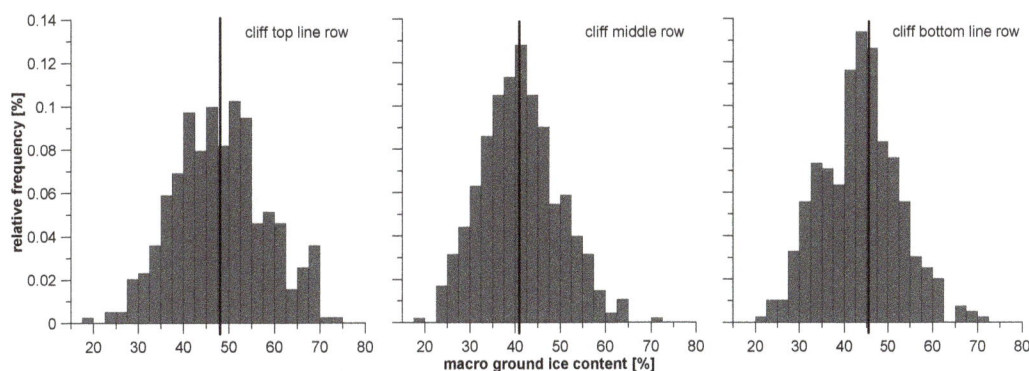

Figure 21. Histograms of macro ground ice content along erosive coastal cliffs on Muostakh Island. Classification of ice-wedge polygons into three different vertical positions shows shift towards higher macro ground ice contents for the lower and upper parts of the subsurface. Vertical lines indicate mean values.

onal form of ice-wedge polygons (Christiansen, 2005). They have a mean edge length of 9 ± 4.8 m, occupy a mean area of 162 m^2, and have a mean diameter of 14.2 ± 2.2 m. Ice-wedge polygon rows along the cliff bottom and top showed slightly higher macro ground ice contents of 43.5 ± 8.7 and 47.9 ± 9.9 vol. %, respectively, compared to 41.1 ± 8.5 vol. % of the middle row (Fig. 21).

5 Discussion

5.1 Permafrost thaw subsidence

The volumetric content of ground ice in unconsolidated permafrost deposits in East Siberian coastal lowlands exceeds their pore volume in the thawed state (Yershov, 2004), and consequently subsidence results when thawing occurs. The mean annual permafrost thaw subsidence rate of -5.8 ± 2.9 cm a^{-1} is not randomly or uniformly distributed across the island but varies with geomorphology. In particular, proximity to erosive coastline segments seem to favour thaw subsidence in the island's interior, where for example rates of $\geq$

$-11\,\mathrm{cm\,a^{-1}}$ were observed close to the northern cape, eroding at $\leq -10\,\mathrm{m\,a^{-1}}$. Similarly, Short et al. (2011) showed -10 to $-15\,\mathrm{cm}$ of terrain displacement during summer 2010 near Collinson Head on Herschel Island, a location of constant rapid coastal erosion rates (Lantuit and Pollard, 2008).

Remote sensing data showed that seasonal thaw settlement in Alaska is in the range of 1–4 cm (Little et al., 2003) and can be up to 12 cm in places (Liu et al., 2014), while in situ long-term observations of permafrost thaw subsidence are 0.8–$1.7\,\mathrm{cm\,a^{-1}}$ (Shiklomanov et al., 2013). Generally, long-term subsidence occurs by thaw at the top of ice-rich permafrost and drainage of meltwater. Yedoma uplands tend to water drainage and deeper thaw compared to alas depressions (Fyodorov-Davydov et al., 2004). Coastal erosion on Muostakh Island maintains a steep hydrological gradient so that rapid drainage is favoured. Increases in PDD sums result in deeper thaw (Burn, 1998) and permafrost warming (Jonsell et al., 2013). In 2012, we saw the highest PDD sum (1010 Kd) for the entire T_air record. Accordingly, ALT on Muostakh in 2012 was on average $47 \pm 19\,\mathrm{cm}$. This agrees with a mean thaw depth of 49 cm at the end of August on Samoylov Island in the Lena Delta (Boike et al., 2013). For the nearby Bykovsky Peninsula, where ALT on yedoma uplands varies by climate zone, Fyodorov-Davydov et al. (2008) document a mean seasonal thaw depth of 32 cm during 2003–2006. During these 4 years, the mean annual PDD sum was 670 Kd, very close to the long-term average of 660 Kd but clearly below the average annual PDD sum for the last decade of 775 and 922 Kd for the last 3 years, when ALT on yedoma surfaces increased by 15 cm.

Our comprehensive observations of intensified thaw subsidence in places of shallow ALT (Fig. 10) indicate the opposite to modelling estimations of ALT as a function of surface subsidence by Liu et al. (2012), but are in accordance with the results of ground-based ALT measurements corrected for thaw subsidence by Shiklomanov et al. (2013), and suggest near-surface occurrence of ground ice in areas of strong subsidence. Where deeper thaw encounters ice-rich basal soil horizons or ice-wedge tops, this results not only in active layer deepening, but also in subsidence. Since the upper Ice Complex is ice-rich, increased heat flow into the ground will cause the island to subside. We therefore suggest that the widespread occurrence of peat mounds particularly in the northern part of the island, which present mainly polygonal centres (Fig. 22), is associated with subsidence of the surrounding terrain, rather than with frost heave.

Comparison of our on-site survey with elevation indications from the literature supports the observation of decreasing height of the island. Generally, backshore elevation along the east coast ranges from 21 m a.s.l. in the north to 13 m a.s.l. in the south. Around the former polar station the mean elevation is 6 m a.s.l. Ivanov and Katasonova (1978) report that the height of the island gradually decreases toward the south from 26 to 6 m. Slagoda (2004) presents profiles of Ice Complex sequences that were sampled in 1982 and reached up to

Figure 22. Photograph of peat mounds (incipient baydzharakhs) in the northern part of Muostakh, where the island's surface is affected by strong permafrost thaw subsidence (person for scale).

25 m a.s.l. Topographic maps showed the highest point of the island was 25 m a.s.l. in 1982. Grigoriev (2008) presents data of continuous on-site visits since the 1990s and takes 22.6 m as reference height for the north cape of Muostakh. Although all former elevation indications refer to the north cape, it is unlikely that only this portion of the island had an elevation ≥ 25 m a.s.l., but subsidence must have occurred mostly during the last 30 years, particularly in view of the fact that positive deviations from mean PDD sums in NE Eurasia occurred since 1988 (Fedorov et al., 2014). However, our 3-D data set spans 62 years, and estimates of annual subsidence were around $-5.8 \pm 2.9\,\mathrm{cm\,a^{-1}}$ over the entire area affected by negative terrain height changes. This is of the same order of magnitude observed elsewhere for ice-rich permafrost (Overduin and Kane, 2006; Fedorov et al., 2012).

5.2   Historical and current erosion development

Recently, TA proceeded at $-3.4 \pm 2.7\,\mathrm{m\,a^{-1}}$ and was therefore 1.9 times more rapid over the past 3 years than the historical record with mean TA of $-1.8 \pm 1.3\,\mathrm{m\,a^{-1}}$. This proportion is consistent with observations made by Günther et al. (2013a), who find recent rates are at least 1.6 times more rapid than historical TD and TA along Ice Complex coastlines throughout the Laptev Sea. They report that for a subset of sites TA accelerated from long-term $-3.3 \pm 1.2\,\mathrm{m\,a^{-1}}$ to $-5.7 \pm 1.2\,\mathrm{m\,a^{-1}}$ over the past few years. When examined over almost 200 km of Ice Complex coastline and the last 42 years, they find long-term TA was $-1.9 \pm 1.5\,\mathrm{m\,a^{-1}}$, almost identical to Muostakh. This suggests that, despite the annually repeating record-breaking erosion rates on the northern cape, coastal thermo-erosion on Muostakh is not exceptional for Ice Complex coasts in the Laptev Sea. Deviations from the mean can be attributed to local variations of macro ground ice content and exposure to environmental parameters such as to offshore or bay–marine environments.

Since its base area shrank by a quarter (24 %) during 62 years, Muostakh Island will disappear within the next 200 years. If the currently observed erosion rates continue, Muostakh is likely to disappear as island earlier, but no later than within the next 100 years. Examples for disappearing Ice Complex islands on the Laptev Sea shelf exist. According to Gavrilov et al. (2003), the former islands Diomede, Semenovsky, and Vasilievsky have become sandbanks with frozen sediments located very close to the seafloor surface, where thermo-abrasion is still active and proceeds with approximate rates of -0.02 to $-0.27\,\mathrm{m\,a}^{-1}$. The subsequent submergence of arctic islands results in shoals on the shallow arctic shelf, grounded sea ice pressure ridges (Reimnitz et al., 1994), and loss of island status, as it happened to Dinkum Sands off the Alaskan Beaufort Sea coast (Reimnitz, 2005). Klyuev et al. (1981) report on the shape of Vasilievsky Island in 1823, when the island had a length of 7.4 km and was 463 m wide, quite similar to Muostakh today. As on Muostakh, erosion on Vasilievsky was most intensive along its major axis. Because of its narrow shape, Vasilievsky broke into two parts and was quickly destroyed afterwards due to the unstable ground-ice-rich composition of the island and to the chaotic erosion with rates of up to $-100\,\mathrm{m\,a}^{-1}$ that occurred from all sides until the island was completely destroyed in 1936 (Klyuev et al., 1981). Vasilievsky Island was located quite far offshore and exposed to larger fetch on all sides. Muostakh Island has a narrow central section where we measured TA on both coasts. On the west-facing side, historical and modern rates do not differ greatly ($-0.8 \pm 0.6\,\mathrm{m\,a}^{-1}$). On the east-facing side, however, modern rates ($-2.3 \pm 1.5\,\mathrm{m\,a}^{-1}$) are more than twice the historical mean ($-1.0 \pm 0.2\,\mathrm{m\,a}^{-1}$). Changes to sea ice cover are probably affecting the east-facing coast more, since it is exposed to Buor Khaya Gulf's comparatively large fetch. Possible future break-up of the island will probably occur at this location. Complete disappearance of Muostakh would take away a protection of the Tiksi port harbour approach, but not enhance erosion of the rocky mainland coast.

5.3 Interannual and seasonal variability of erosion

Very rapid TD rates of $-4.8 \pm 2.3\,\mathrm{m\,a}^{-1}$ in 2010 were followed by slower rates of $-3.4 \pm 1.9\,\mathrm{m\,a}^{-1}$ during the 2011–2012 period. Figure 3 suggests that the coastal exposure undergoes geomorphic changes from a high degree of exposed ground ice to an almost complete coverage of ground ice by thermal denudation debris. Remote sensing derived TD is also in agreement with TD from our on-site repeat surveys. According to this control data set TD was $-2.7 \pm 0.6\,\mathrm{m\,a}^{-1}$ for 2.8 km of the north-eastern coastline from August 2011 to August 2012. For the entire 2010–2013 cycle we therefore suggest that a constantly high positive mean daily T_{air} in 2010 (Fig. 19) resulted in rapid TD, depositing thawed material and obscuring exposed ground ice in the following year (Fig. 3, top photograph).

Figure 23. Photograph of catastrophic collapse through undercutting, reflecting activation of TA during the rising water level at the end of August 2011, when sea water comes in contact with permafrost at the level of thermo-niches. Under calm weather conditions the ground ice block decayed within 2 days.

In 2011, during the very long open water season of 99 days (Fig. 18) TA reworked the coast to predominant steep cliffs (Fig. 23). The main mechanism was block failure due to undercutting by thermo-niches, which probably formed during the strong wind events that occurred in the open water period in 2010 (Fig. 12). However, according to Wobus et al. (2011), thermo-niches develop even under quiet sea conditions. If they have formed during storms, the effect on erosion is combined with a certain delay and consequently not immediately measurable with remote sensing techniques. Maximum thermo-niche development is constrained by ice-wedge polygon size. The continuous presence of thermo-niches was observed in the field exclusively along the north-east and east-facing coast, where backshore height is $\leq 21\,\mathrm{m}$ and syngenetic ice-wedges generally stretch across the whole vertical section, indicating that block failure occurs along the vertical plane according to Hoque and Pollard (2009). Given a mean ice-wedge polygon size of $14.2 \pm 2.2\,\mathrm{m}$ and mean TA of $-4.4 \pm 2.3\,\mathrm{m\,a}^{-1}$ in that area, erosion of one complete polygon block would last for 3 years, and our observations captured this time frame within the 2010–2012 period. Around the northern cape, current TA was $-11.6\,\mathrm{m\,a}^{-1}$ (the most rapid segment in Fig. 11), so that erosion of one complete polygon block takes roughly 1 year there. In general, we suggest that coastal erosion rates are linked to geomorphology particularly through the varying macro ground ice content and its subsurface distribution.

The established dependency of coastal thermo-erosion on T_{air} (Fig. 17) and the seasonally varying intensity of thaw on coastal ice cliffs agrees with other observations. Grigoriev et al. (2006) link coastal dynamics of East Siberian coasts to changes in climatic and permafrost conditions. They find a positive correlation between coastal retreat rate and macro ground ice content, especially when T_{air} exceed

+4 °C, which also seems to be a threshold for TD activity on Muostakh (Fig. 17). In West Siberia, where massive ground ice occurs along the arctic coast, Vasiliev et al. (2006) find that coastal retreat was 2 times higher where ground ice content was 45 %, when compared to places with only 25 %. Lantz and Kokelj (2008) studied retrogressive thaw slumps in the Mackenzie Delta region and find that an increase in mean summer T_{air} by 1.3 °C leads to a 1.4 times more rapid general slump growth and an almost doubling of the slump headwall retreat. Our result of currently 1.9 times more rapid erosion is also consistent with recently observed 1.6 times more rapid coastal erosion rates in the Laptev Sea region by Günther et al. (2013a). Wobus et al. (2011) used time-lapse photography to study thermo-erosion along Alaska's Beaufort Sea coast. They report thaw rates of $1-6 \, cm \, d^{-1}$ during spring, prior to sea ice break-up. During our periods 1 and 2 (15 June–13 July 2010), both lasting for 14 days (Fig. 16, top left and top middle), absolute cliff top line retreat was 0.47 and 0.98 m, and mean daily T_{air} was 6.1 °C and 10.6 °C, respectively (Table 3). With respect to largely exposed ground ice, this corresponds to thaw rates of $3-7 \, cm \, d^{-1}$ or ablation rates of $5.5-6.6 \, mm \, d^{-1} \, °C^{-1}$, according to Braithwaite (1995). Wobus et al. (2011) also observe acceleration after open water season begins, which is consistent with more rapid rates on Muostakh during period 3 (13 July–8 August 2010), the only phase when TD and TA proceeded simultaneously, and thaw was on average $8 \, cm \, d^{-1}$ and, in places, reached $17 \, cm \, d^{-1}$ under a mean daily T_{air} of 12.7 °C. Thaw was slowest during spring 2013 at $0.4 \, cm \, d^{-1}$. The niche and block erosion model of Ravens et al. (2012) turned out to be sensitive primarily to meteorological parameters, not to sea ice extent. Although we did not account for the important energy supply through solar irradiation during sunny and overcast days (Lewkowicz, 1986), nor for sea surface temperature, our TD normalisation efforts also showed TD sensitivity to T_{air} rather than to OWD (Fig. 15). Future studies should extend seasonal analyses to TA, to better account for marine forcing.

5.4 Changes in environmental parameters

The recent lengthening of the sea-ice-free season by 15 days from 81 (1992–2013 mean) to 96 days on average for the 2010–2012 erosion observation period was due to 10 days of prolongation in early summer and 5 days of later freeze-up within the Buor Khaya Gulf. This is in accordance with a general trend of accelerating early ice retreat in the Laptev Sea, probably because of thinner sea ice (Krumpen et al., 2013). Based on our observations, on average, the open water season starts on 20 July. Despite the fact that interannual variations are considerable (35 OWD in 1996 vs. 99 OWD in 2011), Karklin and Karelin (2009) report that negative anomalies of earlier break-up became a rule from 1999 to 2005. Our data show a continuation and strengthening of this trend (Fig. 18), parallel to rising T_{air} (Fig. 13). This also adds

perspective to the trend of OWD increase of $0.5-1 \, da^{-1}$ observed by Barnhart et al. (2014) for coastal cells in the Laptev Sea with significant trends over the entire 1979–2012 SSM/I data set. Overeem et al. (2011) show that open water duration has been increasing along the Alaskan Beaufort Sea coastline from icy to ice-free over decades based on SSM/I ice cover calculations. In our case, most likely, the warm Lena River waters are likely to additionally support local seasonality via early break-up, because of generally thinner sea ice (Spreen et al., 2011) and earlier spring floods (Federova et al., 2009). Even around Muostakh, spatial variations of sea ice conditions are pronounced, where the rapidly eroding east-facing coast can be completely free of ice, while on the west-facing coast, towards the protected Tiksi Bay, sea ice may persist. According to Gukov (2001), the waters west of the island are ice-covered for 12 days longer than in Buor Khaya Gulf. However, depending on either thermal or mechanical sea ice break-up (Petrich et al., 2012), the opposite picture might emerge, as it was in 2013 (Fig. 6).

Wind speeds at Tiksi range between 0 and $25 \, m \, s^{-1}$. Strong winds exceeding $10 \, m \, s^{-1}$ occur almost exclusively during the winter months, with continuous sea ice cover (Fig. 12). Adding to this seasonal protection of the island from wind-driven wave action is the fact that almost all high wind speed events come from the south-west to south direction (Fig. 20), causing water level to fall. Even when ice-free, the maximum fetch south to south-westward of Muostakh is less than 50 km. In terms of constant heavy swells, the earlier start of the open water season might have strong impacts on TA on Muostakh Island, since winds during the sea ice break-up period are almost exclusively from the north and north-western directions, where Muostakh is exposed to nearly unlimited fetch, causing water level to rise.

The very warm summer of 2010 with a positive mean daily T_{air} of 7.7 ±5.3 °C, during 124 days with positive mean T_{air} occurrence, was also accompanied by high net radiation (Boike et al., 2013). In 2011, T_{air} was lower (6.5 ±4.9 °C over 121 days) but winds were stronger, enhancing the convective heat transfer and maintaining melt on ground ice exposures. According to Langer et al. (2011), wind speed in the Lena Delta features a diurnal pattern, with enhanced heat exchange during the day and lowered turbulence during the night. In the very warm summer of 2012, positive mean daily T_{air} was 7.5 ±4.7 °C over the 134-day long TD-active period. Mean July T_{air} of 12.5 °C in 2010, 10.7 °C in 2011, and 11.5 °C in 2012 were higher than the long-term mean July T_{air} of 7.5 °C (1951–2012). Thus, not even the criterion for a subpolar climate, that T_{air} is below 10 °C in the warmest month (Neef, 1956), was met over these 3 years. Besides this warming in the tundra zone, Fedorov et al. (2014) report that the greatest recent summer warming was observed in the adjacent forest tundra and northern taiga.

5.5 Macro ground ice as local controlling factor

Our estimates of 87 % total volumetric ground ice content on Muostakh are slightly above previous specifications of 80 % (Are, 1988a; Slagoda, 2004). Our polygon mapping approach allows sediment centres to touch the polygon's boundary and accordingly mapped ice wedges may have zero width in places. Thus, our geomorphometric method probably overestimates polygon sediment centre size, resulting in conservative rather than excessive assessments of macro ground ice content. Since the sediment on Muostakh is one of the most coarsely grained and poorly sorted examples of Ice Complex (Slagoda, 2004; Schirrmeister et al., 2011), baydzharakhs there are better preserved following thawing, in contrast to those elsewhere which quickly slump or undergo transport (Are et al., 2005).

The shape of syngenetic ice wedges in a vertical plan often deviate from the ideal wedge form (Popov et al., 1985). Generally, 2–4 rows of baydzharakhs could be observed on the slope. Our results imply that macro ground ice distribution has a non-uniform vertical hourglass shape, with higher ice contents and ice wedge widths at the top and bottom of the slope, and probably below sea level. Variable sedimentation and preservation conditions during the Late Pleistocene defined Ice Complex accumulation rates (Wetterich et al., 2011). Thaw unconformities and several ice-wedge generations that are nested in one another also caused varying ice-wedge width across the profile. Field observations also confirm that Pleistocene syngenetic ice wedges on Muostakh Island are thickest at the bottom (rather constant width and not increasing width) and at the top of the coastal exposure, where at the upper 4 m of the profile Holocene ice wedges are often interposed. The higher ice contents thus occur at the positions where we measure coastal erosion and are predisposed to favour intense coastline retreat as a result of warming (TD) and wave action (TA).

6 Conclusions

In this study, we found that continuous coastal erosion on Muostakh Island in the Buor Khaya Gulf of the Laptev Sea during the last 62 years caused the land loss of about 24 % of the island's area, while its volume shrank by 40 %. Muostakh is composed of Ice Complex permafrost deposits, of which up to 87 % of the subsurface is occupied by ground ice. This exposes the island to thermal disturbances from coastal erosion and seasonal thaw and permafrost thaw subsidence, leading to further destruction of the island and its final disappearance, which we expect to take place within the next 100 years under recently changing environmental conditions. Average subsidence of the island's surface was -3.6 ± 1.8 m over 62 years and therefore $-5.8 \pm 2.9 \mathrm{cm\,a}^{-1}$. Average coastal erosion from 1951 to 2012 was $-1.8 \pm 1.3 \mathrm{m\,a}^{-1}$, while recent rates were 1.9 times more rapid at

$-3.4 \pm 2.7 \mathrm{m\,a}^{-1}$. At the highly erosive northern cape of the island, the distance from the mainland has been increasing by $-9.6 \mathrm{m\,a}^{-1}$; this value also accelerated during the recent past to $-17 \mathrm{m\,a}^{-1}$.

Our systematic seasonal analyses demonstrate that the currently higher intensities of the two coastal erosion processes, thermo-abrasion (TA) and thermo-denudation (TD), are controlled at least in part by the increasing open water season and summer air temperatures (T_{air}). The open water season has lengthened from 81 open water days (OWD) on average for the past 2 decades by 15 OWD over the 2010–2012 observation period and, for example, up to 99 OWD in 2011. Annual positive degree day (PDD) sums from 1951 to 2013 in the nearby town of Tiksi were 660 Kd on average and strongly increased to 1010 Kd in 2012. Accordingly, the seasonal duration available for TD has lengthened from 110 days on average to 134 days in 2012.

We show that normalisation of diverse TD rates ($\sigma = 11.5 \mathrm{m\,a}^{-1}$) through PDD sums for each period decreases variability of TD rates across all subperiods to $\sigma = 1.6 \mathrm{m\,a}^{-1}$. We therefore propose that TD rates for short season time periods are corrected by a seasonal factor to make them comparable. Our observations suggest that coastal erosion on this Ice Complex coast increases by $1.2 \mathrm{m\,a}^{-1}$ per 1 °C of positive mean daily T_{air}.

Interannual variations of coastal erosion are also related to local factors, such as macro ground ice content variation with depth, affecting thermo-niche development, which depends on ice-wedge polygon size.

We found a phase shift between TA and TD. TA is only active during the open water season, while TD can proceed throughout the summer. In June, when T_{air} rises and the mainland is already free of snow, mud flows from TD accumulate on top of snow at the cliff bottom, while sea ice is preventing TA. In late August, TD is slowed due to refreezing of the active layer and coastal slopes, while TA may still proceed until land-fast sea ice develops. Currently observed shifts in sea ice duration and T_{air} result primarily in an extension in overlap time of the active seasons for TA and TD – the resulting simultaneity of both processes is more important than the extension of either active season, because the widely varying value domains of TD and TA align and mutually reinforce each other to more effective erosion of thawed permafrost and associated mass displacement.

Acknowledgements. We acknowledge the support of this research through the Potsdam Research Cluster for Georisk Analysis and Sustainability (PROGRESS) and the ERC Starting Grant PETA-CARB. This paper contributes to the Eurasian Arctic Ice 4k project (grant OP 217/2-1 awarded to Thomas Opel by Deutsche Forschungsgemeinschaft). We are thankful for the logistical support of our partners from the Tiksi Hydrobase, the Lena Delta Reserve, and the Arctic and Antarctic Institute, St. Petersburg, Russian Federation. We would like to thank our colleagues H. Meyer, A. S. Makarov, A. N. Sandakov and S. Wetterich for the pleasant collaboration in the field. We kindly thank I. Overeem and an anonymous referee for careful revisions that improved the manuscript. Freely available hydrometeorological data sets (WMO) and sea ice concentrations from Centre de Recherche et d'Exploitation Satellitaire (CERSAT), at IFREMER, Plouzané (France) are gratefully acknowledged.

Edited by: A. Kääb

References

Aguilar, M. A., Aguilar, F. J., Saldaña, M. M., and Fernández, I.: Geopositioning accuracy assessment of GeoEye-1 panchromatic and multispectral imagery, Photogram. Eng. Remote Sens., 78, 247–257, 2012.

Aguirre, A., Tweedie, C. E., Brown, J., and Gaylord, A.: Erosion of the Barrow Environmental Observatory Coastline 2003–2007, Northern Alaska, in: Proceedings of the Ninth International Conference on Permafrost, edited by: Kane, D. L. and Hinkel, K. M., University of Alaska Fairbanks, 29 June–3 July 2008, 1, 7–12, 2008.

Andersen, S., Tonboe, R., Kaleschke, L., Heygster, G., and Pedersen, L.: Intercomparison of passive microwave sea ice concentration retrievals over the high-concentration Arctic sea ice, J. Geophys. Res.-Oceans, 112, C08004, doi:10.1029/2006JC003543, 2007.

Are, F.: The role of coastal retreat for sedimentation in the Laptev Sea, in: Land-Ocean Systems in the Siberian Arctic, edited by: Kassens, H., Bauch, H., Dmitrenko, I., Eicken, H., Hubberten, H.-W., Melles, M., Thiede, J., and Timokohov, L., Springer, Berlin, Heidelberg, Germany, 287–295, 1999.

Are, F., Reimnitz, E., Grigoriev, M., Hubberten, H.-W., and Rachold, V.: The Influence of Cryogenic Processes on the Erosional Arctic Shoreface, J. Coastal Res., 24, 110–121, doi:10.2112/05-0573.1, 2008.

Are, F. E.: Thermal abrasion of sea coasts, Polar Geography and Geology, 12, 1–86, doi:10.1080/10889378809377343, from: Termoabraziya morskikh beregov, Moscow: Nauka, 1980, 158 pp., 1988a.

Are, F. E.: Thermal abrasion of sea coasts, Polar Geography and Geology, 12, 87–157, doi:10.1080/10889378809377352, from: Termoabraziya morskikh beregov, Moscow: Nauka, 1980, 158 pp., 1988b.

Are, F. E.: The contribution of shore thermoabrasion to the Laptev Sea sediment balance, in: Proceedings of the Seventh International Conference on Permafrost, edited by: Lewkowicz, A. G. and Allard, M., Yellowknife, Canada, 25–30, 1998.

Are, F. E.: Razrushenie beregov arkticheskikh primorskikh nizmennostej (Coastal erosion of the Arctic lowlands), Academic publishing house "Geo", Novosibirsk, Russia, 291 pp., 2012.

Are, F. E., Grigoriev, M. N., Hubberten, H.-W., and Rachold, V.: Using thermoterrace dimensions to calculate the coastal erosion rate, Geo-Marine Lett., 25, 121–126, doi:10.1007/s00367-004-0193-y, 2005.

Arp, C. D., Jones, B. M., Schmutz, J. A., Urban, F. E., and Jorgenson, M. T.: Two mechanisms of aquatic and terrestrial habitat change along an Alaskan Arctic coastline, Polar Biol., 33, 1629–1640, doi:10.1007/s00300-010-0800-5, 2010.

Atkinson, D. E.: Observed storminess patterns and trends in the circum-Arctic coastal regime, Geo-Marine Lett., 25, 98–109, doi:10.1007/s00367-004-0191-0, 2005.

Barnhart, K. R., Overeem, I., and Anderson, R. S.: The effect of changing sea ice on the physical vulnerability of Arctic coasts, The Cryosphere, 8, 1777–1799, doi:10.5194/tc-8-1777-2014, 2014.

Bauch, H. A., Mueller-Lupp, T., Taldenkova, E., Spielhagen, R. F., Kassens, H., Grootes, P. M., Thiede, J., Heinemeier, J., and Petryashov, V. V.: Chronology of the Holocene transgression at the North Siberian margin, Global Planet. Change, 31, 125–139, doi:10.1016/S0921-8181(01)00116-3, 2001.

Boike, J., Kattenstroth, B., Abramova, K., Bornemann, N., Chetverova, A., Fedorova, I., Fröb, K., Grigoriev, M., Grüber, M., Kutzbach, L., Langer, M., Minke, M., Muster, S., Piel, K., Pfeiffer, E.-M., Stoof, G., Westermann, S., Wischnewski, K., Wille, C., and Hubberten, H.-W.: Baseline characteristics of climate, permafrost and land cover from a new permafrost observatory in the Lena River Delta, Siberia (1998–2011), Biogeosciences, 10, 2105–2128, doi:10.5194/bg-10-2105-2013, 2013.

Braithwaite, R. J.: Positive degree-day factors for ablation on the Greenland ice sheet studied by energy-balance Modelling, J. Glaciol., 41, 153–160, 1995.

Burn, C. R.: The response (1958–1997) of permafrost and near-surface ground temperatures to forest fire, Takhini River valley, southern Yukon Territory, Can. J. Earth Sci., 35, 184–199, doi:10.1139/e97-105, 1998.

Cheng, P., Toutin, T., Zhang, Y., and Wood, M.: QuickBird: Geometric correction, path and block processing and data fusion, Earth Obs. Magazine, 12, 24–30, 2003.

Christiansen, H. H.: Thermal regime of ice-wedge cracking in Adventdalen, Svalbard, Permafrost Periglac., 16, 87–98, doi:10.1002/ppp.523, 2005.

Dallimore, S. R., Wolfe, S. A., and Solomon, S. M.: Influence of ground ice and permafrost on coastal evolution, Richards Island, Beaufort Sea coast, Can. J. Earth Sci., 33, 664–675, doi:10.1139/e96-050, 1996.

Dmitrenko, I. A., Tyshko, K. N., Kirillov, S. A., Eicken, H., Hölemann, J. A., and Kassens, H.: Impact of flaw polynyas on the hydrography of the Laptev Sea, Global Planet. Change, 48, 9–27, doi:10.1016/j.gloplacha.2004.12.016, 2005.

Dowman, I., Jacobsen, K., Konecny, G., and Sandau, R.: High resolution optical satellite imagery, Whittles Publishing, Dunbeath, UK, 230 pp., 2012.

Ezraty, R., Girard-Ardhuin, F., Piollé, J.-F., Kaleschke, L., and Heygster, G.: Arctic & Antarctic sea ice concentration and arctic sea ice drift estimated from special sensor microwave data – User's manual, Laboratoire d' Océanographie Spatiale, Département d' Océanographie Physique et Spatiale, IFREMER, Brest, France; Institute of Environmental Physics, University of Bremen, Germany, 22 pp., available at: ftp://ftp.ifremer.fr/ifremer/cersat/products/gridded/psi-drift/documentation/ssmi.pdf, 2007.

Federova, I. V., Bolshiyanov, D. Y., Makarov, A. S., Tretyakov, M. N., and Chetverova, A. A.: Sovremennoe gidrologicheskoe sostoyanie delty reki Leny (Current hydrological state of the Lena River Delta), in: Sistema Morya Laptevykh i prilegayushchikh morey Arktiki (System of the Laptev Sea and the adjacent arctic seas), edited by: Kassens, H., Lisitzin, A. P., Thiede, J., Polyakova, Y. I., Timokhov, L. A., and Frolov, I. E., Moscow University Press, Moscow, 278–291, 2009.

Fedorov, A., Gavrilev, P., Konstantinov, P., Hiyama, T., Iijima, Y., and Iwahana, G.: Contribution of Thawing Permafrost and Ground Ice to the Water Balance of Young Thermokarst Lakes in Central Yakutia, in: Proceedings of the Tenth International Conference on Permafrost, Salekhard, Yamal-Nenets Autonomous District, Russia, 25–29 June 2012, 75–80, 2012.

Fedorov, A., Ivanova, R., Park, H., Hiyama, T., and Iijima, Y.: Recent air temperature changes in the permafrost landscapes of northeastern Eurasia, Polar Science, 8, 114–128, doi:10.1016/j.polar.2014.02.001, 2014.

Forbes, D. L. and Hansom, J. D.: Polar Coasts, in: Treatise on Estuarine and Coastal Science, edited by: Wolanski, E. and McLusky, D., Academic Press, Waltham, 245–283, doi:10.1016/B978-0-12-374711-2.00312-0, 2011.

Fraser, C., Dial, G., and Grodecki, J.: Sensor orientation via RPCs, ISPRS J. Photogram. Remote Sens., 60, 182–194, doi:10.1016/j.isprsjprs.2005.11.001, 2006.

Fraser, C. S. and Ravanbakhsh, M.: Georeferencing Accuracy of GeoEye-1 Imagery, Photogram. Eng. Remote Sens., 75, 634–638, 2009.

Fyodorov-Davydov, D. G., Sorokovikov, V. A., Ostroumov, V. E., Kholodov, A. L., Mitroshin, I. A., Mergelov, N. S., Davydov, S. P., Zimov, S. A., and Davydova, A. I.: Spatial and temporal observations of seasonal thaw in the northern Kolyma Lowland, Polar Geography, 28, 308–325, doi:10.1080/789610208, 2004.

Fyodorov-Davydov, D. G., Kholodov, A. L., Ostroumov, V. E., Kraev, G. N., and Sorokovikov, V. A.: Seasonal thaw of soils in the North Yakutian ecosystems, in: Proceedings of the Ninth International Conference on Permafrost, edited by: Kane, D. L. and Hinkel, K. M., University of Alaska Fairbanks, 29 June–3 July 2008, 481–486, 2008.

Gavrilov, A. V., Romanovskii, N. N., Romanovsky, V. E., Hubberten, H.-W., and Tumskoy, V. E.: Reconstruction of Ice Complex Remnants on the Eastern Siberian Arctic Shelf, Permafrost Periglac., 14, 187–198, doi:10.1002/ppp.450, 2003.

Gavrilov, A. V., Romanovskii, N. N., and Hubberten, H.-W.: Paleogeographic scenario of the postglacial transgression on the Laptev Sea shelf, Kriosfera Zemli (Earth Cryosphere), 10, 39–50, available at: http://www.izdatgeo.ru/pdf/krio/2006-1/39.pdf (last access: 16 December 2014), 2006.

Grigoriev, M. N.: Kriomorfogenez Ustevoy Oblasti Reki Leny (Cryomorphogenesis of the Lena River mouth area), Melnikov Permafrost Institute, Russian Academy of Sciences, Siberian Branch, Yakutsk, 176 pp., 1993.

Grigoriev, M. N.: Kriomorfogenez i litodinamika pribrezhno-shelfovoi zony morei Vostochnoi Sibiri (Cryomorhogenesis and lithodynamics of the East Siberian near-shore shelf zone), Habil-
itation thesis, Melnikov Permafrost Institute, Russian Academy of Sciences, Siberian Branch, Yakutsk, 291 pp., 2008.

Grigoriev, M. N., Rachold, V., Schirrmeister, L., and Hubberten, H.-W.: Organic carbon input to the Arctic Seas through coastal erosion, in: The organic carbon cycle in the Arctic Ocean: present and past, edited by: Stein, R. and Macdonald, R., Springer, Berlin, 37–65, 2004.

Grigoriev, M. N., Razumov, S. O., Kunitzky, V. V., and Spektor, V. B.: Dinamika beregov vostochnykh arkticheskikh morey Rossii: Osnovnye faktory, zakonomernosti i tendencii (Dynamics of the Russian East Arctic Sea coast: major factors, regularities and tendencies), Kriosfera Zemli (Earth Cryosphere), 10, 74–94, available at: http://www.izdatgeo.ru/pdf/krio/2006-4/74.pdf last access: 16 December 2014, 2006.

Grigoriev, M. N., Kunitsky, V. V., Chzhan, R. V., and Shepelev, V. V.: On the variation in geocryological, landscape and hydrological conditions in the Arctic zone of East Siberia in connection with climate warming, Geogr. Nat. Res., 30, 101–106, doi:10.1016/j.gnr.2009.06.002, 2009.

Grodecki, J. and Dial, G.: Block Adjustment of High-Resolution Satellite Images Described by Rational Polynomials, Photogram. Eng. Remote Sens., 69, 59–68, 2003.

Grosse, G., Schirrmeister, L., Siegert, C., Kunitsky, V. V., Slagoda, E. A., Andreev, A. A., and Dereviagyn, A. Y.: Geological and geomorphological evolution of a sedimentary periglacial landscape in Northeast Siberia during the Late Quaternary, Geomorphology, 86, 25–51, doi:10.1016/j.geomorph.2006.08.005, 2007.

Grosse, G., Romanovsky, V., Jorgenson, T., Anthony, K. W., Brown, J., and Overduin, P. P.: Vulnerability and feedbacks of permafrost to climate change, Eos, Transactions AGU, 92, 73–74, doi:10.1029/2011EO090001, 2011.

Gukov, A. Y.: Gidrobiologiya Ustevoy Oblasti Reki Leny (Hydrobiology of the Lena River mouth area), Scientific World, Moscow, 288 pp., 2001.

Günther, F., Overduin, P. P., Sandakov, A. V., Grosse, G., and Grigoriev, M. N.: Thermo-erosion along the Yedoma Coast of the Buor Khaya Peninsula, Laptev Sea, East Siberia, in: Proceedings of the Tenth International Conference on Permafrost, edited by: Hinkel, K. M., Salekhard, Yamal-Nenets Autonomous District, Russia, 25–29 June 2012, 1, 137–142, available at: http://epic.awi.de/30828/ (last access: 20 January 2015), 2012.

Günther, F., Overduin, P. P., Sandakov, A. V., Grosse, G., and Grigoriev, M. N.: Short- and long-term thermo-erosion of ice-rich permafrost coasts in the Laptev Sea region, Biogeosciences, 10, 4297–4318, doi:10.5194/bg-10-4297-2013, 2013a.

Günther, F., Sandakov, A., Baranskaya, A., and Overduin, P.: Topographic survey of Ice Complex coasts, in: Russian-German Cooperation System Laptev Sea: The Expeditions Laptev Sea – Mamontov Klyk 2011 & Buor Khaya 2012, Reports on Polar and Marina Research, 664, 16–54, available at: http://epic.awi.de/33371/ (last access: 20 January 2015), 2013b.

Hoque, M. A. and Pollard, W. H.: Arctic coastal retreat through block failure, Can. Geotech. J., 46, 1103–1115, doi:10.1139/T09-058, 2009.

Hussain, M., Chen, D., Cheng, A., Wei, H., and Stanley, D.: Change detection from remotely sensed images: From pixel-based to object-based approaches, ISPRS J. Photogram. Remote Sens., 80, 91–106, doi:10.1016/j.isprsjprs.2013.03.006, 2013.

Hutchinson, M. F. and Gallant, J. C.: Digital elevation models and representation of terrain shape, in: Terrain Analysis: Principles and Applications, edited by: Wilson, J. P. and Gallant, J. C., Wiley, New York, 29–50, 2000.

Ifremer/CERSAT: SSM/I user manual, Ifremer/CERSAT, Plouzane, France, available at: http://www.ifremer.fr/cersat (last access: 12 November 2013), 2000.

Ivanov, M. S. and Katasonova, E. G.: Osobennosti kriolitogennykh otlozhenii ostrova Muostakh (Pecularities of the cryolithogenic deposits on Muostakh), in: Geokriologicheskie i gidrogeologicheskie issledovaniya Yakutii (Geocryological and hydrogeological investigations in Yakutia), Melnikov Permafrost Institute, Yakutsk, 1978.

Ivanov, N. E., Makshtas, A. P., Shutilin, S. V., and Gun, R. M.: Mnogoletnyaya izmenchivost' kharakteristik klimata raiona gidrometeorologicheskoi observatorii Tiksi. (Long-term variability of climate characteristics in the area of Tiksi hydrometeorological observatory), Problemy Arktiki i Antarktiki, 81, 24–40, 2009a.

Ivanov, N. E., Makshtas, A. P., and Shutilin, S. V.: Mnogoletnyaya izmenchivost' kharakteristik klimata raiona gidrometeorologicheskoi observatorii Tiksi. Chast 2. godovoi khod (Long-term variability of climate characteristics in the area of Tiksi hydrometeorological observatory. Part 2. Seasonal variability), Problemy Arktiki i Antarktiki, 83, 97–13, 2009b.

Jacobsen, K.: Direct georeferencing - Exterior orientation parameters, Photogram. Eng. Remote Sens., 67, 1321–1332, 2001.

Jones, B. M., Arp, C. D., Beck, R. A., Grosse, G., Webster, J. M., and Urban, F. E.: Erosional history of Cape Halkett and contemporary monitoring of bluff retreat, Beaufort Sea coast, Alaska, Polar Geography, 32, 129–142, doi:10.1080/10889370903486449, 2009a.

Jones, B. M., Arp, C. D., Jorgenson, M. T., Hinkel, K. M., Schmutz, J. A., and Flint, P. L.: Increase in the rate and uniformity of coastline erosion in Arctic Alaska, Geophys. Res. Lett., 36, L03503, doi:10.1029/2008GL036205, 2009b.

Jonsell, U., Hock, R., and Duguay, M.: Recent air and ground temperature increases at Tarfala Research Station, Sweden, Polar Res., 32, 1–11, doi:10.3402/polar.v32i0.19807, 2013.

Jorgenson, M. T. and Brown, J.: Classification of the Alaskan Beaufort Sea Coast and estimation of carbon and sediment inputs from coastal erosion, Geo-Marine Letters, 25, 69–80, doi:10.1007/s00367-004-0188-8, 2005.

Kääb, A.: Remote sensing of permafrost-related problems and hazards, Permafrost Periglac., 19, 107–136, doi:10.1002/ppp.619, 2008.

Kääb, A., Huggel, C., Fischer, L., Guex, S., Paul, F., Roer, I., Salzmann, N., Schlaefli, S., Schmutz, K., Schneider, D., Strozzi, T., and Weidmann, Y.: Remote sensing of glacier- and permafrost-related hazards in high mountains: an overview, Nat. Hazards Earth Syst. Sci., 5, 527–554, doi:10.5194/nhess-5-527-2005, 2005.

Kaleschke, L., Lüpkes, C., Vihma, T., Haarpaintner, J., Bochert, A., Hartmann, J., and Heygster, G.: SSM/I sea ice remote sensing for mesoscale ocean-atmosphere interaction analysis, Can. J. Remote Sens., 27, 526–537, doi:10.1080/07038992.2001.10854892, 2001.

Karklin, V. P. and Karelin, I. D.: Sezonnaya i mnogoletnyaya izmenchivost kharakteristik ledovogo rezhima morey Laptevykh i Vostochno-Sibirskogo (Seasonal and long-term variability of the ice conditions in the Laptev and East Siberian seas), in: Sistema Morya Laptevykh i prilegayushchikh morey Arktiki (System of the Laptev Sea and the adjacent arctic seas), edited by: Kassens, H., Lisitzin, A. P., Thiede, J., Polyakova, Y. I., Timokhov, L. A., and Frolov, I. E., Moscow University Press, Moscow, 187–201, 2009.

Katasonov, E. M.: Litologiya merzlykh chetvertichnykh otlozhenii (kriolitologiya) Yanskoi Primorskoi Nizmennosti - (Lithology of frozen quaternary deposits (cryolithology) of the Yana Coastal Plain), OAO PNIIIS, 176 pp., 2009.

Klyuev, Y. V., Kotyukh, A. A., and Olenina, N. V.: Kartografo-gidrograficheskaya interpretatsia ischeznoveniya v More Laptevykh ostrovov Semenovskogo i Vasilievskogo (Cartographic-hydrographical interpretation of vanishing of the islands Vasilievsky and Semenovsky in the Laptev Sea), Izvestiya vsesoyuznogo geograficheskogo obshchestva (Bulletin of the All-union Geographical Society), 6, 485–492, 1981.

Knizhnikov, Y. F., Kravtsova, V. I., Baldina, E. E., Gel'man, R. N., Zinchuk, N. N., Zolotarev, E. A., Labutina, I. A., Khar'kovets, E. G., and Kotseruba, A. D.: Tsifrovaya stereoskopicheskaya model' mestnosti: Experimental'nye issledovaniya (Digital stereoscopic terrain model: Experimental investigations), Scientific World, Moscow, 244 pp., 2004.

Konecny, G. and Lehmann, G.: Photogrammetrie, Walter de Gruyter, Berlin, New York, 4 edn., 392 pp., 1984.

Konishchev, V. N.: Paleotemperaturnye usloviya formirovaniya i deformacii sloev Ledovogo Kompleksa (Paleotemperature conditions of formation and deformation of Ice Complex layers), Kriosfera Zemli (Earth Cryosphere), 6, 17–24, 2002.

Krumpen, T., Janout, M., Hodges, K. I., Gerdes, R., Girard-Ardhuin, F., Hölemann, J. A., and Willmes, S.: Variability and trends in Laptev Sea ice outflow between 1992–2011, The Cryosphere, 7, 349–363, doi:10.5194/tc-7-349-2013, 2013.

Kunitsky, V. V.: Kriolitologiya Nizovya Leny (Cryolithology of the Lower Lena), Melnikov Permafrost Institute, USSR Academy of Sciences, Siberian Branch, Yakutsk, 159 pp., 1989.

Langer, M., Westermann, S., Muster, S., Piel, K., and Boike, J.: The surface energy balance of a polygonal tundra site in northern Siberia – Part 1: Spring to fall, The Cryosphere, 5, 151–171, doi:10.5194/tc-5-151-2011, 2011.

Lantuit, H. and Pollard, W. H.: Fifty years of coastal erosion and retrogressive thaw slump activity on Herschel Island, southern Beaufort Sea, Yukon Territory, Canada, Geomorphology, 95, 84–102, doi:10.1016/j.geomorph.2006.07.040, 2008.

Lantuit, H., Overduin, P., Couture, N., Wetterich, S., Aré, F., Atkinson, D., Brown, J., Cherkashov, G., Drozdov, D., Forbes, D., Graves-Gaylord, A., Grigoriev, M., Hubberten, H.-W., Jordan, J., Jorgenson, T., Ødegård, R., Ogorodov, S., Pollard, W., Rachold, V., Sedenko, S., Solomon, S., Steenhuisen, F., Streletskaya, I., and Vasiliev, A.: The Arctic Coastal Dynamics Database: A New Classification Scheme and Statistics on Arctic Permafrost Coastlines, Estuar. Coasts, 35, 383–400, doi:10.1007/s12237-010-9362-6, 2011a.

Lantuit, H., Atkinson, D., Overduin, P. P., Grigoriev, M., Rachold, V., Grosse, G., and Hubberten, H.-W.: Coastal erosion dynamics on the permafrost-dominated Bykovsky Peninsula, north Siberia, 1951–2006, Polar Research, 30, 7341, doi:10.3402/polar.v30i0.7341, 2011b.

Lantuit, H., Overduin, P. P., and Wetterich, S.: Recent Progress Regarding Permafrost Coasts, Permafrost Periglac., 24, 120–130, doi:10.1002/ppp.1777, 2013.

Lantz, T. C. and Kokelj, S. V.: Increasing rates of retrogressive thaw slump activity in the Mackenzie Delta region, Canada, Geophys. Res. Lett., 35, L06502, doi:10.1029/2007GL032433, 2008.

Lewkowicz, A. G.: Rate of short-term ablation of exposed ground ice, Banks Island, Northwest Territories, Canada, J. Glaciol., 32, 511–519, 1986.

Little, J. D., Sandall, H., Walegur, M. T., and Nelson, F. E.: Application of differential global positioning systems to monitor frost heave and thaw settlement in tundra environments, Permafrost Periglac., 14, 349–357, doi:10.1002/ppp.466, 2003.

Liu, L., Schaefer, K., Zhang, T., and Wahr, J.: Estimating 1992–2000 average active layer thickness on the Alaskan North Slope from remotely sensed surface subsidence, J. Geophys. Res.: Earth, 117, F01005, doi:10.1029/2011JF002041, 2012.

Liu, L., Schaefer, K., Gusmeroli, A., Grosse, G., Jones, B. M., Zhang, T., Parsekian, A. D., and Zebker, H. A.: Seasonal thaw settlement at drained thermokarst lake basins, Arctic Alaska, The Cryosphere, 8, 815–826, doi:10.5194/tc-8-815-2014, 2014.

Lomax, A. S., Lubin, D., and Whritner, R. H.: The potential for interpreting total and multiyear ice concentrations in SSM/I 85.5 GHz imagery, Remote Sens. Environ., 54, 13–26, doi:10.1016/0034-4257(95)00082-C, 1995.

Mackay, J. R.: Segregated epigenetic ice and slumps in permafrost, Mackenzie Delta area, Geogr. Bull., 8, 59–80, 1966.

Markus, T., Stroeve, J. C., and Miller, J.: Recent changes in Arctic sea ice melt onset, freezeup, and melt season length, J. Geophys. Res.-Oceans, 114, C12024, doi:10.1029/2009JC005436, 2009.

Meier, W. N. and Stroeve, J.: Comparison of sea-ice extent and ice-edge location estimates from passive microwave and enhanced-resolution scatterometer data, Ann. Glaciol., 48, 65–70, doi:10.3189/172756408784700743, 2008.

Morgenstern, A., Grosse, G., Günther, F., Fedorova, I., and Schirrmeister, L.: Spatial analyses of thermokarst lakes and basins in Yedoma landscapes of the Lena Delta, The Cryosphere, 5, 849–867, doi:10.5194/tc-5-849-2011, 2011.

Mudrov, Y. V., ed.: Merzlotnye yavleniya v kriolitozone ravnin i gor (Permafrost phenomena in mountain and plain cryolithozone - General terms and definitions), Scientific World, Moscow, 316 pp., 2007.

Neef, E.: Das Gesicht der Erde: mit einem ABC, Brockhaus, Leipzig, 980 pp., 1956.

Nuth, C. and Kääb, A.: Co-registration and bias corrections of satellite elevation data sets for quantifying glacier thickness change, The Cryosphere, 5, 271–290, doi:10.5194/tc-5-271-2011, 2011.

Ogorodov, S. A.: Rol' morskikh l'dov v dinamike rel'efa beregovoi zony (The role of sea ice in coastal dynamics), Moscow University Press, Moscow, 173 pp., 2011.

Opel, T.(Ed.): Russian-German cooperation system Laptev Sea: the expedition LENA 2012, Reports on Polar and Marine Research, 684, 104 pp., doi:10.2312/BzPM_0684_2015, 2015.

Overduin, P. P. and Kane, D. L.: Frost Boils and Soil Ice Content: Field Observations, Permafrost Periglac., 17, 291–307, doi:10.1002/ppp.567, 2006.

Overduin, P. P., Hubberten, H.-W., Rachold, V., Romanovskii, N. N., and Grigoriev, M. N. Kasymskaya, M.: The evolution and degradation of coastal and offshore permafrost in the Laptev

and East Siberian Seas during the last climatic cycle, in: Coastline Changes: Interrelation of Climate and geological Processes, edited by: Harff, J., Hay, W. W., and Tetzlaff, D. M., Geol. Soc. Am. Special Paper, 426, 97–111, doi:10.1130/2007.2426(07), 2007.

Overduin, P. P., Strzelecki, M. C., Grigoriev, M. N., Couture, N., Lantuit, H., St-Hilaire-Gravel, D., Günther, F., and Wetterich, S.: Coastal changes in the Arctic, in: Sedimentary Coastal Zones from High to Low Latitudes: Similarities and Differences, edited by: Martini, I. P. and Wanless, H. R., Geological Society, London, Special Publications, 388, 103–129, doi:10.1144/SP388.13, 2014.

Overeem, I., Anderson, R. S., Wobus, C. W., Clow, G. D., Urban, F. E., and Matell, N.: Sea ice loss enhances wave action at the Arctic coast, Geophys. Res. Lett., 38, L17503, doi:10.1029/2011GL048681, 2011.

Petrich, C., Eicken, H., Zhang, J., Krieger, J., Fukamachi, Y., and Ohshima, K. I.: Coastal landfast sea ice decay and breakup in northern Alaska: Key processes and seasonal prediction, J. Geophys. Res.-Oceans, 117, C02003, doi:10.1029/2011JC007339, 2012.

Pizhankova, E. I.: Termodenudatsiya v beregovoi zone Lyakhovskikh Ostrovov - rezultaty deshifrirovaniya aerokosmicheskikh snimkov (Termodenudation in the coastal zone of the Lyakhovsky islands – interpretation of aerospace images), Kriosfera Zemli (Earth Cryosphere), 15, 61–70, http://www.izdatgeo.ru/pdf/krio/2011-3/61.pdf (last access: 16 December 2014), 2011.

Poli, D. and Toutin, T.: Review of developments in geometric modelling for high resolution satellite pushbroom sensors, Photogram. Record, 27, 58–73, doi:10.1111/j.1477-9730.2011.00665.x, 2012.

Popov, A. I., Rozenbaum, G. E., and Tumel, N. V.: Kriolitologiya (Cryolithology), Moscow University Press, Moscow, 1985.

Ravens, T., Jones, B., Zhang, J., Arp, C., and Schmutz, J.: Process-Based Coastal Erosion Modeling for Drew Point, North Slope, Alaska, Journal of Waterway, Port, Coastal Ocean Eng., 138, 122–130, doi:10.1061/(ASCE)WW.1943-5460.0000106, 2012.

Reem, D.: The geometric stability of Voronoi diagrams with respect to small changes of the sites, in: Proceedings of the 27th Annual ACM Symposium on Computational Geometry (SoCG 2011), 254–263, 2010.

Reimnitz, E.: Dinkum Sands - A Recently Foundered Arctic Island, J. Coastal Res., 21, 274–280, doi:10.2112/04-0167.1, 2005.

Reimnitz, E., Dethleff, D., and Nürnberg, D.: Contrasts in Arctic shelf sea-ice regimes and some implications: Beaufort Sea versus Laptev Sea, Marine Geology, 119, 215–225, doi:10.1016/0025-3227(94)90182-1, 1994.

Romanovskii, N. N., Hubberten, H.-W., Gavrilov, A. V., Tumskoy, V. E., Tipenko, G. S., Grigoriev, M. N., and Siegert, C.: Thermokarst and land-ocean interactions, Laptev Sea Region, Russia, Permafrost Periglac., 11, 137–152, 2000.

Romanovsky, V. E., Drozdov, D. S., Oberman, N. G., Malkova, G. V., Kholodov, A. L., Marchenko, S. S., Moskalenko, N. G., Sergeev, D. O., Ukraintseva, N. G., Abramov, A. A., Gilichinsky, D. A., and Vasiliev, A. A.: Thermal state of permafrost in Russia, Permafrost Periglac., 21, 136–155, doi:10.1002/ppp.683, 2010.

Schirrmeister, L., Siegert, C., Kunitsky, V., Grootes, P., and Erlenkeuser, H.: Late Quaternary ice-rich permafrost sequences as a paleoenvironmental archive for the Laptev Sea Region in northern Siberia, Int. J. Earth Sci. (Geologische Rundschau), 91, 154–167, doi:10.1007/s005310100205, 2002.

Schirrmeister, L., Grosse, G., Kunitsky, V., Meyer, H., Derivyagin, A., and Kuznetsova, T.: Permafrost, periglacial and paleoenvironmental studies on New Siberian Islands, in: Russian-German Cooperation System Laptev Sea, The Expedition LENA 2002, edited by: Grigoriev, M. N., Rachold, V., Rachold, V., D. Y., Pfeiffer, E.-M., Schirrmeister, L., Wagner, D., and Hubberten, H.-W., Reports on Polar and Marine Research, 466, 257–260, 10013/epic.10471.d001, 2003.

Schirrmeister, L., Kunitsky, V., Grosse, G., Wetterich, S., Meyer, H., Schwamborn, G., Babiy, O., Derevyagin, A., and Siegert, C.: Sedimentary characteristics and origin of the Late Pleistocene Ice Complex on north-east Siberian Arctic coastal lowlands and islands – A review, Quatern. Int., 241, 3–25, doi:10.1016/j.quaint.2010.04.004, 2011.

Schirrmeister, L., Froese, D., Tumskoy, V., Grosse, G., and Wetterich, S.: Yedoma: Late Pleistocene ice-rich syngenetic permafrost of Beringia, in: The Encyclopedia of Quaternary Science, edited by: Elias, S. A., Amsterdam: Elsevier, iSBN: 9780444536433, 542–552, 2013.

Shcherbakov, Y. E.: Raschet i konstruirovanie aerofotoapparatov (Calculation and construction of air survey cameras), vol. 2, Mashinostroenie, Moskva, 264 pp., 1979.

Shiklomanov, N. I., Streletskiy, D. A., Little, J. D., and Nelson, F. E.: Isotropic thaw subsidence in undisturbed permafrost landscapes, Geophys. Res. Lett., 40, 6356–6361, doi:10.1002/2013GL058295, 2013.

Short, N., Brisco, B., Couture, N., Pollard, W., Murnaghan, K., and Budkewitsch, P.: A comparison of TerraSAR-X, RADARSAT-2 and ALOS-PALSAR interferometry for monitoring permafrost environments, case study from Herschel Island, Canada, Remote Sens. Environ., 115, 3491–3506, doi:10.1016/j.rse.2011.08.012, 2011.

Siegert, C., Schirrmeister, L., and Babiy, O.: The Sedimentological, Mineralogical and Geochemical Composition of Late Pleistocene Deposits from the Ice Complex on the Bykovsky Peninsula, Northern Siberia, Polarforschung, 70, 3–11, available at: http://epic.awi.de/28480/ (last access: 21 January 2015), 2000.

Slagoda, E. A.: Kriolitogennye otlozheniya primorskoi ravniny morya Laptevykh: litologiya i mikromorfologiya (poluostrov Bykovskiy i ostrov Muostakh) - Cryolitogenic sediments of the Laptev Sea coastal lowland: lithology and micromorphology (Bykovsky Peninsula and Muostakh Island), Ekspress, Tyumen, 122 pp., 2004.

Solomatin, V. I.: O strukture poligonal'no-zhilnogo l'da (On the structure of polygonal veined ice), in: Podzemnyi led (Underground ice), edited by: Popov, A. I., Moscow University Press, Moscow, 46–72, 1965.

Spreen, G., Kwok, R., and Menemenlis, D.: Trends in Arctic sea ice drift and role of wind forcing: 1992–2009, Geophys. Res. Lett., 38, L19501, doi:10.1029/2011GL048970, 2011.

Strauss, J., Schirrmeister, L., Wetterich, S., Borchers, A., and Davydov, S. P.: Grain-size properties and organic-carbon stock of Yedoma Ice Complex permafrost from the Kolyma lowland, northeastern Siberia, Global Biogeochem. Cycl., 26, GB3003, doi:10.1029/2011GB004104, 2012.

Toutin, T.: Spatiotriangulation With Multisensor VIR/SAR Images, IEEE Trans. Geosci. Remote Sens., 42, 2096–2103, doi:10.1109/TGRS.2004.834638, 2004.

Toutin, T.: State-of-the-art of geometric correction of remote sensing data: A data fusion perspective, Int. J. Image Data Fusion, 2, 3–35, doi:10.1080/19479832.2010.539188, 2011.

Tweedie, C. E., Aguirre, A., Cody, R., Vargas, S., and Brown, J.: Spatial and Temporal Dynamics of Erosion along the Elson Lagoon Coastline near Barrow, Alaska (2002–2011), in: Proceedings of the Tenth International Conference on Permafrost, edited by: Hinkel, K. M., Salekhard, Yamal-Nenets Autonomous District, Russia, 25–29 June 2012, 1, 425–430, 2012.

Vasiliev, A. A.: Permafrost controls of coastal dynamics at the Marre-Sale key site, western Yamal, in: Permafrost: proceedings of the 8th International Conference on Permafrost, edited by: Philips, M., Springman, S., and Arenson, L., Zürich, Switzerland, 21–25 July 2003, A. A. Balkema Publishers, Rotterdam, 1173–1178, 2003.

Vasiliev, A. A., Streletskaya, I. D., Cherkashev, G. A., and Vanshtein, B. G.: Coastal dynamics of the Kara Sea, Kriosfera Zemli (Earth Cryosphere), 10, 56–67, available at: http://www.izdatgeo.ru/pdf/krio/2006-2/56.pdf (last access: 16 December 2014), 2006.

Wetterich, S., Rudaya, N., Tumskoy, V., Andreev, A. A., Opel, T., Schirrmeister, L., and Meyer, H.: Last Glacial Maximum records in permafrost of the East Siberian Arctic, Quat. Sci. Rev., 30, 3139–3151, doi:10.1016/j.quascirev.2011.07.020, 2011.

Winterfeld, M., Schirrmeister, L., Grigoriev, M. N., Kunitsky, V. V., Andreev, A., Murray, A., and Overduin, P. P.: Coastal permafrost landscape development since the Late Pleistocene in the western Laptev Sea, Siberia, Boreas, 40, 697–713, doi:10.1111/j.1502-3885.2011.00203.x, 2011.

Wobus, C., Anderson, R., Overeem, I., Matell, N., Clow, G., and Urban, F.: Thermal erosion of a permafrost coastline: Improving process-based models using time-lapse photography, Arct. Antarct. Alp. Res., 43, 474–484, doi:10.1657/1938-4246-43.3.474, 2011.

Yershov, E. D.: General Geocryology, Studies in Polar Research, Cambridge University Press, Cambridge, UK, 580 pp., 2004.

Zhang, Y.: Understanding image fusion, Photogram. Eng. Remote Sens., 70, 657–661, 2004.

Zhigarev, L. A.: Osobennosti dinamiki beregovoi kriolitozony arkticheskikh morei (Pecularities of coastal cryolithozone dynamics of arctic seas), in: Dinamika arkticheskikh poberezhii Rossii (Dynamics of russian arctic shores), edited by: Solomartin, V. I., Sovershaev, V. A., and Mazur, I. I., Moscow State University publishing house, Moscow, 19–34, 1998.

Boundary conditions of an active West Antarctic subglacial lake: implications for storage of water beneath the ice sheet

M. J. Siegert[1], N. Ross[2], H. Corr[3], B. Smith[4], T. Jordan[3], R. G. Bingham[5], F. Ferraccioli[3], D. M. Rippin[6], and A. Le Brocq[7]

[1]Bristol Glaciology Centre, School of Geographical Sciences, University of Bristol, Bristol, BS8 1SS, UK
[2]School of Geography, Politics and Sociology, Newcastle University, Newcastle upon Tyne, NE1 7RU, UK
[3]British Antarctic Survey, Cambridge CB3 0ET, UK
[4]Applied Physics Lab, Polar Science Center, University of Washington, Seattle, WA 98105, USA
[5]School of GeoSciences, University of Edinburgh, Edinburgh EH8 9XP, UK
[6]Environment Department, University of York, York YO10 5DD, UK
[7]Geography, College of Life and Environmental Sciences, University of Exeter, Exeter EX4 4RJ, UK

Correspondence to: M. J. Siegert (m.j.siegert@bristol.ac.uk)

Abstract. Repeat-pass ICESat altimetry has revealed 124 discrete surface height changes across the Antarctic Ice Sheet, interpreted to be caused by subglacial lake discharges (surface lowering) and inputs (surface uplift). Few of these active lakes have been confirmed by radio-echo sounding (RES) despite several attempts (notable exceptions are Lake Whillans and three in the Adventure Subglacial Trench). Here we present targeted RES and radar altimeter data from an "active lake" location within the upstream Institute Ice Stream, into which at least 0.12 km^3 of water was previously calculated to have flowed between October 2003 and February 2008. We use a series of transects to establish an accurate depiction of the influences of bed topography and ice surface elevation on water storage potential. The location of surface height change is downstream of a subglacial hill on the flank of a distinct topographic hollow, where RES reveals no obvious evidence for deep (> 10 m) water. The regional hydropotential reveals a sink coincident with the surface change, however. Governed by the location of the hydrological sink, basal water will likely "drape" over topography in a manner dissimilar to subglacial lakes where flat strong specular RES reflections are measured. The inability of RES to detect the active lake means that more of the Antarctic ice sheet bed may contain stored water than is currently appreciated. Variation in ice surface elevation data sets leads to significant alteration in calculations of the local flow of basal water indicating the value of, and need for, high-resolution altimetry data in both space and time to establish and characterise subglacial hydrological processes.

1 Introduction

Over the last ten years, our appreciation of basal hydrology in Antarctica has changed significantly. In the most recent inventory, Wright and Siegert (2012) collated evidence for 379 Antarctic subglacial lakes. Most of these ($\sim$250) are evidenced solely by discrete and distinct radio wave reflections from flat ice–water interfaces, using ice penetrating radio-echo sounding (RES). A few have been interpreted from flat ice sheet surfaces derived from a large region of floating ice (Bell et al., 2006, 2007). The remainder (numbering 124) have been established from repeat-pass satellite altimetric measurements of ice surface changes interpreted as evidence of water flowing into, and discharge from, "active" subglacial lakes (Smith et al., 2009). In some instances, RES data from a subglacial lake have been combined with repeat-pass altimetry, revealing compelling evidence for a substantial and active basal hydrological system in Antarctic ice streams (Christianson et al., 2012; Horgan et al., 2012). In one case, subglacial water discharge beneath Byrd Glacier, inferred from satellite altimetry, has been shown to coincide with an increase in measured ice surface velocity (Stearns et al., 2008). In addition, Scambos et al. (2011) used ICESat altimetry to reveal how rapid localised surface elevation change on the Antarctic Peninsula caused the drainage of a previously unknown subglacial lake. While most surface elevation changes have been measured close to the ice sheet margin, satellite altimetric evidence of basal water flow in the centre of East Antarctica (Wingham et al., 2006; McMillan et al., 2013) points to a potentially highly connected basal hydrological system linking large stores of water beneath the ice divide through fast flowing ice streams to the ice margin (Wright et al., 2012).

Several altimetrically derived subglacial lakes have numerous repeat-pass transects, which define accurately both lake outline and the loss/gain of water (e.g. Lake Whillans, Fricker et al., 2007). However, around half of the "active lake" inventory (Smith et al., 2009) is comprised of evidence for losses/gains of subglacial water from fewer than five transects, often just two and occasionally only one. In such cases, interpretation of discrete ice surface height changes as "active lakes" beneath the ice is plausible yet, given the paucity of data, inconclusive. Other explanations could include migration of "packets" of basal water (or sediment), as postulated by Gray et al. (2005) rather than volume loss/gain in distinct lake locations, or even the surface manifestation in changes to basal drag (Sergienko et al., 2007). As it is not known whether proposed "active lakes" drain completely before refilling (see e.g. Evatt et al., 2006; Wingham et al., 2006; Pattyn, 2008; Fowler, 2009), they are potentially ephemeral features. In both the Aurora and southern Wilkes subglacial basins, airborne RES data were acquired at the locations of proposed "active lakes" (Wright et al., 2012, 2014). Based on available data, no classic bright, flat and strong radio-wave reflections characteristic of deepwater lakes have been found to coincide with the locations of the "active lakes". In a few cases, "active lakes" were found to coincide with radio reflections showing one or two of these traits (e.g. Lake Mercer in the Siple Coast of West Antarctica; Fricker et al., 2007; Carter et al., 2007). Explanations for this mismatch include both the inability of radar to detect subglacial lakes under certain glaciological conditions (e.g. surface crevasses, or warm, $> -10\,°C$, ice), and a lack of water during the time of RES data acquisition. Furthermore, for all "active lakes", surrounding physiographical information that might help to understand how water flows into and out of a region cannot be detected using satellite altimetry alone.

Here we present targeted RES and radar altimeter data from an 'active lake' within the upstream Institute Ice Stream in West Antarctica, named "Institute E2". These data are coupled with ICESat ice surface measurements to ascertain evidence for basal water, local and regional basal topography and hydrological potential. In doing so, we reveal how the subglacial system can store and route basal water without the build-up of deep-water lakes, and demonstrate the ability

of (and necessity for) high-resolution surface altimetry and RES to detect and understand such a system.

2 Geophysical methods and data

Institute E2 was defined by two tracks of ICESat repeat-pass altimeter data (October 2003 and February 2008), during which the surface elevation rose by up to $\sim 5.5\,m$. Notwithstanding alternative explanations for ice surface height changes (noted above), the available data reveal a discrete monopole dome of uplift explicable as a consequence of the isolated build-up of basal water. Based on the ICESat data, Smith et al. (2009) defined a simple polygon of the lake outline. According to calculations by Smith et al. (2009), the lake experienced a gain in water of $0.12\,km^3$ and, according to the most recent data, was still gaining water at the end of 2008 (Figs. 2 and 3). The centre of the lake (as defined by the ICESat-inferred polygon) was used to define the location of a deep field camp on the Institute Ice Stream where, in the austral summer of 2010/2011, a RES campaign to measure basal topography and conditions of the Institute and Möller Ice Streams was undertaken (Ross et al., 2012). As the camp was positioned just a few km from the central lake coordinate, input/output survey flights were organised such that numerous RES transects were recovered from the lake at a variety of orientations. The survey geometry also led to RES data being acquired from the immediately surrounding regions, at a flight line spacing of $\sim 7.5\,km$ (Fig. 1).

The RES equipment used was a coherent system with a carrier frequency of 150 MHz, a bandwidth of 12 MHz and a pulse-coded waveform acquisition rate of 312.5 Hz. Aircraft position was obtained from differential GPS. Surface elevation of the ice sheet was derived from radar altimeter terrain-clearance measurements, with an accuracy of $\pm 1\,m$. Doppler processing was used to migrate radar-scattering hyperbola at the bed in the along-track direction. The onset of the received bed echo was picked in a semi-automatic manner using PROMAX seismic processing software. The post-processed data rate was 6.5 Hz, giving a spatial sampling interval of $\sim 10\,m$. The travel time in the near-surface firn layer is taken as the sum of two components; solid ice and an air gap. When calculating ice thickness we used a nominal value of 10 m to correct for the firn layer. A spatial variation in density affects the equivalent air gap, however, and this could account for variations across the survey area in the order of $\pm 3\,m$. This error is small relative to the overall error budget, however, which is dominated by the uncertainty in the overall ice thickness, estimated to be in the order of $\pm 1\,\%$ (see methods section in Ross et al., 2012, for details of the calculation).

To calculate basal power returns, the RES data were processed by unfocused SAR; a coherent integration over the returns from a section of the first Fresnel zone. This processing workflow is distinct from the Doppler processing described above, and was necessary to maintain the true

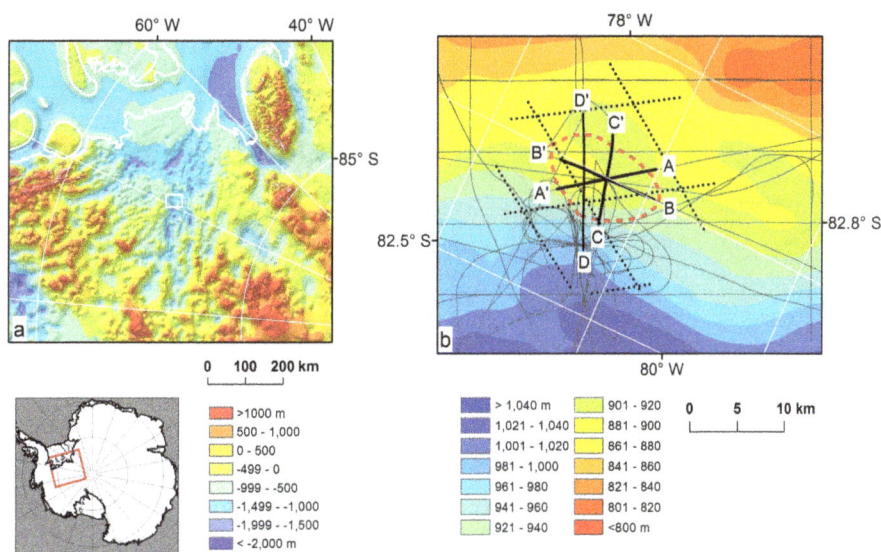

Fig. 1. (a) BEDMAP2 subglacial topography of the Institute Ice Stream region (with insert for study region in West Antarctica) (Fretwell et al., 2013). (b) Ice surface elevation with RES (grey lines) (Ross et al., 2012) and ICESat transects (black dotted lines) over and around Institute E2 (red dashed line), as delineated by Smith et al. (2009). RES lines in black, and labelled A–A', B–B', C–C' and D–D' refer to RES transects provided in Figs. 7 and 8. To assist comparison with Fig. 2, ICESat lines are oriented such that those closest to the 80° W longitude are also closest to the Easting's axis in Fig. 2.

return power of the signal. The window length was chosen to be ∼ 50 m within which the peak value of the basal reflector power was picked. The signal power was then compensated for differences in path length, transmitted output, and dielectric absorption. A nominal average two-way absorption of 3.0 dB/100 m was applied. This value was based on predictions of radar attenuation in Antarctica (Matsuoka et al., 2012), previous radar studies undertaken using the same equipment, and approximate calibration against the flat ice–water interface radar returns from Ellsworth Subglacial Lake (Woodward et al., 2010). Reducing the rate of dielectric absorption increases the relative basal powers in shallower ice, but does not adversely affect the general distribution of received RES strength (i.e. calculations of regions with relatively high reflectivity are insensitive to variation around the likely absorption rate). Finally, an estimated system performance figure of 200 dB was subtracted from the result.

Measurement of ice thickness, subtracted from surface elevation (Fig. 1) yields subglacial topography (Fig. 4a). Assuming water is driven by gravity and overburden pressure of ice according to Shreve (1972), the bed and ice surface elevations can be used to calculate the hydrological potential (Fig. 4b), which through definition of the hydro-potential surface can deliver hydrological pathways (assuming the bed is wet everywhere). Finally, the basal radio-echo power returns, when normalised (i.e. corrected for englacial absorption and geometry spreading) to account for ice thickness, can provide information on basal conditions (Fig. 5). The radar altimeter data (and indeed RES data for bed elevations) were gridded, using the ArcGIS topo-2-raster function, and used

to define the ice surface elevation around Institute E2. The ice surface elevations were compared with the most recent digital elevation model compilation of Antarctica (Bamber et al., 2009), and revealed significant (10–20 m) differences in elevation at a number of locations, most notably over the centre of Institute E2 (as defined by the lake outline polygon) (Fig. 6). We believe our surface elevation is more accurate over and around Institute E2 than Bamber et al.'s (2009) satellite-derived ice surface because of the greater density of airborne versus satellite tracks in this region and the fact that the airborne data were acquired within a short survey window of no more than 3 weeks (as opposed to several years).

The absolute elevations of ICESat (up to late 2008) and aircraft radar altimeter (December/January 2010/2011) data were compared to confirm the differences observed above were not due to significant basal water exchanges between the data acquisition dates. Because the aircraft radar has a coarser vertical resolution than ICESat, and because a fraction of the radar energy penetrates the snow surface before returning to the transmitter, there are potential offsets between the radar-estimated and ICESat elevations. Where radar tracks cross ICESat profiles, both inside and outside the boundary of Institute E2, the differences between the ICESat and radar elevations are typically on the order of 0.5 m, although in a few places excursions of up to 2–3 m were measured. Given that the ice surface elevation uplift measured by ICESat was ∼ 6 m (Fig. 2), this suggests that while the lake did not drain or fill substantially between late 2008 and the 2010–2011 measurements, we cannot use the radar altimeter

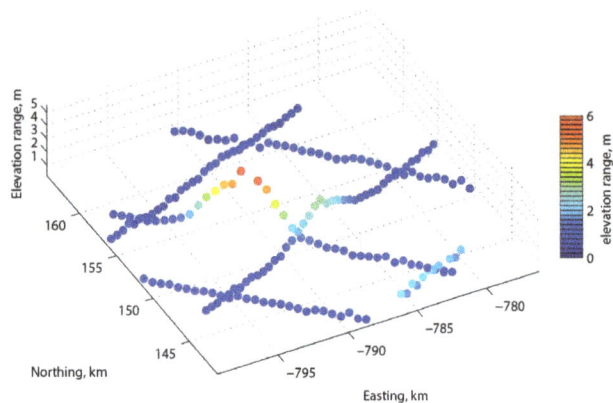

Fig. 2. Three-dimensional view of ICESat tracks (as noted in Fig. 1b) and the uplift detected between October 2003 and February 2008 (Smith et al., 2009). The ICESat lines are oriented such that those closest to the Easting's axis are also closest to the 80° W longitude in Figs. 1b and 6b.

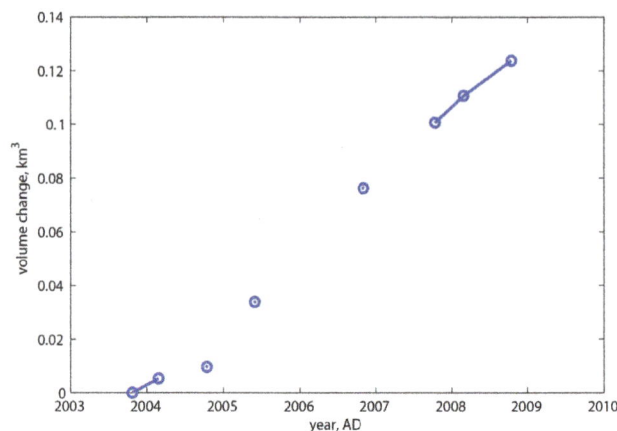

Fig. 3. Time-dependent change in the volume of Institute E2, using the method detailed in Smith et al. (2009) and ICESat surface elevation data between 2003 and 2009. Note that the data points are connected when intervening campaigns gave valid volume estimates. The data point at the end of 2008 is new to this paper (others being available in Smith et al., 2009).

data to form a precise estimate of the 2009–2010/2011 volume change.

3 Analysis

3.1 Ice surface elevation, bed topography and lake sensitivity

Some of Institute E2 (as defined by the lake outline polygon) lies beneath a hydrological potential minima, and on the downslope flank of a topographic hill close to a sizeable valley, ~ 2 km below the ice surface, at least 15 km in length and ~ 1–2 km wide (Fig. 4; Supplement video). In many places, the influence of local topography on the flow of basal water is sufficiently large that subglacial basins are coincident with hydrological sinks and, hence, are topographically controlled subglacial lakes (Bell et al., 2006; Siegert et al., 2007). While it is possible for the hydrological sink to be displaced from obvious topographic basins (Christianson et al., 2012), what makes Institute E2's topographic situation unusual (compared with RES-defined subglacial lakes) is that its bed slope is aligned along the approximate direction of ice flow. This makes Institute E2 controlled as much by surface slope rather as bed topography (see Sect. 4) and, therefore, potentially susceptible to change if the ice surface elevation is adjusted. Obviously, one way of adjusting the ice surface is to fill the lake, thus the stability of Institute E2 may be related to its own growth. Since Institute E2 is not confined by topography, any discharge will only cease once the surface-slope-driven hydrological sink (Figs. 5, 7 and 8) is re-established. According to Wingham et al. (2006) this might not occur until the lake has completely discharged (although others point to restricted discharges; e.g. Fowler, 2009). Hence, Institute E2 may be ephemeral.

The ice sheet in this region has only relatively recently relaxed from its Last Glacial Maximum form (perhaps within the last 500–1000 yr; Livingstone et al., 2013; Siegert et al., 2013). Given Institute E2's sensitivity to ice surface elevation, it is unlikely to predate the most recent glaciological setting, making it probably relatively young compared with some topographically controlled subglacial lakes (such as Ellsworth Subglacial Lake, Woodward et al., 2010).

3.2 RES evidence of basal water

We focus analyses on transects flown directly over the central coordinate of Institute E2 (Smith et al., 2009) in different orientations (Fig. 7), and along the approximate line of ice flow (Fig. 8). At the centre of Institute E2 (as defined by the lake outline polygon), there is little obvious qualitative evidence from RES for the presence of a deep-water subglacial lake (Figs. 5, 7 and 8). Subglacial lakes are traditionally characterised by RES as strong reflections (10–20 dB greater than surrounding bed) that are specular and flat, provided they are sufficiently deep (> 10 m) (Oswald and Robin, 1973; Carter et al., 2007). RES from Institute E2 reveals bed reflections that are non-specular and non-flat, however. According to Gorman and Siegert (1999), VHF radio waves can penetrate through shallow pure-water bodies and, in six examples, they demonstrated how RES reflections from the ceiling and floor of a subglacial lake can interfere providing the water depth is < 10 m. This can lead to RES reflections that are non-characteristic of deep subglacial lakes, and provides one explanation for why a sharp, mirror like interface is not observed in Institute E2 (in accordance with observations elsewhere, e.g. Langley et al., 2011; Welch et al., 2009).

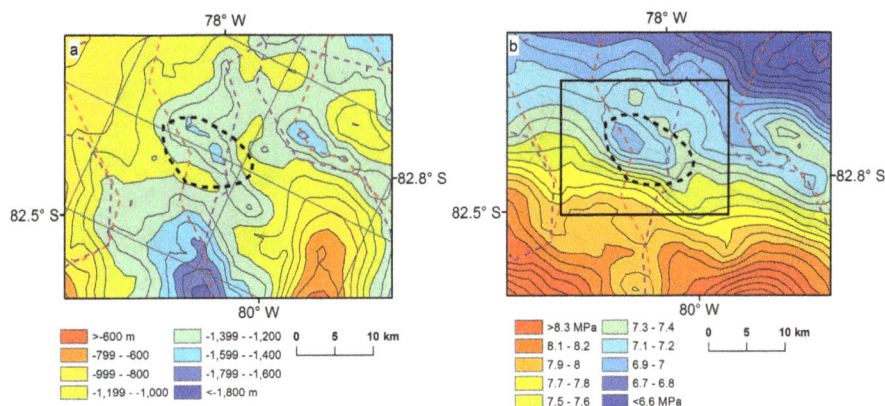

Fig. 4. Hydrological pathways (dashed lines) superimposed on **(a)** bed topography and **(b)** hydrological potential of the region surrounding Institute E2. The black box in **(b)** denotes the region depicted in Fig. 5. Two sets of hydrological pathways are given: red dashed lines, calculated using our aircraft ice surface altimeter data (Ross et al., 2012); and purple dashed lines, calculated using surface elevations from Bamber et al. (2009). Black dashed line delineates the approximate location of Institute E2, according to Smith et al. (2009). The direction of ice flow is predominantly from the bottom of the figures to the top.

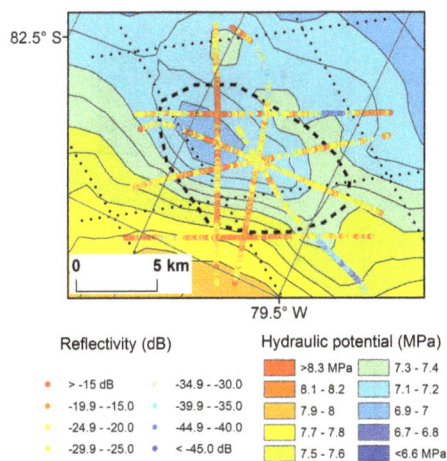

Fig. 5. Received RES strength (in dB) from basal reflections along RES lines defined in Fig. 1 over and around Institute E2 (see Fig. 4b for location). The RES lines are superimposed onto the ice sheet hydro-potential (as in Fig. 4b). Also shown is the outline of Institute E2, as proposed by Smith et al. (2009) (black dashed line).

Relative RES power returns from Institute E2 have a noticeable spatial variability (Fig. 5). Over the centre coordinate of Institute E2 (Smith et al., 2009) they do not appear to be anomalously large. In both the centre of the topographic basin, and the centre of the hydrological potential sink (Fig. 4b) (which is offset to one side of the previously proposed area of Institute E2), relative basal power returns are generally ~ 10–20 dB greater than from surrounding regions, however, although there is a noticeable along-track scatter (Fig. 5). Similarly strong RES returns, and therefore probably basal water, are also recorded from other regions in the survey area (Fig. 5). Conceptually, even if the water depths are very low (of the order of centimetres), enhanced

RES reflections should still be observed compared with a dry bed. Hence, it is possible that the distribution of water in Institute E2 is discontinuous. This opens the possibility for bed roughness to further reduce basal power returns, which may also help to explain the spatial complexity observed (Fig. 5). Noting the spatial correspondence between the surface elevation anomaly identified in the along-track ICESat data, enhanced reflectivity of the along-track RES data, and a minima in the gridded hydro-potential surface (Figs. 2 and 5), we suggest that the spatial extent of the lake is restricted to the region(s) of relatively high basal reflectivity. While assigning a lake outline is not possible with certainty, it is clear that the lake outline defined in Smith et al. (2009) is over-simplistic and, probably, too large.

An alternative explanation for the distribution of basal power returns is that they reflect defects in the RES system (either in data acquisition or processing). This is unlikely, given the system's ability to detect well-defined subglacial lake features in other regions of the wider survey, however (e.g. Ellsworth Subglacial Lake; Woodward et al., 2010).

3.3 Flow of basal water

According to Shreve (1972), based on our RES depiction of bed topography and radar measurement of ice surface elevation, water will be driven from the Institute E2 region toward the main trunk of the Institute Ice Stream (Fig. 4), where a larger, deeper subglacial basin exists, which (considering the thickness of ice within it) is also very likely to contain subglacial water (Fig. 1). This larger basin is part of a significant, topographically complex valley that routes water northwards toward the trunk of the Institute Ice Stream and, from there, to the Filchner Ronne Ice Shelf (Le Brocq et al., 2013). Differences between our radar measurements of ice surface elevation (acquired in less than 3 weeks) and Bamber et al.'s

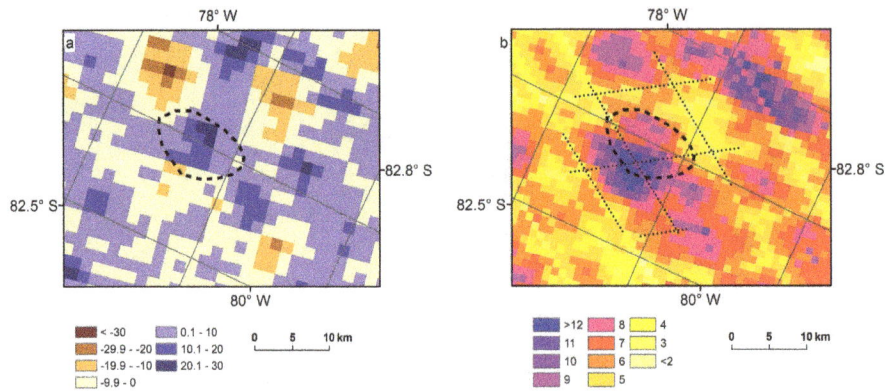

Fig. 6. Quantification (in meters) of errors in ice sheet surface elevation measurements. (**a**) Difference between Griggs and Bamber (2009) and this paper's (Fig. 1b) radar-altimeter surface elevations. (**b**) Map of the distribution of RMSE (root mean square error) in the Bamber et al. (2009) ice surface digital elevation model (DEM) (derived in this area exclusively from ICESat data) calculated using a multiple regression based on airborne validation data (adapted from Fig. 11 of Griggs and Bamber, 2009). Also shown is the outline of Institute E2, as proposed by Smith et al. (2009) (black dashed line), and ICESat lines (in **b**), which are oriented such that those closest to the 80° W longitude are also closest to the Easting's axis in Fig. 2.

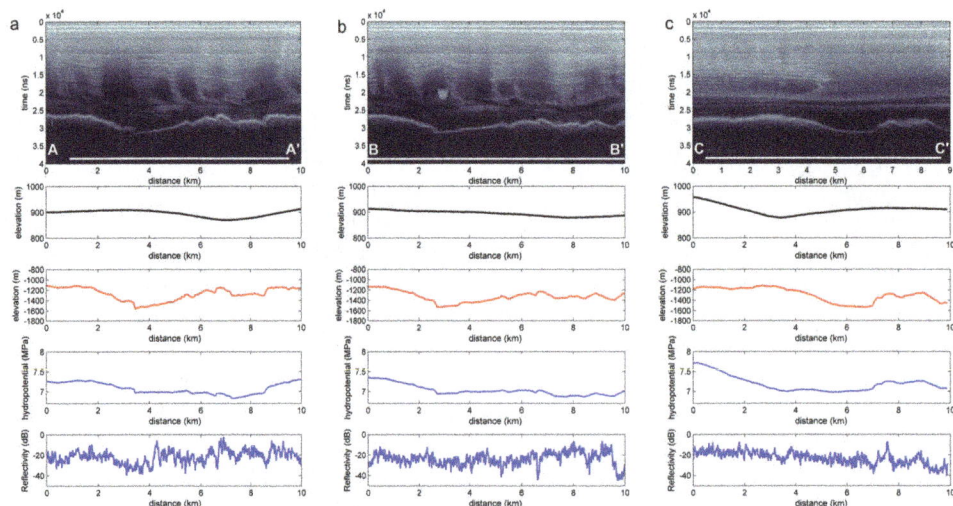

Fig. 7. RES transects centred on the "active" subglacial lake Institute E2. The locations of the transects are provided in Fig. 1b. (**a**) Transect A–A'. (**b**) Transect B–B'. (**c**) Transect C–C'. For each transect, the coverage of the ICESat-derived lake extent (after Smith et al., 2009) is shown as a white bar on the radargram. Beneath the radargrams graphs of ice surface elevation (black line), bed elevation (red line), basal hydropotential (upper blue line) and basal reflectivity (lower blue line) are provided.

(2009) DEM (collected over years), which are of the order of 10–20 m in places, result in local changes to the calculation of the expected route of basal water flow (Fig. 4b), demonstrating a high level of sensitivity in basal hydrology to ice surface elevation in this region. It is interesting to note that a 20 m error in ice surface elevation has the equivalent effect of a ~ 220 m error in the bed elevation; underlining the importance of accurate surface elevation data to quantify the flow of basal water (Wright et al., 2008).

The calculated routing of the subglacial water is also highly dependent upon the requirement to fill the hydropotential lows (i.e. "hydrological sinks") to make a hydro-

logically workable surface as input for the flow routing algorithm. As the hydro-potential lows in Fig. 4b comprise a significant part of the study area they have a consequentially greater influence on our ability to determine the localised routing of the subglacial water than in studies of wider survey areas.

4   Basal processes and data quality

The measurement of a subglacial hydrological sink coincident with the ICESat-derived "active lake" is crucial to appreciating how water may both pond and discharge at the

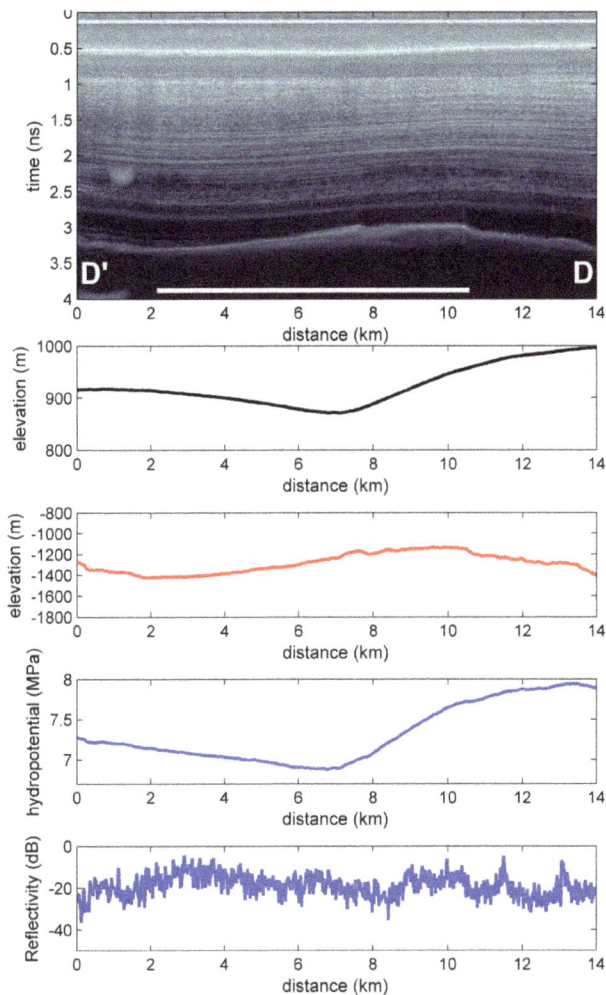

Fig. 8. RES transect D–D' (see Fig. 1b for location), along the approximate line of ice flow (from D to D'). The coverage of the ICESat-derived lake extent (after Smith et al., 2009) is shown as a white line on the radargram. Beneath the radargram graphs of ice surface elevation (black line), bed elevation (red line), basal hydropotential (upper blue line) and basal reflectivity (lower blue line) are provided.

site. From RES alone, it is likely that the existence of the subglacial lake would have been missed due to its possible thickness (< 10 m) and/or its surface shape (i.e. not flat). Calculating the hydrological sink requires RES data, however, as does an appreciation of subglacial hydrological pathways. The example presented in this paper shows that RES alone may be both inadequate to fully comprehend the locations of "active lakes" yet crucial to appreciating the flow of water beneath the ice sheet. Currently, RES data are too sparse in Antarctica to accurately measure subglacial hydrological sinks other than those that are the consequence of large scale topography. Sub-km resolution in RES data are needed, and yet although such data allow enhanced appreciation of the subglacial environment (e.g. Bo et al., 2009; Woodward et

al., 2010; King et al., 2009) the vast majority of the ice sheet base is sampled with transect separations normally > 5 km, and often far greater. Consequently, although the broad potential subglacial flow of water is understood (e.g. Siegert et al., 2007), information on the flow of basal water is currently spatially insufficient to fully comprehend how "active" subglacial lakes exchange water and, ultimately, influence ice sheet dynamics.

While we think it unlikely that Institute E2 discharged all of its water in the short period (~ 1 yr) between the end of the ICESat time series and when the RES data were acquired (see Sect. 2), we cannot rule out entirely the possibility that the lake was relatively dry of water when the airborne data were collected, thus explaining the lack of evidence for a lake in the RES data. Whether a subglacial lake is able to drain completely has been the subject of several studies (Evatt et al., 2006; Wingham et al., 2006; Fowler, 2009). The pertinent issue being that a subglacial lake's ice seal theoretically becomes stronger as water is evacuated, thus potentially restricting discharges. It is also possible that the broader ice surface elevation may have changed recently, thus affecting the flow of basal water and our ability to calculate it correctly. These are serious data-related issues for the detection and understanding of basal water flow from active subglacial lakes, and their solution requires a time series of high-resolution data that is currently unavailable.

Institute E2 was chosen for investigation because of its location to a planned RES survey. Through this opportunity we have been able to reveal that the lake is unlikely to comprise deep (> 10 m) water, and that it exists as a consequence of the interrelation between basal topography, ice flow and hydrology. Sergienko and Hulbe (2011) indicate that subglacial lakes may accumulate downstream of ice sheet "sticky spots", which may provide an explanation for why the lake grew in its present location. They show how water will infill on the lee side of subglacial obstacles, and the volume of water stored will be regulated by the rates of both water supply and subglacial freezing. Figure 8, in which RES data were collected along the approximate line of ice flow, reveals that the hydro-potential minimum is positioned downstream of a subglacial hill, suggesting Sergionko and Hulbe's theory may be valid in Institute E2. Hence, although Institute E2 does not occupy a subglacial valley, it still may be controlled by subglacial topography to some degree.

As RES data from Institute E2 appear similar to many of the other 124 lakes identified using satellite altimetric changes (Wright et al., 2014), we should be confident that the mismatch between intuitive identification of subglacial lakes from RES and inferred subglacial hydrological changes from ice surface altimetry may be repeated in some (possibly many) other regions of Antarctica. The implication of the inability of RES to independently detect the locations of "active" subglacial lakes leads to the likelihood that much more of the Antarctic bed contains water than is currently appreciated from RES alone.

5 Summary and discussion

Between October 2003 and November 2008, ICESat surface altimetry revealed a discrete zone of ice sheet uplift within the Institute Ice Stream, interpreted by Smith et al. (2009) to be due to the filling of a subglacial lake by at least $0.12\,\mathrm{km}^3$ of water (named Institute E2). In the austral summer of 2010–2011, an airborne geophysical survey of the Weddell Sea sector of West Antarctica targeted Institute E2 by flying a series of transects centred on the middle of the uplifted region (using a coordinate given in Smith et al., 2009). RES was used to measure subglacial topography and evidence for basal water. Airborne radar altimetry was used to measure ice sheet surface elevation. Institute E2 lies downstream of a subglacial hill, on the upstream flank of a major subglacial hollow. Sergienko and Hulbe (2011) predict basal water accumulation on the lee side of subglacial obstacles. Thus, although Institute E2 does not occupy a topographic basin, its location may still be influenced by the interrelation between basal water (production and flow) and topography. Using altimetry information to calculate how water may flow beneath the ice sheet reveals a hydropotential minimum coincident with the ICESat-derived anomaly of Institute E2 (but not the polygon as defined by Smith). RES also reveals relatively stronger (10–20 dB) basal reflections from beneath the hydropotential minimum than in adjacent regions, suggesting the presence of water, although considerable along-track scatter is also recorded. No classic RES reflections from a "deep-water" (> 10 m) subglacial lake were observed, however. Based on these data, we consider Institute E2 is controlled as much by ice sheet surface slopes as by basal topography, making it potentially susceptible to small ice surface changes. Such changes may occur as a consequence of ponding by basal water, making the lake's stability potentially influenced by its own growth. As such, Institute E2 is likely to be ephemeral and shallow; we believe it unlikely that the water depth is greater than 10 m.

Institute E2 is one of more than 100 active subglacial lakes detected through ice surface altimetric changes. They are scattered across the Antarctic ice sheet mostly, but not exclusively, located further toward the ice sheet margin than the divide. We have shown that RES alone is incapable of independently detecting Institute E2. Similar observations have been found for "active" lakes in both the Aurora (Wright et al., 2012) and southern Wilkes (Wright et al., 2014) subglacial basins. We therefore propose that considerably more basal water may be present in Antarctica than currently appreciated from RES. Given that the satellite altimetry is restricted to a short time series ($\sim 10\,\mathrm{yr}$) and that evidence for significant "active" lake discharges are being discovered from analysis of new altimetric data sets (e.g. Flament et al., 2013; McMillan et al., 2013), our appreciation of the volume of stored basal water (Dowdeswell and Siegert, 1999; Wright and Siegert, 2011) is in need of upwards revision. Indeed, given recent RES evidence of widespread subglacial hydro-logical regimes in both East (Wright et al., 2012) and West (Schroeder et al., 2013) Antarctica, the level of stored water within these and comparable systems may be significantly underestimated.

Acknowledgements. Funding was provided by UK NERC AFI grant NE/G013071/1. C. Robinson (Airborne Survey engineer), I. Potten and D. Cochrane (pilots) and M. Oostlander (air mechanic) are thanked for their invaluable assistance in the field. We also thank three referees, S. Carter, M. Wolovick and B. Csatho for their insightful and constructive reviews.

Edited by: E. Larour

References

Bamber, J. L., Gomez-Dans, J. L., and Griggs, J. A.: A new 1 km digital elevation model of the Antarctic derived from combined satellite radar and laser data – Part 1: Data and methods, The Cryosphere, 3, 101–111, doi:10.5194/tc-3-101-2009, 2009.

Bell, R. E., Studinger, M., Fahnestock, M. A., and Shuman, C. A.: Tectonically controlled subglacial lakes on the flanks of the Gamburtsev Subglacial Mountains, East Antarctica, Geophys. Res. Lett., 33, L02504, doi:10.1029/2005GL025207, 2006.

Bell, R. E., Studinger, M., Shuman, C. A., Fahnestock, M. A., and Joughin, I.: Large subglacial lakes in East Antarctica at the onset of fast-flowing ice streams, Nature, 445, 904–907, 2007.

Bo, S., Siegert, M. J., Mudd, S. M., Sugden, D. E., Fujita, S., Xiangbin, C., Yunyun, J., Xueyuan, T., and Yuansheng, L.: The Gamburtsev Mountains and the origin and early evolution of the Antarctic Ice Sheet, Nature, 459, 690–693, 2009.

Carter, S. P., Blankenship, D. D., Peters, M. E., Young, D. A., Holt, J. W., and Morse, D. L.: Radar-based subglacial lake classification in Antarctica, Geochem. Geophy. Geosy., 8, Q03016, doi:10.1029/2006GC001408, 2007.

Christianson, K., Jacobel, R. W., Horgan, H. J., Anandakrishnan, S., and Alley, R. B.: Subglacial Lake Whillans – Ice penetrating radar and GPS observations of a shallow active reservoir beneath a West Antarctic ice stream, Earth Planet. Sc. Lett., 331–332, 237–245, 2012.

Dowdeswell, J. A. and Siegert, M. J.: The dimensions and topographic setting of Antarctic subglacial lakes and implications for large-scale water storage beneath continental ice sheets, Geol. Soc. Am. Bull., 111, 254–263, 1999.

Evatt, G. W., Fowler, A. C., Clark, C. D., and Hulton, N. R. J.: Subglacial floods beneath ice sheets, Philos. T. R. Soc. A, 364, 1769–1794, 2006.

Fowler, A. C.: Dynamics of subglacial floods, P. Roy. Soc. A-Math. Phy., 465, 1809–1828, 2009.

Flament, T., Berthier, E., and Rémy, F.: Cascading water underneath Wilkes Land, East Antarctic Ice Sheet, observed using altimetry and digital elevation models, The Cryosphere Discuss., 7, 841–871, doi:10.5194/tcd-7-841-2013, 2013.

Fretwell, P., Pritchard, H. D., Vaughan, D. G., Bamber, J. L., Barrand, N. E., Bell, R., Bianchi, C., Bingham, R. G., Blankenship, D. D., Casassa, G., Catania, G., Callens, D., Conway, H., Cook, A. J., Corr, H. F. J., Damaske, D., Damm, V., Ferraccioli, F., Forsberg, R., Fujita, S., Gim, Y., Gogineni, P., Griggs, J. A., Hindmarsh, R. C. A., Holmlund, P., Holt, J. W., Jacobel, R. W., Jenkins, A., Jokat, W., Jordan, T., King, E. C., Kohler, J., Krabill, W., Riger-Kusk, M., Langley, K. A., Leitchenkov, G., Leuschen, C., Luyendyk, B. P., Matsuoka, K., Mouginot, J., Nitsche, F. O., Nogi, Y., Nost, O. A., Popov, S. V., Rignot, E., Rippin, D. M., Rivera, A., Roberts, J., Ross, N., Siegert, M. J., Smith, A. M., Steinhage, D., Studinger, M., Sun, B., Tinto, B. K., Welch, B. C., Wilson, D., Young, D. A., Xiangbin, C., and Zirizzotti, A.: Bedmap2: improved ice bed, surface and thickness datasets for Antarctica, The Cryosphere, 7, 375–393, doi:10.5194/tc-7-375-2013, 2013.

Fricker, H. A., Scambos, T., Bindshadler, R., and Padman, L.: An active subglacial water system in West Antarctica mapped from space, Science, 315, 1544–1548, 2007.

Gorman, M. R. and Siegert, M. J.: Penetration of Antarctic subglacial lakes by VHF electromagnetic pulses: Information on the depth and conductivity of subglacial water bodies, J. Geophys. Res., 104, 29311–29320, 1999.

Gray, L., Joughin, I., Tulaczyk, S., Spikes, V. B., Bindshadler R., and Jezek, K.: Evidence for subglacial water transport in the West Antarctic Ice Sheet through three-dimensional satellite radar interferometry, Geophys. Res. Lett., 32, L03501, doi:10.1029/2004GL021387, 2005.

Griggs, J. A. and Bamber, J. L.: A new 1 km digital elevation model of Antarctica derived from combined radar and laser data – Part 2: Validation and error estimates, The Cryosphere, 3, 113–123, doi:10.5194/tc-3-113-2009, 2009.

Horgan, H. J., Anandakrishnan, S., Jacobel, R. W., Christianson, K., Alley, R. B., Heeszel, D. S., Picotti, S., and Walter, J. I.: Subglacial Lake Whillans – Seismic observations of a shallow active reservoir beneath a West Antarctic ice stream, Earth Planet. Sc. Lett., 331–332, 201–209, 2012.

King, E., Hindmarsh, R., and Stokes, C.: Formation of mega-scale glacial lineations observed beneath a West Antarctic ice stream, Nat. Geosci., 2, 585–588, 2009.

Langley, K., Kohler, J., Matsuoka, K., Sinisalo, T., Scambos, T., Neumann, T., Muto, A., Winther, J.-G., and Albert, M.: Recovery Lakes, East Antarctica: Radar assessment of sub-glacial water extent, Geophys. Res. Lett., 38, L05501, doi:10.1029/2010GL046094, 2011.

Le Brocq, A., Ross, N., Griggs, J., Bingham, R., Corr, H., Ferroccioli, F., Jenkins, A., Jordan, T., Payne, A., Rippin, D., and Siegert, M. J.: Ice shelves record the history of channelised flow beneath the Antarctic ice sheet, Nat. Geosci., 6, 945–948, doi:10.1038/ngeo1977, 2013.

Livingstone, S. J., Clark, C. D., and Woodward, J.: Predicting subglacial lakes and meltwater drainage pathways beneath the Antarctic and Greenland ice sheets, The Cryosphere Discuss., 7, 1177–1213, doi:10.5194/tcd-7-1177-2013, 2013.

Matsuoka, K., MacGregor, J. A., and Pattyn, F.: Predicting radar attenuation within the Antarctic ice sheet, Earth Planet. Sc. Lett., 359–360, 173–183 doi:10.1016/j.epsl.2012.10.018, 2012

McMillan, M., Corr, H., Shepherd, A., Ridout, A., Laxon, S., and Cullen, R.: Three-dimensional mapping by CryoSat-2 of subglacial lake volume changes, Geophys. Res. Lett., 40, 4321–4327, doi:10.1002/grl.50689, 2013.

Oswald, G. K. A. and Robin, G. de Q.: Lakes beneath the Antarctic Ice Sheet, Nature, 245, 251–254, 1973.

Pattyn, F.: Investigating the stability of subglacial lakes with a full Stokes ice-sheet model, J. Glaciol., 54, 353–361, 2008.

Ross, N., Bingham, R. G., Corr, H., Ferraccioli, F., Jordan, T. A., Le Brocq, A., Rippin, D. M., Young, D., Blankenship, D. D., and Siegert, M. J.: Steep reverse bed slope at the grounding line of the Weddell Sea sector in West Antarctica, Nat. Geosci., 5, 393–396, doi:10.1038/ngeo1468, 2012.

Scambos, T. A., Berthier, E. and Shuman, C. A.: The triggering of sub-glacial lake drainage during the rapid glacier drawdown: Crane Glacier, Antarctic Peninsula, Ann. Glaciol., 52, 74–82, 2011.

Schroeder, D. M., Blankenship, D. D., and Young, D. A.: Evidence for a water system transition beneath Thwaites Glacier, West Antarctica, P. Natl. Acad. Sci., 110, 12225–12228, doi:10.1073/pnas.1302828110, 2013.

Sergienko, O. V. and Hulbe, C. L.: "Sticky spots" and subglacial lakes under ice streams of the Siple Coast, Antarctica, Ann. Glaciol., 52, 18–24, 2011.

Sergienko, O. V., MacAyeal, D. R., and Bindschadler, R. A.: Causes of sudden, short term changes in ice-stream surface elevation, Geophys. Res. Lett., 34, L22504, doi:10.1029/2007GL031775, 2007.

Shreve, R. L.: Movement of water in glaciers, J. Glaciol., 11, 205–214, 1972.

Siegert, M. J., Le Brocq, A., and Payne A.: Hydrological connections between Antarctic subglacial lakes and the flow of water beneath the East Antarctic Ice Sheet, in: Glacial Sedimentary Processes and Products, edited by: Hambrey, M. J., Christoffersen, P., Glasser, N. F. and Hubbard, B. P, Special Publication #39, International Association of Sedimentologists, 3–10, 2007.

Siegert, M. J., Ross, N., Corr, H., Kingslake, J., and Hindmarsh, R.: Late Holocene ice-flow reconfiguration in the Weddell Sea sector of West Antarctica, Quaternary Sci. Rev., 78, 98–107, 2013.

Smith, B. E., Fricker, H. A., Joughin, I. R., and Tulaczyk, S.: An inventory of active subglacial lakes in Antarctica detected by ICESat (2003–2008), J. Glaciol., 55, 573–595, 2009.

Stearns, L. A., Smith, B. E., and Hamilton, G. S.: Increased flow speed on a large East Antarctic outlet glacier caused by subglacial floods, Nat. Geosci., 1, 827–831, 2008.

Wingham, D. J., Siegert, M. J., Shepherd, A., and Muir, A. S.: Rapid discharge connects Antarctic subglacial lakes, Nature, 440, 1033–1036, 2006.

Woodward, J., Smith, A., Ross, N., Thoma, M., Grosfeld, C., Corr, H., King, E., King, M., Tranter, M., and Siegert, M. J.: Location for direct access to subglacial Lake Ellsworth: An assessment of geophysical data and modelling, Geophys. Res. Lett., 37, L11501, doi:10.1029/2010GL042884, 2010.

Welch, B. C., Jacobel, R. W., and Arcone, S.: First results from radar profiles collected along the US-ITASE traverse from Taylor Dome to South Pole (2006–2008), Ann. Glaciol., 50, 35–41, 2009.

Wright, A. and Siegert, M. J.: The identification and physiographical setting of Antarctic subglacial lakes: an update based on recent geophysical data, in: Subglacial Antarctic Aquatic Environments, edited by: Siegert, M., Kennicutt, C., Bindschadler, B., AGU Geophysical Monograph 192, Washington D.C., 9–26, 2011.

Wright, A. P. and Siegert, M. J.: A fourth inventory of Antarctic subglacial lakes, Antarct. Sci., 24, 659–664, doi:10.1017/S095410201200048X, 2012.

Wright A. P., Siegert, M. J., Le Brocq, A., and Gore, D.: High sensitivity of subglacial hydrological pathways in Antarctica to small ice sheet changes, Geophys. Res. Lett., 35, L17504, doi:10.1029/2008GL034937, 2008.

Wright, A. P., Young, D. A., Roberts, J. L., Dowdeswell, J. A., Bamber, J. L., Young, N., Le Brocq, A. M., Warner, R. C., Payne, A. J., Blankenship, D. D., van Ommen, T., and Siegert, M. J.: Evidence for a hydrological connection between the ice divide and ice sheet margin in the Aurora Subglacial Basin sector of East Antarctica, J. Geophys. Res.-Earth, 117, F01033, doi:10.1029/2011JF002066, 2012.

Wright, A. P., Young, D. A., Bamber, J. A., Dowdeswell, J. A., Payne, A. J., Blankenship, D. D., and Siegert, M. J.: Subglacial hydrological connectivity within the Byrd Glacier catchment, J. Glaciol., in press, 2014.

3

Low below-ground organic carbon storage in a subarctic Alpine permafrost environment

M. Fuchs[1,*]**, P. Kuhry**[1]**, and G. Hugelius**[1]

[1]Department of Physical Geography, Stockholm University, 106 91 Stockholm, Sweden
[*]now at: Alfred Wegener Institute Helmholtz Centre for Polar and Marine Research, Telegrafenberg A43, 14473 Potsdam, Germany

Correspondence to: M. Fuchs (matthias.fuchs@awi.de)

Abstract. This study investigates the soil organic carbon (SOC) storage in Tarfala Valley, northern Sweden. Field inventories, upscaled based on land cover, show that this alpine permafrost environment does not store large amounts of SOC, with an estimate mean of $0.9 \pm 0.2 \, \text{kg C m}^{-2}$ for the upper meter of soil. This is 1 to 2 orders of magnitude lower than what has been reported for lowland permafrost terrain. The SOC storage varies for different land cover classes and ranges from $0.05 \, \text{kg C m}^{-2}$ for stone-dominated to $8.4 \, \text{kg C m}^{-2}$ for grass-dominated areas. No signs of organic matter burial through cryoturbation or slope processes were found, and radiocarbon-dated SOC is generally of recent origin ($< 2000 \, \text{cal yr BP}$). An inventory of permafrost distribution in Tarfala Valley, based on the bottom temperature of snow measurements and a logistic regression model, showed that at an altitude where permafrost is probable the SOC storage is very low. In the high-altitude permafrost zones (above 1500 m), soils store only ca. $0.1 \, \text{kg C m}^{-2}$. Under future climate warming, an upward shift of vegetation zones may lead to a net ecosystem C uptake from increased biomass and soil development. As a consequence, alpine permafrost environments could act as a net carbon sink in the future, as there is no loss of older or deeper SOC from thawing permafrost.

1 Introduction

The permafrost-affected soil area in the northern circumpolar region is widespread, occupying about 17.8 million km^2 (Hugelius et al., 2014). The soils in the northern permafrost region store large amounts of soil organic carbon (SOC), which are vulnerable to climate change. With a warming climate, which is expected to be most pronounced in northern high latitudes, thawing permafrost soils may cause remobilization of soil organic matter (SOM) previously protected in permafrost (Gruber et al., 2004; Schuur et al., 2008). This can lead to an increased microbial decomposition of SOM and a release of carbon dioxide (CO_2) and methane (CH_4) into the atmosphere. As a consequence, permafrost soils may act as a future carbon (C) source and lead to a positive climate feedback. However, the total storage of SOC within the northern permafrost region and the amount of greenhouse gases that can be released into the atmosphere and trigger accelerated climate warming are still uncertain (Schuur et al., 2009, 2013; Kuhry et al., 2010; McGuire et al., 2010).

Several local-to-regional-scale studies have been carried out to investigate stocks of SOC in northern permafrost environments (e.g., Michaelson et al., 1996; Kuhry et al., 2002; Zimov et al., 2006a; Ping et al., 2008; Horwath Burnham and Sletten, 2010; Hugelius et al., 2010, 2011). Based on the Northern Circumpolar Soil Carbon Database (NCSCD), Hugelius et al. (2014) estimated the 0–300 cm SOC stock in the northern permafrost region to be $1035 \pm 150 \, \text{Pg C}$ ($\pm 95\%$ confidence interval). However, many regions in the NCSCD are underrepresented and contain few sampled pedons, leading to a more generalized estimation of the C stocks for some remote areas (Mishra et al., 2013). Especially in regions of thin sedimentary overburden, including highlands and alpine terrain, estimates are based on limited data and associated with wide uncertainty ranges (Hugelius et al., 2014).

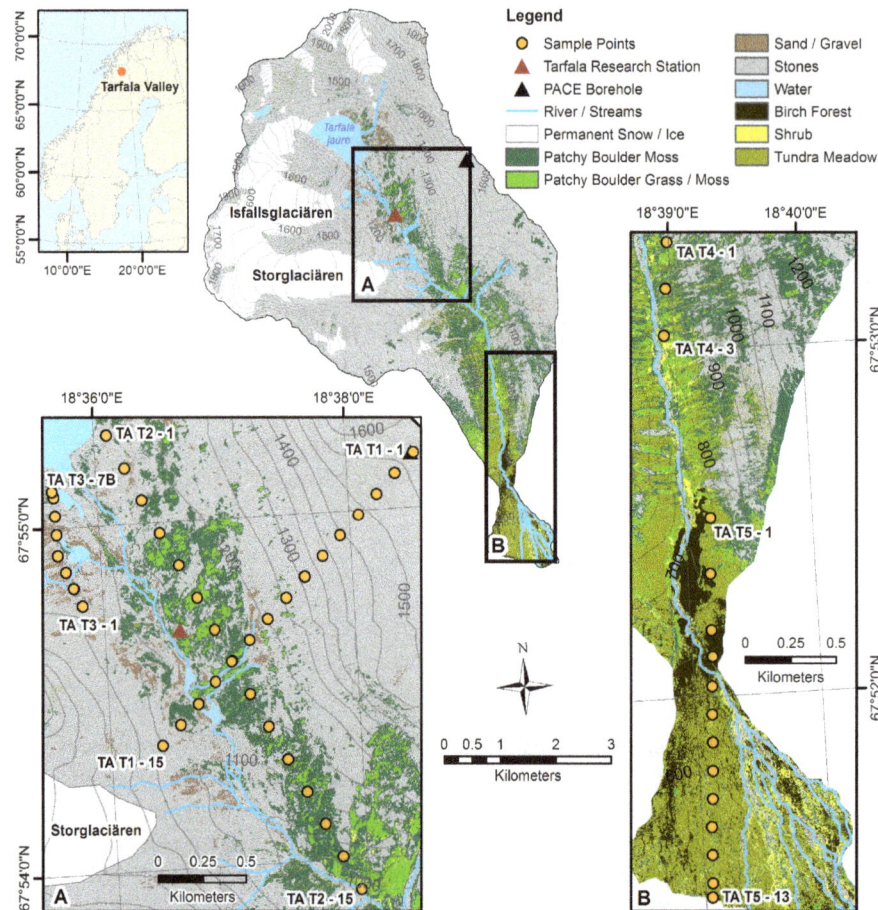

Figure 1. The Tarfala Valley study area, including an overview location map, a map of the whole study area with land cover classification and detailed maps showing transect and sample point locations in the central (**a**) and lower (**b**) parts of the valley.

This study presents a detailed SOC inventory for a subarctic alpine permafrost environment by investigating the C stocks in soils of Tarfala Valley, northern Sweden. It is essential to establish to what extent these type of environments contribute to the large SOC storage in the northern permafrost region. Mountain areas and alpine permafrost are sensitive to climate change due to steep ecoclimatic gradients. The aim of this study is to assess the permafrost extent and SOC pools in a subarctic alpine environment and evaluate their potential fate under conditions of future global warming.

2   Study area

Tarfala Valley is located in the Scandes mountains of northern Sweden, at ca. 67°55′ N, 18°37′ E. The study area (31.2 km^2) is delineated based on the catchment of Tarfala River (Tarfalajåkk), which drains into the broader Ladtjovagge. It includes the alluvial fan of Tarfala River to encompass the entire altitudinal gradient from the source to

the outlet of Tarfala River. The area ranges between 550 and 2100 m a.p.s.l. (above present sea level) and is characterized in the upper part by six glaciers that drain into Tarfala River (Fig. 1).

The mean annual air temperature (MAAT) at the Tarfala research station is −3.4 °C (1965–2009) and the mean annual precipitation for the Tarfala River catchment is 1997 mm (Dahlke et al., 2012). The MAAT in Tarfala increased by 0.54 °C per decade for the period 1969–2009, whereas the mean annual precipitation did not change significantly (Dahlke et al., 2012). The mean altitudinal lapse rate between the Tarfala research station (1135 m a.p.s.l.) and the mountain saddle (Tarfalaryggen) along the eastern border of the study area (1540 m a.p.s.l.) is ca. 4.5 °C km^{-1}; however, the lapse rate in the summer months (JJA) of around 5.8 °C km^{-1} is significantly higher than the winter (DJF) lapse rate of around 2.7 °C km^{-1} (Jonsell et al., 2013).

The vegetation cover in the study area is generally sparse. In high-elevation areas there is mostly barren ground. The middle part of the valley, around the Tarfala research station (1135 m a.p.s.l.), is characterized by patchy boulder fields and shallow soils with a mix of bare rocks, grasses, mosses and lichen. Further down the valley, dwarf shrubs (mainly *Salix* species and *Empetrum hermaphroditum*) appear up to 1000 m a.p.s.l. and the mountain birch forest (*Betula pubescens* ssp. *czerepanovii*) reaches up to ca. 750 m a.p.s.l. On the alluvial fan, in the lowest part of the study area, the vegetation consists of a mix of deciduous and evergreen shrubs, graminoids and herbs.

Tarfala Valley is characterized by little and very shallow soil development. The predominant soils in the study area are characterized by very limited soil formation with poorly developed soil genetic horizons, high stone content and shallow regolith. These soils are classified as Leptosols and Regosols (IUSS Working Group WRB, 2006). On Tarfalaryggen, soil movement caused by frost–thaw cycles (cryoturbation) has led to patterned ground formation, and there is permafrost in the upper 2 m of soil; these soils are classified as Turbic Cryosols. In riverbed deposits of glacial streams (e.g., in the glacier forefield of Isfallsglaciären), soils are classified as Fluvisols.

Extensive research has been carried out in Tarfala Valley, focusing mainly on glaciology and permafrost. Glaciers are the main subject of studies, with Storglaciären having the longest ongoing glacier mass balance measurements in the world (Holmlund et al., 2005). According to Brown et al. (1997), Tarfala Valley is located in the discontinuous permafrost zone. A permafrost borehole installed by the PACE (Permafrost and Climate in Europe) project is situated 1540 m a.p.s.l. on Tarfalaryggen (Harris et al., 2001). The borehole measures the soil temperature down to 100 m every 6 h (Sollid et al., 2000). Mean annual ground temperature at the depth of zero annual amplitude is $-2.8\,^{\circ}\mathrm{C}$, with a mean active layer depth of 1.5–1.6 m. Permafrost is currently not present in a 15 m deep borehole located at an elevation of 1135 m a.p.s.l. near Tarfala research station (Bolin Centre for Climate Research, 2013). King (1984) reports an active layer depth of 2.5–4 m in the valley floor around 1200 m a.p.s.l.

Even though many scientific studies have been carried out in Tarfala Valley (e.g., Stork, 1963, on vegetation cover; King, 1984, and Isaksen et al., 2007, on permafrost; Holmlund et al., 2005, and Jansson and Pettersson, 2007, on glaciology; Dahlke et al., 2012, on hydrology), there are no previous studies on SOC storage from this area.

3 Methods

3.1 Soil sampling

In August 2012, a stratified random sampling program was executed in Tarfala Valley, during which soil profiles were collected along five transects. Transects were chosen to represent the altitudinal zones and vegetation types in the valley. Strict equidistant individual profiles were placed along the transects to introduce a degree of randomness in the sampling. Near-surface organic layers were collected from pits dug into the soils by cutting out samples of known volume. Deeper soil layers were sampled by hammering a steel pipe of ca. 4 cm diameter into the soil vertically in 5–10 cm depth increments. Coring was pursued so that the whole profile was collected to a depth of 1 m, if possible. Most of the collected soil profiles were shallow because the stony soils did in no single case enable a sampling to the full reference depth of 100 cm. The soil was mostly of uniform nature and, during collection of soil samples, no indication of soil organic matter buried through cryoturbation could visually be detected. Furthermore, permafrost was never encountered during coring, even at high elevations, indicating generally deep active layers in Tarfala Valley. In total, 56 profile sites were sampled and described and 295 individual soil samples collected.

3.2 Land cover classification

A description of the vegetation cover in a ground-truth plot (diameter 10 m) was made around each profile site, with special attention paid to the occurrence of stones and boulders (see description of SOC mass calculation below). For upscaling purposes, a land cover classification (LCC) was compiled from remotely sensed data. For this LCC, an orthophoto (compiled with ERDAS Imagine LPS from CIR aerial photographs with 0.5 m spatial resolution) (Lantmäteriet, 2008), a WorldView2 satellite image (European Space Imaging GMBH, 2012) and a Landsat 5 (TM) (US Geological Survey, 2011) satellite image were used. The remote mountainous area as well as cloud and snow cover in the images made a usage of different data sets unavoidable to cover the whole valley. The LCC includes nine different classes which have been separated by a combination of a 3-D stereo analysis and supervised classification (maximum likelihood). The requirements for a supervised classification in general and the training areas in particular followed Campbell (2011). To verify the classification, the kappa index of agreement was calculated based on the 56 ground-truth plots. Nine dominant land cover classes were recognized in Tarfala Valley and form the basis for establishing a land cover classification based on field and remotely sensed data. The classes are presented in the Supplement (Table S1).

3.3 Geochemical analyses

Soil samples of known volume were weighed in the laboratory after oven drying at $60\,^{\circ}\mathrm{C}$ (for 48 h) to calculate dry bulk density (DBD, $\mathrm{g\,cm^{-3}}$). For loss on ignition (LOI), samples were burned at $550\,^{\circ}\mathrm{C}$ for 6 h to determine the organic carbon content and at $950\,^{\circ}\mathrm{C}$ for 2 h to determine the carbonate content (Dean, 1974; Heiri et al., 2001). In addition, a subset of 96 samples was further homogenized, freeze-dried and analyzed, first with a CarloErba NC 2500 elemental analyzer to determine C/N (weight) ratios and second with a coupled mass spectrometer (Finnigan DeltaV Advantage) to determine the stable isotope composition of δ^{13}C and δ^{15}N. Four bulk soil samples were submitted to the Radiocarbon Laboratory in Poznan, Poland, for dating with the accelerator mass spectrometry (AMS) approach (Walker, 2005). After the analysis, radiocarbon dates were calibrated into calendar years, cal yr BP (1950), and expressed as the mean age of the highest 68 % probability interval using the software OxCal 4.1.7 (Bronk Ramsey, 2010).

3.4 SOC storage calculations

The organic C values obtained from the elemental analysis for 96 samples were used to estimate the C percentage (C %) of the remaining 199 samples for which only LOI results were available. Rather than using a constant conversion factor, this is based on a third order polynomial regression between the C % and LOI for those samples where both parameters were measured ($n = 96$, $r^2 = 0.95$):

$$C(\%) = 0.000004 \cdot (LOI_{550})^3 - 0.000352 \cdot (LOI_{550})^2 + 0.481602 \cdot (LOI_{550}). \tag{1}$$

SOC mass (kg C m^{-2}) was calculated for each sample with the DBD (g cm^{-3}), the percentage organic C, the coarse fragment fraction (> 2 mm) (CF, %) and the sample depth interval with the following equation:

$$SOC \left(kg\,C\,m^{-2} \right) = DBD \cdot C \cdot (1 - CF) \cdot depth \cdot 10. \tag{2}$$

The SOC storage (kg C m^{-2}) in each soil profile was calculated by adding up the SOC mass of all samples (5–10 cm depth increments) for the reference depths of 0–30 and 0–100 cm. It should be noted, however, that in all cases it was not possible to reach a full depth of 100 cm due to the occurrence of large stones, boulders or bedrock (these are assumed to contain no SOC). Storage was calculated separately for the organic-rich top soil layer and the underlying mineral soil layer. The division between these layers was made based on field observations. The mean SOC storage for each of the recognized land cover types is calculated as the arithmetic mean of all soil profiles representing those land cover types. To avoid overestimation of the C content, each LCC mean SOC kg C m^{-2} value was weighted by the mean percentage of large stones (> 4 cm diameter) visible at the surface. These areas were considered to have no soil development and to contain no SOC. The coverage of large stones was derived by field observations at every sample spot within a radius of 5 m. Thereafter, the mean SOC storage in Tarfala Valley was calculated based on the proportions of the land cover classes in the LCC. These calculations were performed for all land cover classes together (including glaciers, barren grounds and lakes) and for the vegetated classes only.

3.5 Statistical methods

The results from the geochemical analyses and the upscaling were further analyzed with statistical methods. All statistical analyses were carried out with the open-source statistical analysis package PAST 2.17 (Hammer et al., 2001). Three main statistical analyses were carried out: (1) confidence intervals (CI) for the mean C estimates of the total study area were calculated according to Hugelius (2012); (2) linear correlations (Pearson's correlation) between soil depth and the different geochemical parameters (DBD, C %, LOI, C/N-ratio, δ^{13}C, δ^{15}N) were calculated to examine whether the different parameters decrease or increase significantly with increasing depth; (3) the Student's t test was applied to examine if there is a statistically significant difference between the organic-rich top soil and the underlying mineral samples for all the different geochemical parameters. In all cases, the probability limit of $p \leq 0.01$ was chosen for statistical significance.

3.6 Permafrost mapping

In addition to the SOC inventory, the permafrost distribution in Tarfala Valley was mapped. Bottom temperature of snow (BTS) measurements were carried out in March 2013, with the precision temperature-measuring instrument, Series P400 (Dostmann Electronic, 2013). This handheld thermometer has an accuracy of $\pm 0.3\,^{\circ}$C and a resolution of $0.1\,^{\circ}$C. The temperature probe was calibrated in ice water to $0\,^{\circ}$C before every field day. The BTS method is a simple and cost-effective approach to get a first impression on the distribution of permafrost by measuring the temperature at the snow–ground surface interface. For this method a snow cover of a minimum of 80 cm is required to provide sufficient insulation from variable air temperatures above the snow pack (Haeberli, 1973; King, 1983). With the BTS values, a logistic regression with altitude as the single independent variable was used to map the probability of permafrost occurrence. For the logistic regression, BTS values were classified into permafrost likely and non-permafrost likely. The threshold values for permafrost-likely BTS values vary dependent on snow depth and range from -2.5 to $-4.5\,^{\circ}$C (King, 1984). Altitude was chosen as single independent variable because other possibly important parameters for permafrost occurrence (slope, aspect, solar radiation, etc.) showed no significant correlation with measured BTS values. Using the permafrost probability map, the amount of SOC stored in probable permafrost areas could be estimated.

4 Results

4.1 Land cover classification

The LCC presented in Fig. 1 has an overall accuracy of 72.2 % and a kappa index of agreement of 0.68. The rather low kappa index can be explained by snow cover at higher elevations in the orthophoto, which needed to be corrected by a Landsat 5 (TM) image with a coarser spatial resolution. The LCC shows that Tarfala Valley is dominated by rocks and stones that cover almost 60 % of the area and permanent snow and ice that cover more than 18 %. The largest vegetated land cover class is "patchy boulder moss" which covers almost 10 % of the landscape, but this class is defined as a mix of moss and stones that on average has more than 40 % stones. All land cover classes include a certain amount of stones, which ranges from 4 % in the class "birch forest" to 47 % in the class "sand/gravel" (for more details, see Table S1).

4.2 SOC quantity

The mean SOC storage of the study area including all land cover classes is 0.7 ± 0.2 and $0.9 \pm 0.2 \, \mathrm{kg\,C\,m^{-2}}$ for 0–30 and 0–100 cm soil depths, respectively (mean $\pm 95 \%$ CI) (Table 1). Calculations have also been made for the vegetated area only. This area excludes the low SOC land cover classes "stone", "sand/gravel", "water" and "permanent snow/ice" and, therefore, the mean C storage is considerably higher than for the entire study area. The mean SOC for the vegetated area only is 3.7 ± 0.8 and $4.6 \pm 1.2 \, \mathrm{kg\,C\,m^{-2}}$ for 0–30 and 0–100 cm soil depths, respectively (mean $\pm 95 \%$ CI).

A detailed analysis of the different land cover classes shows the partitioning of C stored in Tarfala (Fig. 2). Most of the SOC in Tarfala Valley is stored in the class "tundra meadow" (35 % of SOC) even though it only covers 4.3 % of the total study area. However, the highest mean value occurs in the class "patchy boulder grass/moss", which stores on average $8.4 \pm 5.4 \, \mathrm{kg\,C\,m^{-2}}$ (Table 1) and accounts for 24 % of the total SOC storage in Tarfala Valley.

The coefficient of variation of the mean SOC values of the land cover classes is high (near 1 in many cases), which is an effect of the high within-class variability in depth of the fine-grained deposits overlying coarse regolith or bedrock (also reflected in the standard deviation of the mean profile depth, see Table 1). Therefore the variability of profile depth within the different land cover classes is reflected in the variability of organic carbon for single classes. Additionally, the coarse fragment fraction ($> 2 \, \mathrm{mm}$) varied within classes (data not shown). Besides the variability in fine-soil depth, the results show that most of the organic C is stored in near-surface layers. On average, more than 80 % of the SOC is stored within the upper 30 cm of soil and a third of the SOC is stored in the organic-rich top soil layer. This also allows an estimation of the SOC stored within the permafrost layer. As the active layer in Tarfala Valley seems to be on the order of 1.5–4 m thick (King, 1984; Isaksen et al., 2007), it can be assumed that only a very minor-to-negligible amount of organic C is stored within the permafrost layer. It should be noted that permafrost was never reached during field coring due to the occurrence of bare rock and stones.

Table 1. Mean soil organic carbon (SOC) storage and sample site characteristics for the different land cover classes in Tarfala Valley.

Land cover classes	Mean SOC storage				Mean profile depth			Area (km²)	Area (%)	Mean altitude of profiles incl. range (m a.p.s.l.)		Profile distribution	
	0–30 cm (kg C m⁻² ± std)	0–100 cm (kg C m⁻² ± std)	Organic-rich top soil layer (kg C m⁻² ± std)	Mineral layer (kg C m⁻² ± std)	Profile site (cm ± std)	Organic-rich top soil layer (cm ± std)	Mineral layer (cm ± std)					Number of sites	Sites in permafrost[c] (cont./discont./sporad./isol. or none)
Birch forest	5.7 ± 3.5	6.6 ± 5.0	2.0 ± 0.6	4.6 ± 4.6	25 ± 22	3.4 ± 0.9	22 ± 22	0.3	1.0	656	(636–675)	3	–/–/–/3
Tundra meadow	6.0 ± 3.0	7.2 ± 5.5	2.7 ± 1.4	4.5 ± 4.7	24 ± 24	4.3 ± 2.1	20 ± 23	1.3	4.3	652	(562–864)	8	–/–/–/8
Shrub	4.6 ± 4.3	4.6 ± 4.3	1.5 ± 0.8	3.1 ± 3.6	16 ± 11	4.0 ± 1.8	12 ± 10	0.3	0.9	688	(581–824)	5	–/–/–/5
Pat. bould. grass/moss	6.2 ± 4.0	8.4 ± 5.4	1.4 ± 0.8	7.0 ± 5.2	38 ± 25	4.8 ± 1.8	33 ± 25	0.8	2.6	1129	(993–1202)	11	–/–/11/–
Patchy boulder moss	1.8 ± 1.7	2.3 ± 1.9	0.6 ± 0.6	1.8 ± 1.5	40 ± 23	2.9 ± 1.5	37 ± 24	3.1	9.9	1181	(1076–1485)	12	–/1/11/–
Sand/gravel	0.7 ± 0.8	1.0 ± 0.9	0.1 ± 0.1	0.9 ± 0.9	53 ± 27	1.3 ± 1.0	52 ± 27	0.3	1.0	1293	(1122–1559)	9	–/3/6/–
Stones	0.05 ± 0.1	0.05 ± 0.1	0.05 ± 0.1	0.0	3 ± 2	3.0 ± 1.6	0	18.6	59.8	1304	(1154–1542)	8	–/5/3/–
Water	–	–	–	–	–	–	–	0.7	2.1	1194[d]	(–)	0	–
Permanent snow/ice	–	–	–	–	–	–	–	5.8	18.5	1530[d]	(–)	0	–
Study area[a]	0.7 ± 0.2	0.9 ± 0.2	0.3 ± 0.1	0.6 ± 0.2	–	2.4	6.2	31.2	100	1059	(562–1559)	56	–/9/31/16
Vegetated area[a,b]	3.7 ± 0.8	4.6 ± 1.2	1.3 ± 0.3	3.3 ± 1.0	–	3.5	30.2	5.8	18.7	954	(562–1485)	39	–/1/22/16

[a] Mean SOC storage is based on the land cover classification upscaling. The second number in each column is not the standard deviation like in the land cover classes but the 95 %-confidence interval (calculated according to Hugelius, 2012) which is based on the SOC variance and areal extent of each LCC. [b] Only the vegetated area is considered. The following classes have been excluded from the calculations: "sand/gravel", "stones", "water", "permanent snow/ice". [c] The permafrost table was not reached during sampling at any of the sample sites. [d] The mean altitude of the classes "water" and "permanent snow/ice" is based on the land cover classification and not on profile sites.

Table 2. Statistics of the geochemical analyses of soil samples.

Geochemical analysis	All samples, mean ± std	Organic-rich top soil layer samples, mean ± std	Mineral layer samples, mean ± std	Significant difference between organic and mineral samples, Student's t test	Correlation with increasing depth, Pearson's correlation
DBD (g cm^{-3})*	0.9 ± 0.8	0.4 ± 0.3	1.6 ± 0.7	yes ($p < 0.01$)	0.71 ($p < 0.01$)
LOI$_{550}$ (%)*	21.6 ± 27.0	40.3 ± 28.5	4.8 ± 8.1	yes ($p < 0.01$)	−0.47 ($p < 0.01$)
LOI$_{950}$ (%)*	0.4 ± 0.4	0.4 ± 0.4	0.3 ± 0.4	no ($p = 0.06$)	−0.11 ($p = 0.05$)
% C	11.4 ± 13.8	25.8 ± 13.7	3.8 ± 5.2	yes ($p < 0.01$)	−0.54 ($p < 0.01$)
C/N ratio (−)	17.6 ± 8.5	23.3 ± 11.4	14.6 ± 4.1	yes ($p < 0.01$)	−0.38 ($p < 0.01$)
$\delta^{13}C_{tot}$ vs. PDB (‰)	−26.1 ± 1.2	−26.8 ± 1.0	−25.6 ± 1.0	yes ($p < 0.01$)	0.42 ($p < 0.01$)
$\delta^{15}N$ vs. air (‰)	1.8 ± 2.6	−0.54 ± 2.0	3.2 ± 1.8	yes ($p < 0.01$)	0.53 ($p < 0.01$)

* Calculations carried out with all 295 samples; other calculations based on 96 samples from elemental analysis.

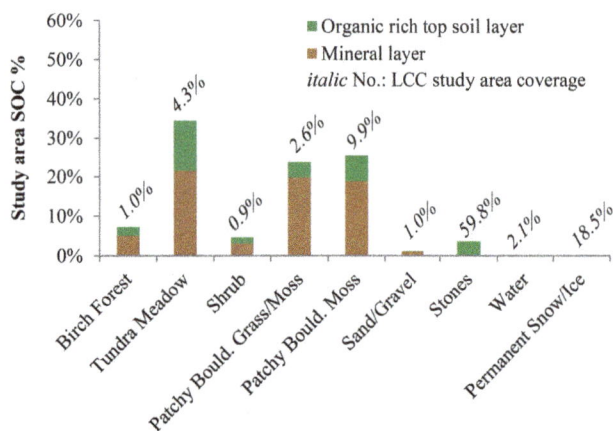

Figure 2. Partitioning of total SOC storage and proportional area coverage of land cover classes in Tarfala Valley (31.2 km^2).

The soils of Tarfala Valley display no signs of cryoturbation of the organic-rich top soil layer into the deeper mineral soil horizons. Likewise, no burial of the organic-rich layer due to solifluction processes on slopes was observed.

4.3 SOM quality and age

The soils in Tarfala Valley are characterized by a steady, statistically significant ($p < 0.01$) increase in bulk density with depth (Fig. 3a; Table 2). However, LOI (550 °C) and C % show strong, statistically significant ($p < 0.01$) negative correlations with depth (Fig. 3b; Table 2). As a result, there is less SOM with greater depth in the soil. There is also a statistically significant (t test, $p < 0.01$) difference in the C content of the organic-rich top soil layer and the underlying mineral layer (Table 2). During field sampling there were no observations of buried SOM through, for example, cryoturbation or solifluction. Similarly, the laboratory results showed no single value or outlier with high C % below the top organic layer. Therefore, there are neither visual nor laboratory

results indicating burial of organic carbon to depth in the investigated soil profiles.

Besides C content, other geochemical analyses of the soil samples also show a coherent picture. The C/N ratio and stable isotopic composition of SOM reflect its relative state of decomposition (e.g., Mariotti and Balesdent, 1990; Kuhry and Vitt, 1996; Ping et al., 1998; Hugelius et al., 2012). There is a statistically significant (t test, $p < 0.01$) difference between the mean C/N ratio of the organic-rich top soil layer (23.3 ± 11.4) and that of the mineral layer (14.6 ± 4.05). The C/N ratio decreases with increasing depth ($p < 0.01$), indicating progressively more decomposed SOM (Fig. 3c). Ping et al. (1998) pointed out that the C/N ratio is dependent on vegetation cover and that trends need to be interpreted carefully. In Tarfala Valley, the decrease of C/N ratio with depth is consistent across all land cover classes. However, these trends are not statistically significant for the separate land cover classes, probably due to the limited number of replicates within each class (data not shown). The stable isotope composition of $\delta^{13}C$ vs. PDB and $\delta^{15}N$ vs. air shows statistical significant ($p < 0.01$) enrichment of stable isotopes with increasing soil depth (Fig. 3d; Table 2). The enrichment of $\delta^{13}C$ and $\delta^{15}N$ with depth can be considered an indication of SOM degradation through microbial respiration (Mariotti and Balesdent, 1990; Ping et al., 1998).

Four bulk soil samples (living roots removed) from two profiles belonging to the class "patchy boulder grass/moss", located close to the floor of the central Tarfala Valley, have been radiocarbon-dated (Table 3). These profiles were selected because they had the thickest organic-rich top soil layer among the collected profiles in the study area and displayed a slight, but highly unusual for this area, C enrichment in the underlying mineral soil (weak B-horizon development). Results indicate that the SOM close to the surface is recent in age (< 100 years old), whereas the mineral soil at greater depths contains slightly older SOM, with ages of 1269 and 1919 cal yr BP (Table 3). Considering the fact that the two dated profiles are among the most well-developed

Table 3. Results from the radiocarbon analysis.

Site	Depth (cm)	Lab. no.	Site and sample description	Age ^{14}C	Age cal yr BP
TA T1-9B	19–20	Poz-51853	Grass/moss patch, base of top organics	123.48 ± 0.4 pMC	modern
TA T1-9B	50–60	Poz-51854	Grass/moss patch, silty sand and stones	2035 ± 35 BP	1919
TA T2-11	10–15	Poz-51856	Grass/moss patch, base of top organics	95 ± 30 BP	20
TA T2-11	33–37	Poz-51857	Grass/moss patch, silty sand and small stones	1380 ± 30 BP	1269

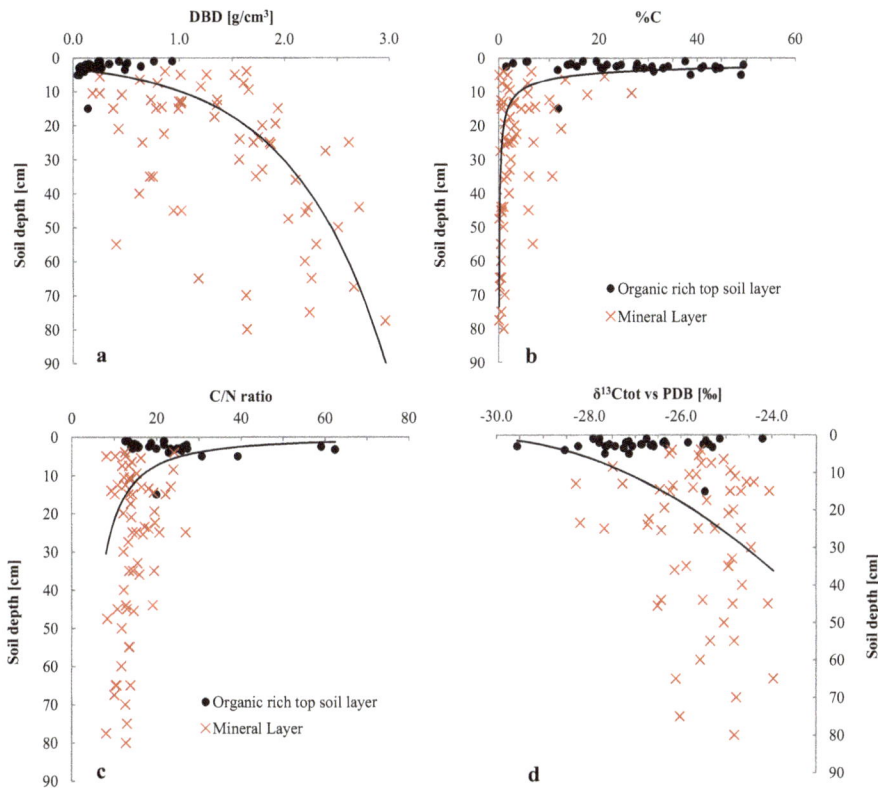

Figure 3. Results of the geochemical analyses of the soils samples of Tarfala Valley. DBD is Dry bulk density (**a**); % C is percentage C (**b**); C/N is the weight ratio (**c**); $\delta^{13}C_{tot}$ vs. PDB is the stable isotope δ^{13}C analyzed to the international standard PeeDeeBelemnite (**d**). Lines are best-fit power, polynomial or exponential regressions, shown for graphic representation of mean trends only. Some high bulk density values (up to 3.0 g cm^{-3}) in sandy profile sites are probably the result of errors in field volume estimates due to difficulties in collecting these loose materials.

soils in the study area, it is most likely that most of the SOM in Tarfala Valley is of a very young age. The geochemistry of these two dated profiles, which reflect the general trends described for the whole data set, is presented in the Supplement (Fig. S2).

4.4 Permafrost mapping

Permafrost zones are commonly separated into the classes continuous, discontinuous, sporadic and isolated patches (e.g., Brown et al., 1997). However, with the logistic regression approach the probability for the occurrence, rather than the areal extent, of permafrost was used to map the per-

mafrost distribution into the conventional classes (Fig. 4). This was already applied by Lewkowicz and Ednie (2004) in their study in Yukon Territory, Canada. However, with this approach the permafrost distribution has to be interpreted carefully, especially in a highly heterogeneous alpine environment like Tarfala Valley. Areas with a > 90 % probability for the occurrence of permafrost are considered as continuous, which in Tarfala Valley includes all areas above 1561 m a.p.s.l. The discontinuous permafrost zone (probability between 50 and 90 %) occurs at an altitude between 1218 and 1561 m a.p.s.l., while the sporadic permafrost zone commences at an altitude above 875 m a.p.s.l. (probability > 10 %). The altitudinal zonation

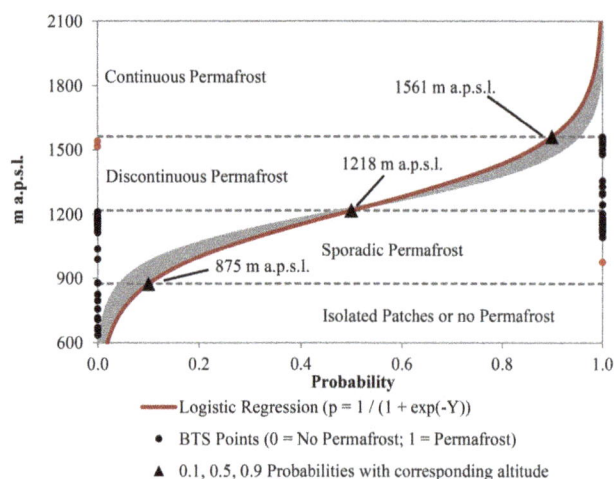

Figure 4. Permafrost probability in relation to altitude: the probability is based on a logistic regression model with the altitude as single independent variable. The grey corridor shows the range of the permafrost probability if outliers (red dots) are removed from the model.

of permafrost as depicted in Fig. 4 is very similar to those proposed by King (1983) and Marklund (2011), particularly if some outliers are removed from our analysis. The lowermost site where BTS values suggest permafrost is located at 976 m a.p.s.l.; measurements at two high-elevation sites (ca. 1500 m a.p.s.l.) suggest an absence of permafrost. While there are no technical reasons to reject these results, these outliers should be considered with caution due to the inherently large uncertainty range in the BTS method.

5 Discussion

5.1 Current SOC quantity and SOM composition

The results presented for Tarfala Valley show very low SOC storage compared to inventories from lowland areas in the northern permafrost region (e.g., Michaelson et al., 1996; Kuhry et al., 2002; Hugelius et al., 2010). However, the mean value of $0.9 \, \mathrm{kg \, C \, m^{-2}}$ (0–100 cm) is quite close to values reported for other mountainous environments. Kuhry et al. (2002) estimated a mean value of $0.3 \, \mathrm{kg \, C \, m^{-2}}$ for the land cover class "natural barelands" and $1.3 \, \mathrm{kg \, C \, m^{-2}}$ for the land cover class "alpine sparse tundra", which together represent ca. 8 % of the total catchment area of the Usa basin (northeast European Russia); Ping et al. (2008) estimated a value of $3.8 \, \mathrm{kg \, C \, m^{-2}}$ for "mountain soils" in the North American Arctic region. The number of pedons in both these studies is very low ($n = 1$ to 4).

Considering values from only the vegetated area in Tarfala Valley, the mean SOC values are $3.7 \, \mathrm{kg \, C \, m^{-2}}$ for 0–30 cm and $4.6 \, \mathrm{kg \, C \, m^{-2}}$ for 0–100 cm soil-depth intervals. Similar SOC inventories on vegetated patches have been carried out

in the Tibetan plateau. Doerfer et al. (2013) measured the SOC content in the Huashixia and Wudaoliang region, which resulted in mean values of 10.4 and $3.4 \, \mathrm{kg \, C \, m^{-2}}$, respectively, for 0–30 cm. The land cover was in both cases classified as "alpine meadow". Our mean SOC value for the class "tundra meadow" and the corresponding depth interval is $6.0 \, \mathrm{kg \, C \, m^{-2}}$. Other SOC inventories on the Tibetan plateau showed similar results. Ohtsuka et al. (2008) measured a mean SOC content of $1.0–13.7 \, \mathrm{kg \, C \, m^{-2}}$ for 0–30 cm in "alpine meadow"; Yang et al. (2008) measured $9.6 \, \mathrm{kg \, C \, m^{-2}}$ in "alpine meadow" and $3.1 \, \mathrm{kg \, C \, m^{-2}}$ for "alpine steppe"; and Wang et al. (2008) measured $9.3–10.7 \, \mathrm{kg \, C \, m^{-2}}$ for "alpine grasslands" (our corresponding value for the "patchy boulder grass/moss" class is $6.2 \, \mathrm{kg \, C \, m^{-2}}$). A SOC inventory from the Swiss Alps showed higher values than Tarfala Valley. Zollinger et al. (2013) investigated "alpine grassland" (at 2700 m a.p.s.l.) and "subalpine forest" (at 1800 m a.p.s.l.) soils and estimated the C stocks down to the C-horizon at ca. $10 \, \mathrm{kg \, C \, m^{-2}}$ for permafrost and ca. $15 \, \mathrm{kg \, C \, m^{-2}}$ for non-permafrost sites. It has to be emphasized that these values represent only the mean of the vegetated sites and are not based on a landscape upscaling to include all mountainous terrain. Nonetheless, in all these studies, the high SOC content often reported from lowland permafrost areas, ranging between ca. 25 and $50 \, \mathrm{kg \, C \, m^{-2}}$ (e.g., Michaelson et al., 1996; Kuhry et al., 2002; Hugelius et al., 2010), is never achieved.

Several reasons for the low SOC values in Tarfala Valley seem obvious. There is a high amount of bare ground and glaciated terrain in the study area (almost ca. 80 %) which leads to very limited in situ production of organic plant matter in the system. Even the vegetated classes have abundant stone cover which diminishes the landscape fraction with fine-soil development. The fraction of stone coverage in the different land cover classes varied between 4 and 47 % (the soil volume occupied by stones was considered devoid of SOC in stock calculations). Furthermore, no signs of SOM burial by cryoturbation or solifluction processes were observed in any investigated soil profile. Burial of SOM through cryoturbation or slope processes are important mechanisms explaining high SOC stocks in other permafrost environments (Palmtag et al., 2015). The active layer in Tarfala Valley is significantly deeper than the depth of active soil formation, which means that organic carbon decomposition is not impeded by sub-zero temperatures during the warm season. The steep topography and coarse sediments favor rapid drainage and aerated soils. No peat formation or peaty soils were observed in Tarfala Valley. Finally, the soils are rather shallow; in most cases they do not reach a depth of 1 m and sometimes not even 30 cm. As a consequence of all these factors, the soils in Tarfala Valley are not characterized by any of the pedogenic processes that often lead to the accumulation of high stocks of SOC in permafrost region soils (Tarnocai et al., 2009).

The mean value for Tarfala soils down to 1 m depth ($0.9\,\mathrm{kg\,C\,m^{-2}}$) is considerably lower than the one reported for the Swedish mountains ($26.1\,\mathrm{kg\,C\,m^{-2}}$) in the Northern Circumpolar Soil Carbon Database (Hugelius et al., 2013). The high value in the NCSCD can be explained by the highly generalized soil map on which these estimates are based. The NCSCD soil polygon that overlaps with the Tarfala Valley study area has an area of ca. $2900\,\mathrm{km^2}$ and includes adjacent lowland terrain with peatland (Histosols) and forested (Podsols) areas.

Geochemical indicators, such as C/N ratios and stable isotopes ($\delta^{13}C$ and $\delta^{15}N$), indicate that the SOM in Tarfala soils becomes gradually more decomposed with depth and age. Cryoturbation of C-enriched material is one of the mechanisms that significantly increases SOC storage in permafrost soils (e.g., Ping et al., 2008). In Tarfala we did not find evidence for burial of relatively non-decomposed SOM from the organic-rich top soil layer deeper into the profiles. The two dated soil profiles are exceptional for Tarfala Valley as they have the thickest organic-rich top soil layer and relatively high carbon values in greater depths. However, basal dates for even these thickest organic-rich top soil layers are recent and the SOC at greater depth is also quite young ($< 2000\,\mathrm{cal\,yr\,BP}$). Therefore, much of the SOM in Tarfala Valley seems to be cycled within 100 years or less and does not accumulate at greater depths. This is in stark contrast to permafrost soils from lowland regions, which are reported to have extensive cryoturbation of relatively non-decomposed SOM that has been preserved at greater soil depths for thousands of years (e.g., Bockheim, 2007; Hugelius et al., 2010).

5.2 Future developments

Our results indicate that there is not a large amount of SOC stored in the soils of Tarfala Valley. The relatively highest mean SOC storage is found in vegetated ground at lower elevations. A further analysis that takes into account the permafrost zonation shows that the potential for SOC storage in permafrost-affected soils is very small (Fig. 5). The mean SOC value at an elevation of 1250 m a.p.s.l., where the probability for permafrost is just above 50 %, is $0.7\,\mathrm{kg\,C\,m^{-2}}$ (for 0–100 cm) and at an altitude of 1500 m a.p.s.l. (permafrost probability 85 %) it is only $0.1\,\mathrm{kg\,C\,m^{-2}}$. Therefore, most of the SOC in Tarfala Valley is stored at lower elevations where the probability for permafrost-affected soils is low. Taking into account that the active layer is 2.5–4 m thick in the valley floor around 1200 m a.p.s.l. (King, 1984) and the fine soil is only rarely deeper than 1 m, the amount of SOC stored in the permafrost layer is assumed to be negligible.

The vegetation and SOC distribution in Tarfala Valley allow some considerations about future total ecosystem C storage in the area under conditions of global climate change. Climate warming will result in an upwards shift of vegetation zones with the corresponding initiation of soil development in currently high-alpine barren areas. Upwards altitudinal

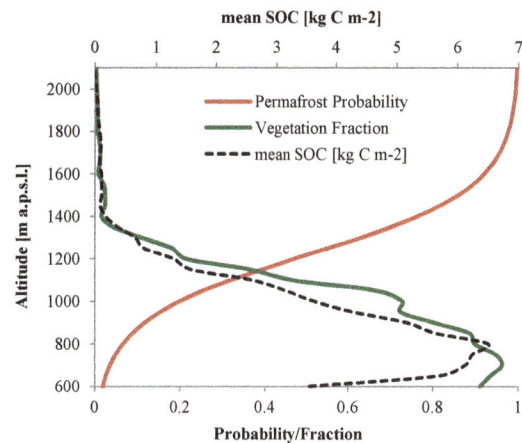

Figure 5. Fraction of vegetation cover and probability for permafrost presence in relation to altitude in Tarfala Valley, including the mean SOC storage per altitude (calculated in 50 m altitudinal intervals). The permafrost probability is based on the BTS measurements and a logistic regression with the altitude as single independent variable. The vegetation fraction is based on the altitudinal distribution of vegetated classes in the land cover classification. Slightly lower SOC values at elevations below 700 m are related to exposed streambeds in the Tarfala river alluvial fan.

shifts of plants due to increased temperatures have been observed in alpine regions (e.g., Walther et al., 2005), including the Scandinavian mountain range (e.g., Klanderud and Birks, 2003; Kullman, 2002, 2010). Kullman and Öberg (2009) report an altitudinal upward shift of trees of about 200 m in the past 100 years in the Swedish Scandes, in accordance with observed temperature increases. For a first rough estimation of potential upwards shifts of vegetation zones, the mean summer temperature change was taken as a first indicator even though many other factors will affect the vegetation (e.g., winter temperatures, precipitation, wind exposure; Kullman, 2010). The projected mean summer (JJA) temperature increase for the Tarfala mountain region until 2100 is $2.8\,^{\circ}\mathrm{C}$ (SRES A1B scenario, SMHI, 2013). Considering a summer lapse rate of $5.8\,^{\circ}\mathrm{C\,km^{-1}}$ (Jonsell et al., 2013), the potential altitudinal upward shift for the vegetation cover is ca. 500 m. Grace et al. (2002) and Kullman (2010) calculated a similar potential treeline shift in the region by the end of this century. However, not the entire Tarfala Valley will be suitable for plant colonization, because of steep slopes, a lack of fine-soil matrix, and wind-exposed ridges.

Schuur et al. (2009) showed that in the Alaskan tundra, increased plant productivity is eventually outweighed by increased decomposition of deeper and older SOM following permafrost thaw. For projections of permafrost degradation in Tarfala Valley, the mean annual air temperature has to be considered. A climate scenario for the Tarfala mountain region estimates a mean annual temperature increase of ca. $4.6\,^{\circ}\mathrm{C}$ until 2100 (SRES A1B scenario, SMHI, 2013). Taking into account a mean annual lapse rate of $4.5\,^{\circ}\mathrm{C\,km^{-1}}$

(Jonsell et al., 2013), the 0 °C air temperature isotherm could rise with ca. 1000 m, which would greatly affect permafrost occurrence in the area. Data from the PACE borehole at Tarfalaryggen show that the permafrost temperature at the zero annual amplitude depth of 20 m has already experienced a warming of 0.047 °C yr^{-1} (Jonsell et al., 2013). Even though future permafrost degradation is highly plausible for most of the upper Tarfala Valley, only a negligible amount of SOC is currently stored in the area and could be affected by thaw. Under future climate warming and permafrost thawing, little or no SOC will be remobilized from permafrost soils in Tarfala Valley. On the contrary, increased temperatures will lead to an upward vegetation shift, phytomass production and soil development, with the result of an increased C uptake in Tarfala Valley in the future. The only way that projected permafrost thaw might negatively affect C uptake is through an initial increased slope instability in steep terrain (Gregory and Goudie, 2011; French, 2007), which could prevent vegetation establishment and soil development.

Compared to lowland permafrost regions in the northern circumpolar region (see e.g., Gruber et al., 2004; Zimov et al., 2006b; Schuur et al., 2009), a subarctic high-alpine permafrost environment like the upper Tarfala Valley cannot be considered a future source of C to the atmosphere. In general, alpine permafrost environments above the contiguous vegetation limit have the potential of becoming a C sink in the future and therefore stand out as an exception in the general assessment of thawing permafrost soils representing an important positive feedback to future climate warming (e.g., Schuur et al., 2013).

6   Conclusions

The SOC inventory in Tarfala Valley, with a mean storage of 0.9 kg C m^{-2} for the upper meter of soil, shows that this area cannot be considered a C-rich permafrost environment. This low value is a result of the high amount of barren ground and stony surfaces in the study area, low plant productivity, shallow soils and lack of SOM burial through cryoturbation or slope processes. The low SOC storage leads to the conclusion that environments like Tarfala Valley cannot become significant sources of C with future permafrost thawing. Instead, they could act as net C sinks following an upward shift of vegetation zones causing increased phytomass production, soil development and SOM accumulation. The potential magnitude of an increased C uptake in this type of mountainous permafrost region remains to be addressed by further studies. Nevertheless, this study shows that there is a need to include alpine environments to estimate the total SOC stock in permafrost soils of the northern circumpolar region and to fully assess the permafrost thaw–C feedback.

Acknowledgements. We wish to thank Eva-Lisa and Juri Palmtag, Niels Weiss and Malin Brandel for help with the fieldwork, Andreas Bergström and his team for the support at Tarfala research station and two anonymous reviewers for their constructive comments. The study was supported by the Nordic Center of Excellence DEFROST project.

Edited by: P. Marsh

References

Bockheim, J. G.: Importance of cryoturbation in redistributing organic carbon in permafrost-affected soils, Soil Sci. Soc. Am. J., 71, 1335–1342, doi:10.2136/sssaj2006.0414N, 2007.

Bolin Centre for Climate Research: Bolin Centre Database, Tarfala data, permafrost monitoring data, http://bolin.su.se/data/tarfala/permafrost.php, last access: 30 October 2013.

Bronk Ramsey, C.: OxCal v4.1.7, Radiocarbon Calibration Software, Research Lab for Archaeology, Oxford, UK, available at: http://c14.arch.ox.ac.uk (last access: 13 February 2013), 2010.

Brown, J., Ferrians Jr., O. J., Heginbottom, J. A., and Melnikov, E. S.: Circum-Arctic map of permafrost and ground-ice conditions, 1 : 10 000 000, Map CP-45, United States Geological Survey, International Permafrost Association, Washington, D.C., 1997.

Campbell, J. B.: Introduction to remote sensing, 5th Edn., Guilford Press, New York, 667 pp., 2011.

Dahlke, H. E., Lyon, S. W., Stedinger, J. R., Rosqvist, G., and Jansson, P.: Contrasting trends in floods for two sub-arctic catchments in northern Sweden – does glacier presence matter?, Hydrol. Earth Syst. Sci., 16, 2123–2141, doi:10.5194/hess-16-2123-2012, 2012.

Dean, W. E.: Determination of carbonate and organic matter in calcareous sediments and sedimentary rocks by loss on ignition: Comparison with other methods, J. Sediment. Petrol., 44, 242–248, 1974.

Doerfer, C., Kuehn, P., Baumann, F., He, J.-S., and Sholten, T.: Soil organic carbon pools and stocks in permafrost-affected soils on the Tibetan Plateau, PLOS ONE, 8, e57024, doi:10.1371/journal.pone.0057024, 2013.

Dostmann Electronic: Temperature Probe P400, http://www.dostmann-electronic.de/englisch/Produkte/produkt.php?pid=205&lang=eng, last access: 6 May 2013.

European Space Imaging GMBH: WorldView2 satellite image, date of acquisition: 4 July 2012, http://www.euspaceimaging.com/ (last access: 6 May 2013), 2012.

French, H. M.: The Periglacial Environment, 3rd Edn., John Wiley & Sons Ltd, West Sussex, England, 480 pp., doi:10.1002/9781118684931, 2007.

Grace, J., Berninger, F., and Nagy, L.: Impacts of climate change on tree line, Ann. Bot.-London, 90, 537–544, 2002.

Gregory, K. J. and Goudie, S. A.: The SAGE Handbook of Geomorphology, SAGE Publications Ltd, London, England, 648 pp., 2011.

Gruber, N., Friedlingstein, P., Field, C. B., Valentini, R., Heimann, M., Richey, J. E., Romero-Lankao, P., Schulze, D., and Chen, C. T. A.: The vulnerability of the carbon cycle in the 21st century: An assessment of carbon-climate-human interactions, in: The Global Carbon Cycle: Integrating Humans, Climate, and the Natural World, edited by: Field, C. and Raupach, M., Island Press, Washington, D.C., 45–76, 2004.

Haeberli, W.: Die Basis-Temperatur der winterlichen Schneedecke als möglicher Indikator für die Verbreitung von Permafrost in den Alpen, Z. Gletscherk. Glaziol., 9, 221–227, 1973.

Hammer, Ø., Harper, D. A. T., and Ryan, P. D.: PAST: Paleontological statistics software package for education and data analysis, Palaeontol. Electron., 4, available at: http://palaeo-electronica.org/2001_1/past/issue101.htm (last access: 6 March 2013), 2001.

Harris, C., Haeberli, W., Vonder Muehll, D., and King, L.: Permafrost monitoring in the high mountains of Europe: the PACE project in its global context, Permafrost Periglac., 12, 3–11, doi:10.1002/ppp.337, 2001.

Heiri, O., Lotter, A. F., and Lemcke, G.: Loss on ignition as a method for estimating organic and carbonate content in sediments: reproducibility and comparability of results, J. Paleolimnol., 25, 101–110, doi:10.1023/A:1008119611481, 2001.

Holmlund, P., Jansson, P., and Pettersson, R.: A re-analysis of the 58 year mass-balance record of Storglaciären, Sweden, Ann. Glaciol., 42, 389–394, 2005.

Horwath Burnham, J. and Sletten, R. S.: Spatial distribution of soil organic carbon in northwest Greenland and underestimates of high arctic carbon stores, Global Biogeochem. Cy., 24, GB3012, doi:10.1029/2009GB003660, 2010.

Hugelius, G.: Spatial upsacling using thematic maps: An analysis of uncertainties in permafrost soil carbon estimates, Global Biogeochem. Cy., 26, GB2026, doi:10.1029/2011GB004154, 2012.

Hugelius, G., Kuhry, P., Tarnocai, C., and Virtanen, T.: Soil organic carbon pools in a periglacial landscape: A case study from the central Canadian Arctic, Permafrost Periglac., 21, 16–29, doi:10.1002/ppp.677, 2010.

Hugelius, G., Virtanen, T., Kaverin, D., Pastukhov, A., Rivkin, F., Marchenko, S., Romanovsky, V., and Kuhry, P.: High-resolution mapping of ecosystem carbon storage and potential effects of permafrost thaw in periglacial terrain, European Russian Arctic, J. Geophys. Res., 116, G03024, doi:10.1029/2010JG001606, 2011.

Hugelius, G., Routh, J., Kuhry, P., and Crill, P.: Mapping the degree of decomposition and thaw remobilization potential of soil organic matter in discontinuous permafrost terrain, J. Geophys. Res., 117, G02030, doi:10.1029/2011JG001873, 2012.

Hugelius, G., Tarnocai, C., Broll, G., Canadell, J. G., Kuhry, P., and Swanson, D. K.: The Northern Circumpolar Soil Carbon Database: spatially distributed datasets of soil coverage and soil carbon storage in the northern permafrost regions, Earth Syst. Sci. Data, 5, 3–13, doi:10.5194/essd-5-3-2013, 2013.

Hugelius, G., Strauss, J., Zubrzycki, S., Harden, J. W., Schuur, E. A. G., Ping, C.-L., Schirrmeister, L., Grosse, G., Michaelson, G. J., Koven, C. D., O'Donnell, J. A., Elberling, B., Mishra, U., Camill, P., Yu, Z., Palmtag, J., and Kuhry, P.: Estimated stocks of circumpolar permafrost carbon with quantified uncertainty ranges and identified data gaps, Biogeosciences, 11, 6573–6593, doi:10.5194/bg-11-6573-2014, 2014.

Isaksen, K., Sollid, J. L., Holmlund, P., and Harris, C.: Recent warming of mountain permafrost in Svalbard and Scandinavia, J. Geophys. Res., 112, F02S04, doi:10.1029/2006JF000522, 2007.

IUSS Working Group WRB: World reference base for soil resources, 2nd Edn., World Soil Resources Reports 103, FAO, Rome, 143 pp., 2006.

Jansson, P. and Pettersson, R.: Spatial and temporal characteristics of a long mass balance record, Storglaciären, Sweden, Arct. Antarct. Alp. Res., 39, 432–437, 2007.

Jonsell, U., Hock, R., and Duguay, M.: Recent air and ground temperature increases at Tarfala Research Station, Sweden, Polar Res., 32, 19807, doi:10.3402/polar.v32i0.19807, 2013.

King, L.: High mountain permafrost in Scandinavia, Permafrost: Fourth international conference, Proceedings, Nat. Acad. Press, Washington, D.C., 612–617, 1983.

King, L.: Permafrost in Skandinavien, Untersuchungsergebnisse aus Lappland, Jotunheimen und Dovre/Rondane, Heidelberger Geographische Arbeiten 76, University of Heidelberg, Department of Geography, Heidelberg, 174 pp., 1984.

Klanderud, K. and Birks, H. J. B.: Recent increases in species richness and shifts in altitudinal distribution of Norwegian mountain plants, Holocene, 13, 1–6, 2003.

Kuhry, P. and Vitt, D. H.: Fossil carbon/nitrogen ratios as a measure of peat decomposition, Ecology, 77, 271–275, doi:10.2307/2265676, 1996.

Kuhry, P., Mazhitova, G. G., Forest, P.-A., Deneva, S. V., Virtanen, T., and Kultti, S.: Upscaling soil organic carbon estimates for the Usa Basin (Northeast European Russia) using GIS-based landcover and soil classification schemes, Dan. J. Geogr., 102, 11–25, 2002.

Kuhry, P., Dorrepaal, E., Hugelius, G., Schuur, E. A. G., and Tarnocai, C.: Potential remobilization of belowground permafrost carbon under future global warming, Permafrost Periglac., 21, 208–214, doi:10.1002/ppp.684, 2010.

Kullman, L.: Rapid recent range-margin rise of tree and shrub species in the Swedish Scandes, J. Ecol., 90, 68–77, 2002.

Kullman, L.: A richer, greener and smaller alpine world: review and projection of warming-induced plant cover change in the Swedish Scandes, Ambio, 39, 159–169, doi:10.1007/s13280-010-0021-8, 2010.

Kullman, L. and Öberg, L.: Post-little ice age tree line rise and climate warming in the Swedish Scandes: a landscape ecological perspective, J. Ecol., 97, 415–429, doi:10.1111/j.1365-2745.2009.01488.x, 2009.

Lantmäteriet: CIR aerial images 08o48zu50_31 ~ 2008-09-10_105654_98–08o48zu50_31_2008-09-10_105654_103, http://www.lantmateriet.se/ (last access: 6 May 2013), 2008.

Lewkowicz, A. G. and Ednie, M.: Probability mapping of mountain permafrost using the BTS Method, Wolf Creek, Yukon Territory, Canada, Permafrost Periglac., 15, 67–80, doi:10.1002/ppp.480, 2004.

Mariotti, A. and Balesdent, J.: ^{13}C natural abundance as a tracer of soil organic matter turnover and paleoenvironment dynamics, Chem. Geol., 84, 217–219, 1990.

Marklund, P.: Alpin permafrost i Kebnekaisefjällen: Modellering med logistic regression och BTS-data, Självständigt arbete i geovetenskap, Nr. 22, Uppsala University, Disciplinary Domain of Science and Technology, Earth Sciences, Department of Earth Sciences, LUVAL, Uppsala, 2011.

McGuire, A. D., Macdonald, R. W., Schuur, E. A. G., Harden, J. W., Kuhry, P., Hayes, D. J., Christensen, T. R., and Heimann, M.: The carbon budget of the northern cryosphere region, Curr. Opin. Environ. Sustain., 2, 231–236, 2010.

Michaelson, G. J., Ping, C. L., and Kimble, J. M.: Carbon storage and distribution in tundra soils of Arctic Alaska, U.S.A., Arctic Alpine Res., 28, 414–424, 1996.

Mishra, U., Jastrow, J. D., Matamala, R., Hugelius, G., Koven, C. D., Harden, J. W., Ping, C. L., Michaelson, G. J., Fan, Z., Miller, R. M., McGuire, A. D., Tarnocai, C., Kuhry, P., Riley, W. J., Schaefer, K., Schuur, E. A. G., Jorgenson, M. T., and Hinzman, L. D.: Empirical estimates to reduce modelling uncertainties of soil organic carbon in permafrost regions: a review of recent progress and remaining challenges, Environ. Res. Lett., 8, 035020, doi:10.1088/1748-9326/8/3/035020, 2013.

Ohtsuka, T., Hirota, M., Zhang, X., Shimono, A., Senga, Y., Du, M., Yonemura, S., Kawashima, S., and Tang, Y.: Soil organic carbon pools in alpine to nival zones along an altitudinal gradient (4400–5300 m) on the Tibetan Plateau, Polar Science, 2, 227–285, 2008.

Palmtag J., Hugelius, G., Lashchinskiy, N. Tamstorf, M. P., Richter, A., Elberling, B., and Kuhry, P.: Storage, landscape distribution, and burial history of soil organic matter in contrasting areas of continuous permafrost, Arct. Antart. Alp. Res., 47, 71–88, doi:10.1657/AAAR0014-027, 2015.

Ping, C. L., Bockheim, J. G., Kimble, J. M., Michaelson, G. J., and Walker, D. A.: Characteristics of cryogenic soils along a latitudinal transect in Arctic Alaska, J. Geophys. Res., 103, 28917–28928, doi:10.1029/98JD02024, 1998.

Ping, C. L., Michaelson, G. J., Jorgenson, M. T., Kimble, J. M., Epstein, H., Romanovsky, V. E., and Walker, D. A.: High stocks of soil organic carbon in the North American arctic region, Nat. Geosci., 1, 615–619, doi:10.1038/ngeo284, 2008.

Schuur, E. A. G., Bockheim, J., Canadell, J. G., Euskirchen, E., Field, C. B., Goryachkin, S. V., Hagemann, S., Kuhry, P., Fafleur, P. M., Lee, H., Mazhitova, G., Nelson, F. E., Rinke, A., Romanovsky, V. E., Shiklomanov, N., Tarnocai, C., Venevsky, S., Vogel, J. G., and Zimov, S. A.: Vulnerability of permafrost carbon to climate change: Implications for the global carbon cycle, BioScience, 58, 701–714, doi:10.1641/B580807, 2008.

Schuur, E. A. G., Vogel, J. G., Crummer, K. G., Lee, H., Sickman, J. O., and Osterkamp, T. E.: The effect of permafrost thaw on old carbon release and net carbon exchange from tundra, Nature, 459, 556–559, doi:10.1038/nature08031, 2009.

Schuur, E. A. G., Abbott, B. W., Bowden, W. B., Brovkin, V., Camill, P., Canadell, J. G., Chanton, J. P., Chapin III, F. S., Christensen, T. R., Ciais, P., Crosby, B. T., Czimczik, C. I., Grosse, G., Harden, J., Hayes, D. J., Hugelius, G., Jastrow, J. D., Jones, J. B., Kleinen, T., Koven, C. D., Krinner, G., Kuhry, P., Lawrence, D. M., McGuire, A. D., Natali, S. M., O'Donnell, J. A., Ping, C. L., Riley, W. J., Rinke, A., Romanovsky, V. E., Sannel, A. B. K., Schädel, C., Schaefer, K., Sky, J., Subin, Z. M., Tarnocai, C., Turetsky, M. R., Waldrop, M. P., Walter Anthony, K. M., Wickland, K. P., Wilson, C. J., and Zimov, S. A.: Expert assessment of vulnerability of permafrost carbon to climate change, Climatic Change, 119, 359–374, doi:10.1007/s10584-013-0730-7, 2013.

SMHI, Sveriges meteorologiska och hydrologiska institut, Klimatscenarier: http://www.smhi.se/klimatdata/ Framtidens-klimat/Klimatscenarier/2.2252/2.2264 (last access: 8 November 2013), 2013.

Sollid, J. L., Holmlund, P., Isaksen, K., and Harris C.: Deep permafrost boreholes in western Svalbard, northern Sweden, and southern Norway, Norsk. Geogr. Tidsskr., 54, 186–191, 2000.

Stork, A.: Plant immigration in front of retreating glaciers, with examples from the Kebnekajse area, Northern Sweden, Geograf. Ann., 45, 1–22, 1963.

Tarnocai, C., Canadell, J. G., Mazhitova, G., Schuur, E. A. G., Kuhry, P., and Zimov, S.: Soil organic carbon pools in the northern circumpolar permafrost region, Global Biogeochem. Cy., 23, GB2023, doi:10.1029/2008GB003327, 2009.

US Geological Survey: Earth Resources Observation and Science Center (EROS), USGS Global Visualization Viewer, Landsat 5 TM satellite image, acquisition date: 18 September 2011, http://glovis.usgs.gov/ (last access: 25 October 2012), 2011.

Walker, M.: Quaternary dating methods, John Wiley & Sons Ltd, England, 306 pp., 2005.

Walther, G.-R., Beissner, S., and Burga, C. A.: Trends in the upward shift of alpine plants, J. Veg. Sci., 16, 541–548, 2005.

Wang, G., Li, Y., Wang, Y., and Wu, Q.: Effects of permafrost warming on vegetation and soil carbon pool losses on the Qinghai-Tibet Plateau, China, Geoderma, 143, 143–152, 2008.

Yang, Y., Fang, J., Tang, Y., Ji, C., Zheng, C., He, J., and Zhu, B.: Storage, patterns and controls of soil organic carbon in the Tibetan grasslands, Global Change Biol., 14, 1592–1299, doi:10.1111/j.1365-2486.2008.01591.x, 2008.

Zimov, S. A., Schuur, E. A. G., and Chapin, F. S.: Permafrost and the global carbon budget, Science, 312, 1612–1613, doi:10.1126/science.1128908, 2006a.

Zimov, S. A., Davydov, S. P., Zimova, G. M., Davydova, A. I., Schuur, E. A. G., Dutta, K., and Chapin III, F. S.: Permafrost carbon: Stock and decomposability of a globally significant carbon pool, Geophys. Res. Lett., 33, L20502, doi:10.1029/2006GL027484, 2006b.

Zollinger, B., Alewell, C., Kneisel, C., Meusburger, K., Gärtner, H., Brandová, D., Ivy-Ochs, S., Schmidt, M. W. I., and Egli, M.: Effect of permafrost on the formation of soil organic carbon pools and their physical-chemical properties in the Eastern Swiss Alps, Catena, 11, 70–85, doi:10.1016/j.catena.2013.06.010, 2013.

4

The impact of snow depth, snow density and ice density on sea ice thickness retrieval from satellite radar altimetry: results from the ESA-CCI Sea Ice ECV Project Round Robin Exercise

S. Kern[1], **K. Khvorostovsky**[2], **H. Skourup**[3], **E. Rinne**[4], **Z. S. Parsakhoo**[1,*], **V. Djepa**[5], **P. Wadhams**[5], and **S. Sandven**[2]

[1]Center for Climate System Analysis and Prediction CliSAP, University of Hamburg, Hamburg, Germany
[2]Nansen Environmental and Remote Sensing Center NERSC, Bergen, Norway
[3]Danish Technical University-Space, Copenhagen, Denmark
[4]Finnish Meteorological Institute FMI, Helsinki, Finland
[5]University of Cambridge, Cambridge, UK
[*]now at: Institute for Meteorology and Geophysics, University of Cologne, Cologne, Germany

Correspondence to: S. Kern (stefan.kern@zmaw.de)

Abstract. We assess different methods and input parameters, namely snow depth, snow density and ice density, used in freeboard-to-thickness conversion of Arctic sea ice. This conversion is an important part of sea ice thickness retrieval from spaceborne altimetry. A data base is created comprising sea ice freeboard derived from satellite radar altimetry between 1993 and 2012 and co-locate observations of total (sea ice + snow) and sea ice freeboard from the Operation Ice Bridge (OIB) and CryoSat Validation Experiment (CryoVEx) airborne campaigns, of sea ice draft from moored and submarine upward looking sonar (ULS), and of snow depth from OIB campaigns, Advanced Microwave Scanning Radiometer (AMSR-E) and the Warren climatology (Warren et al., 1999). We compare the different data sets in spatiotemporal scales where satellite radar altimetry yields meaningful results. An inter-comparison of the snow depth data sets emphasizes the limited usefulness of Warren climatology snow depth for freeboard-to-thickness conversion under current Arctic Ocean conditions reported in other studies. We test different freeboard-to-thickness and freeboard-to-draft conversion approaches. The mean observed ULS sea ice draft agrees with the mean sea ice draft derived from radar altimetry within the uncertainty bounds of the data sets involved. However, none of the approaches are able to reproduce the seasonal cycle in sea ice draft observed by moored ULS. A sensitivity analysis of the freeboard-to-thickness conversion suggests that sea ice density is as important as snow depth.

1 Introduction

As part of the European Space Agency (ESA) Climate Change Initiative (CCI) Sea Ice Essential Climate Variable (ECV) project (SICCI project), quality-controlled long-term data sets of sea ice thickness and concentration will be derived from Earth observation data. The product of sea ice thickness and sea ice area is the sea ice volume which is considered to be among the most sensitive indicators of the amplification of climate change in the Arctic (Schweiger et al., 2011; Zhang et al., 2012; Krinner et al., 2010; Stranne and Björk, 2012; Wadhams et al., 2012).

The main data source for hemispheric sea ice thickness distribution is satellite radar altimetry. Laxon et al. (2003) used European Remote Sensing Satellites (ERS1/2) radar altimeter (RA) data to obtain a first estimate of the sea ice thickness distribution in the Arctic Ocean south of 81.5° N. More recently, Envisat and CryoSat-2 RA data has been used to compute sea ice thickness (Giles et al., 2008; Laxon et al., 2013); the northern limit for Envisat RA data is also 81.5° N, while CryoSat-2 allows sea ice thickness retrieval up to 88° N. In a number of studies, the retrieved sea ice freeboard and its derived thickness product were evaluated (e.g. Laxon et al., 2003; Giles and Hvidegaard, 2006; Giles et al., 2007; Connor et al., 2009). Yet to be calculated and evaluated is the sea ice thickness using the combined time series of ERS-1/2 RA data and Environmental Satellite (Envisat) radar altimeter-2 (RA-2) data of the period 1993 to 2012.

Sea ice thickness can be obtained with other methods than radar altimetry. The first Ice Cloud and Elevation Satellite (ICESat-1) with its Geoscience Laser Altimeter System (GLAS) allowed computing sea ice thickness from laser altimetry for up to three periods each year of about 1 month duration for years 2003 to 2009 (Kwok et al., 2009). Methods using spaceborne active or passive microwave sensor data (e.g. Kwok et al., 1995; Martin et al., 2004; Kaleschke et al., 2012) or using spaceborne infrared sensor data (e.g. Yu and Rothrock, 1996) do not allow computation of an Arctic-wide sea ice thickness distribution. These methods are limited in the maximum thickness to be retrieved, which is less than a metre, and can additionally be hampered by clouds. Also, satellite laser altimetry is influenced by clouds.

Ground-based, submarine-based, moored and airborne sensors provide sea ice thickness information via measurement of sea ice freeboard or thicknessor total (sea ice plus snow) freeboard or sea ice draft. Such data form the basis of our current understanding of Arctic Ocean sea ice volume loss (Rothrock et al., 2008; Lindsay, 2010; Haas et al., 2008, 2010; Schweiger et al., 2011, Wadhams et al., 2011). On the one hand this data has limited spatio-temporal coverage in contrast to satellite remote sensing data. On the other hand this data is extremely valuable for validation of sea ice thickness products obtained from satellite observations.

In order to derive sea ice thickness for all methods mentioned in the previous three paragraphs, assumptions need to be made about, e.g. ice and snow density, vertical sea ice structure, location of the dynamic sea surface height, and snow depth distribution. In addition to these, the RA method must also assume the penetration depth of radar waves into the snow. The only direct sea ice thickness measurement is a drill hole. Therefore it is important to keep in mind that products of the above-mentioned sources might have a bias and do have a finite uncertainty.

Within the SICCI project, a selection of the most suitable retrieval methods and the most appropriate input data sets for freeboard-to-thickness conversion using RA data is carried out in the so-called Round Robin Exercise (RRE). The RRE is based on analysis of data compiled in the Round Robin Data Package (RRDP). The RRDP comprises ERS-1/2 and Envisat RA sea ice freeboard data, input data for the freeboard-to-thickness conversion and validation data of sea ice thickness, freeboard, draft, snow depth and total freeboard. The main goal is to find an optimal set of assumptions and input data for the freeboard-to-thickness conversion – assuming that the RA sea ice freeboard is correct. To do this, we investigate the quality of the data used and estimate the

sensitivity of the methods used to the input parameters. Validation of RA sea ice freeboard and thickness data will be carried out at a later stage of the SICCI project. This is the reason why a number of data sets one would expect to be used in this study are not used. The amount of sea ice thickness data is limited and we could not use the same data in algorithm selection and validation. We chose to save the sea ice thickness derived from ICESat-1 measurements (Kwok et al., 2009), the total (sea ice + snow) thickness derived from electromagnetic (EM) induction sounding (Haas et al., 2008, 2010) and data from recent (2011 to the present) Operation Ice Bridge (OIB) campaigns for the validation exercise.

The paper is organized as follows. Section 2 describes the RRDP. Section 3 describes the methods used. In Sect. 4, we present the results of our analyses. These are discussed in Sect. 5 and concluded in Sect. 6. We note that the results presented reflect the work of the SICCI project consortium and have been carried out at the respective institutions.

2   Data

The RRDP comprises satellite data: ERS-1/2 RA and Envisat RA-2 sea ice freeboard and snow depth from Advanced Microwave Scanning Radiometer aboard Earth Observation Satellite (AMSR-E). The RRDP includes snow depth and density data from the Warren climatology (Warren et al., 1999), henceforth abbreviated with W99, and it includes a variety of sea ice data from other platforms. These are basically data from moored, submarine and airborne sensors as listed in Table 1. All data will be described in the following paragraphs. Figure 1 shows a sample Envisat RA-2 sea ice freeboard map for March 2010 together with the locations where these other data are taken from. The majority of RA-2 sea ice freeboard values are in a reasonable range (between 0.1 m and 0.4 m).

Sea ice freeboard data as used in the RRDP are derived from ERS-1/2 RA and Envisat RA-2 data using the methodology introduced by Laxon et al. (2003) and Giles et al. (2008) and described in detail in the SICCI ATBD (ESA SICCI project consortium, 2013). To shortly recap, elevation measurements from leads and ice floes are distinguished based on the pulse peakiness of the waveform. After re-tracking the range and applying necessary corrections (namely the Doppler range and delta Doppler, the ionospheric, the dry tropospheric and the modelled wet tropospheric, ocean tide, long-period tide, loading tide, earth tide, pole tide and inverse barometer corrections) and filters (removal of complex waveforms, failed re-tracking and echoes that yielded elevations more than 2 m from the mean dynamic sea surface height), the local sea level at ice floe locations is interpolated from nearby lead elevations. Freeboard is then calculated as the difference of radar-altimetry-measured ice floe elevation and the local sea level. Individual radar altimeter freeboard measurements are present in the RRDP data

Table 1. Validation data used in the RRDP for sea ice thickness.

Date	Location	Parameter	Source	Acronym
2003–2008	Beaufort Sea	Ice draft, snow depth	BGEP moored ULS, AMSR-E	BGEP
Apr 1994 Oct 1996	Beaufort Sea	Ice draft	NSIDC US submarine ULS	BS
Mar 2007	Fram Strait, Beaufort Sea	Ice draft, snow depth	UCAM UK submarine ULS, AMSR-E	BSS
May 2011 Apr 2008	Fram Strait	Ice freeboard, thickness, snow depth	DTU ALS, ASIRAS, AMSR-E	FS
Oct 2009	Western Arctic	Ice freeboard, thickness, snow depth	NSIDC IceBridge	OIB

base. These measurements correspond to the freeboard of ice within the surface footprint of the altimeter. The size of the footprint, i.e. the spatial resolution of the instrument, depends on the target surface properties and is of the order of 2 to 10 km (Connor et al., 2009).

The net uncertainty of the gridded RA-derived freeboards is unknown. The factors contributing to the freeboard uncertainty include sub-footprint surface roughness, ambiguities in radar penetration into snow, bias due to wave shape from leads and floes, tides, the uncertainty in satellite position and radar speckle. Due to the speckle a large number of RA freeboard estimates must be averaged to get a meaningful estimate. In this work individual RA freeboard estimates are averaged according to the collocation areas defined in Sect. 2 further below, or into a 2° longitude × 0.5° latitude grid (approximately 60 km grid cell size). Averaging is always done over 1 calendar month. Depending on latitude and number of leads identified this results hardly in more than 200 measurements per grid cell to be averaged for the gridded product. This is illustrated in Fig. 2 showing for months October to March the average number N of single orbit Envisat RA-2 sea ice freeboard data used per month per 100 km grid cell – which is the grid resolution of the SICCI project SIT prototype product. Averaging is done over the entire Envisat RA-2 period, i.e. winters 2002/03 to 2011/12. Note the decline in areas with $N > 200$ over the season (compare November to March) in the northern Beaufort and Chukchi Seas. This can be most likely attributed to a smaller number of leads as shown by Bröhan and Kaleschke (2014).

In this paper we do not discuss the uncertainty of RA freeboards. This will be done later as part of the Sea Ice CCI validation exercise. Instead we take the freeboard estimates as accurate and study the effect of using different assumptions about the sea ice and snow density as well as different sources of snow depth estimates.

W99 snow depth and density data is available as climatological monthly values for a given location of the Arctic Ocean. Because the W99 climatology is a second-degree polynomial decreasing rapidly outside the central Arctic Ocean (Warren et al., 1999), extrapolated estimates, e.g. in

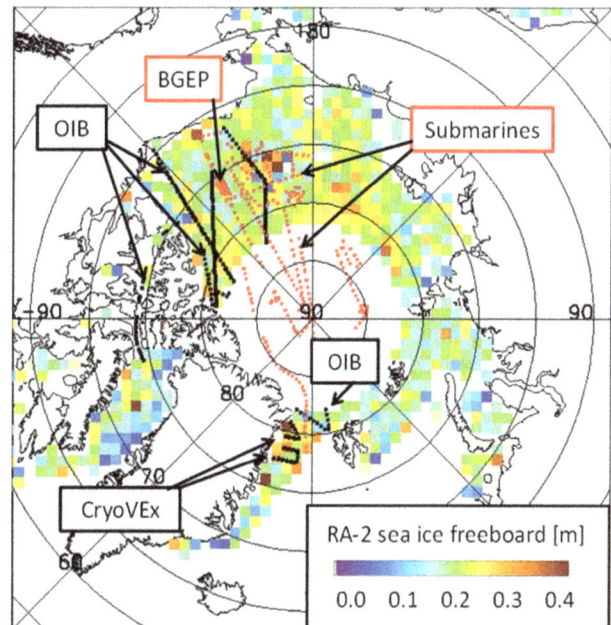

Figure 1. Envisat RA-2 sea ice freeboard distribution for March 2010 superposed with locations of campaigns used for our intercomparison study: airborne campaigns (in black): CryoVEx, OIB; moored and submarine upward looking sonar (ULS) in red: BGEP, Submarines. Grid resolution is 100 km. The white circular area around the pole indicates the region north of the 81.5° N parallel with no Envisat RA-2 data.

the Hudson Bay or the Bering Sea, should not be taken as real snow depth values. W99 data can be considered reliable up to the coasts on the Pacific and Eurasian side of the Arctic Ocean. Towards the Atlantic side the approximate southern limit of useful W99 data is 80° N (Warren et al., 1999); south of this latitude no or only few observations contributed to the climatology. W99 snow depth and density data are collocated individually for each single RA freeboard estimate and averaged over the same area and time as the freeboard (see above paragraph and Sect. 2 further below).

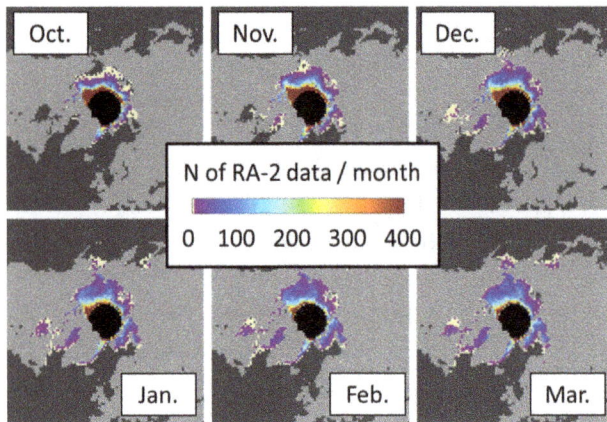

Figure 2. Average number N of Envisat RA-2 data per 100 km grid cell per month for the period 2002/03 to 2011/12.

AMSR-E snow depth on sea ice is taken for the Arctic from the AMSR-E/Aqua Daily L3 12.5 km Brightness Temperature, Sea Ice Concentration, & Snow Depth Polar Grids product (http://nsidc.org/data/docs/daac/ae_si12_12km_tb_sea_ice_and_snow.gd.html, Cavalieri et al., 2004) available from NSIDC. These data are provided daily at 12.5 km grid resolution as running 5-day means and are limited to snow depths below 0.45 m on seasonal ice (Markus and Cavalieri, 1998; Comiso et al., 2003). The algorithm is sensitive to sea ice roughness (Worby et al., 2008, Ozsoy-Cicek et al., 2011; Kern et al., 2011) as well as snow wetness and grain size (Maksym and Markus, 2008; Markus and Cavalieri, 1998). Recently, the quality of AMSR-E snow depth was assessed for the Arctic (Cavalieri et al., 2012; Brucker and Markus, 2013). A comparison between OIB and AMSR-E snow depths for about six hundred 12.5 km grid cells from the years 2009 to 2011 (Brucker and Markus, 2013) indicated a basin average bias of up to 0.07 m and RMSD values between 0.03 m and 0.15 m. Under ideal conditions, i.e. for high concentration (> 90 %) level first-year ice (FYI) thicker than 0.5 m, the RMSD is below 0.06 m for, on average, 0.2 m thick snow (Brucker and Markus, 2013). For our study, AMSR-E snow depth is collocated with RA sea ice freeboard by averaging data over a calendar month over a disc of 100 km radius centred at each RA sea ice freeboard grid cell.

The combination of a laser scanner and snow radar or a radar altimeter provides simultaneous collocated snow depth, total (sea ice + snow) freeboard and sea ice freeboard data. The laser scanner senses the snow surface and is used to derive the total freeboard – similar to the ICESat-1 GLAS instrument – if the instantaneous sea surface height (SSH) is known. The snow radar directly measures snow depth on top of sea ice using the range difference between reflections at the two interfaces, ice–snow and snow–air. For a radar altimeter operating at Ku-band frequencies it is assumed that it provides the height of the ice–snow interface above the SSH:

the sea ice freeboard, under dry snow and/or freezing conditions.

The RRDP includes a combination of CryoVEx laser scanner (ALS) and radar altimeter data (ASIRAS). ALS and ASIRAS data are taken from DTU Space, National Space Institute (ftp://ftp2.spacecenter.dk/pub/ESACCI-SI/) and are averaged over 50 km transects of flight line (see Fig. 1 for location). We use CryoVEx data from campaigns at the end of April 2008 and beginning of May 2011. The collocated RA-2 data are averages for April of the respective year of observation from all orbits within a disc of 100 km radius centred at each ALS 50 km transect centre. ALS data are used to derive total freeboard (Hvidegaard and Forsberg, 2002) with accuracy and precision of independent measurements of about 0.1 m to 0.15 m. ASIRAS sea ice freeboard data are derived using a method similar to Ricker et al. (2012) and have an accuracy of 0.15 m to 0.2 m for independent measurements. As measurements are averaged along 50 km transects located in an area of frequent lead occurrence the accuracy relevant for this study is of the order of 0.01 m for the ALS data. For the same reason it can be expected that the accuracy of the ASIRAS data is better than the numbers given above and has a magnitude of 0.05 m to 0.1 m.

We note that the radius of 100 km seems to be quite large. We have demonstrated, though, that a month of averaging over single orbit RA-2 sea ice freeboard data and hence using a large number of data points per grid cell (Fig. 2) is required for a sufficient reduction of particularly speckle noise. Using a smaller radius of, for example, 50 km would reduce the number of data points per averaging area substantially. In addition, airborne campaign data are usually for only a few days and are therefore a snapshot compared to the RA-2 data averaging period of a calendar month. The sea ice sensed during the airborne campaign might have drifted out of the collocation area around the transect centre used if a too small collocation area had been chosen. Hence, for all collocations with airborne or submarine-based data, we used a collocation area radius of 100 km.

The RRDP includes OIB laser scanner (Airborne Thematic Mapper, ATM)-measured and snow radar-measured total freeboard, snow depth, and ice thickness (Panzer et al., 2013; Kurtz et al., 2013). OIB data are taken from the NSIDC (http://nsidc.org/data/icebridge/index.html) and are averaged over 50 km transects along track. The collocated RA-2 data are monthly averages of observations from all orbits within a disc of 100 km radius centred at each OIB 50 km transect centre. We used data from OIB campaigns in April 2009 and March and April 2010 (see Fig. 1 for location). Kurtz et al. (2013) summarize the uncertainty sources of OIB snow depth retrieval. They point out that the results of Farrell et al. (2012) are a bit too optimistic (0.01 m uncertainty in snow depth) and instead suggest a snow depth uncertainty of 0.06 m in agreement with Kwok et al. (2011): 0.03 m to 0.05 m for snow depths between 0.1 m and 0.7 m. Lowest retrievable snow depth is of the magnitude 0.05 m (see also Kwok and Maksym, 2014).

In addition to snow depth, the OIB freeboards are shown to be accurate. Past problems identified with the automatic SSH retrieval from ATM data alone for 2009 (Nathan Kurtz, personal communication, 2013) were mitigated starting with the 2010 OIB data by including contemporary digital imagery (Onana et al., 2013). For the bulk of total freeboard obtained from OIB ATM measurements, the bias can be expected to be close to zero, with a precision of between 0.05 m and 0.1 m (Farrell et al., 2012; Kurtz et al., 2013). This is confirmed by a study of Kwok et al. (2012), who found agreement between ICESat-1 and OIB-ATM freeboards of within 0.01 m and a measurement repeatability of about 0.04 m.

Upward looking sonar (ULS) observes sea ice draft which can be converted into sea ice thickness in a similar way as the sea ice freeboard. In the RRDP we use data from the Beaufort Gyre Exploration Project (BGEP) where three, sometimes four, moored ULS measured sea ice draft. The approximate location of these moorings is denoted by the red triangles in Fig. 1. BGEP ULS data are taken for years 2003 to 2008 from WHOI (http://www.whoi.edu/page.do?pid=66559). Accuracy of the data is between 0.05 m and 0.1 m (Krishfield and Proshutinsky, 2006). This data provides an independent measure of the seasonal cycle of sea ice draft and thus sea ice thickness. The collocated data are monthly averages of observations from all single orbit RA-2 sea ice freeboards which fall into a box centred at the BGEP mooring location (see Fig. 1) extending over 12 degree latitude and 30 degree longitude. Snow depth data are averaged over the same area. This box may be oversized. The rationale behind using such a large co-location area was to maximize the number of valid RA freeboard estimates and to minimize the effect of sea ice motion changing ice type composition in that area.

Another source of ULS data in the RRDP are those carried on board submarines. Submarine ULS draft data were successfully used by Laxon et al. (2003) for a first assessment of Arctic Ocean sea ice thickness distribution obtained from ERS-1/2 data. The RRDP contains submarine ULS data from three cruises (red dots in Fig. 1). Data from two of the cruises from US submarines (April 1994 and October 1996) are available from NSIDC (http://nsidc.org/data/g01360.html). Data from the third cruise by a UK submarine (March/April 2007) are available from University of Cambridge (UCAM), see also (Wadhams et al., 2011). Submarine ULS data are in general less accurate than the BGEP data but are the only information about draft distribution over a larger region. Rothrock and Wensnahan (2007) report a bias of 0.29 m and a standard deviation of 0.25 m. An assessment of the UK submarine ULS data used reveals a standard deviation of 0.29 m and a bias of 0.4 m; these numbers are worse compared to the US submarine data due to classified submarine positions. The collocated RA-2 data are monthly averages of observations from all orbits within a disc of 100 km radius centred at each

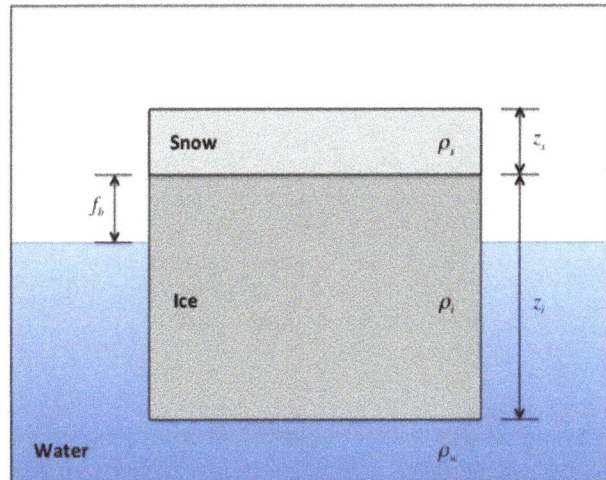

Figure 3. Illustration of the parameters involved in sea ice thickness computation using sea ice freeboard.

submarine ULS 50 km transect centre. A transect length of 50 km is recommended by Rothrock and Wensnahan (2007).

3　Methods

It is assumed that satellite radar altimetry measures the sea ice freeboard. By assuming isostasy, sea ice freeboard can be used to compute sea ice thickness z_i:

$$z_i = \frac{z_s \rho_s + f_b \rho_w}{\rho_w - \rho_i} \tag{1}$$

and also sea ice draft D

$$D = \frac{z_s \rho_s + f_b \rho_i}{\rho_w - \rho_i}, \tag{2}$$

with snow depth z_s, sea ice freeboard f_b, and the densities of sea water, sea ice and snow: ρ_w, ρ_i, and ρ_s, respectively. Figure 3 illustrates the parameters used in Eq. (1).

The main objectives of the Round Robin Exercise are

- to select the best snow depth (product) for freeboard-to-thickness conversion

- to investigate the validity and influence of retrieval assumptions, such as using constant sea ice density, on the sea ice thickness retrieval

In order to achieve these goals, the following investigations are carried out:

1. Snow depth data of the different data sets involved are inter-compared.

2. RA-2 sea ice freeboard is converted to total freeboard by adding snow depth information and compared with OIB and CryoVEx total freeboard.

3. RA and RA-2 sea ice freeboard is used to compute sea ice draft D using Eq. (2) with different input data and compared to ULS sea ice draft data. This is done using a "standard set of densities" (see below). For BGEP mooring ULS data, we additionally compute sea ice draft separately for multiyear ice (MYI) and FYI densities and two different fixed snow densities.

4. RA-2 sea ice freeboard is used to compute sea ice thickness combining the standard set of densities with various snow depth information; the results are compared to OIB sea ice thickness.

The standard set of densities is $\rho_i = 900 \, \text{kg m}^{-3}$, which is the average density of MYI and FYI, and $\rho_w = 1030 \, \text{kg m}^{-3}$ (Wadhams et al., 1992). The snow density is taken from W99 and varies over space and time. In order to account for the effect of different densities for MYI and FYI (in investigation 3, see above), we use sea ice densities published elsewhere (e.g. Timco and Frederking, 1996; Alexandrov et al., 2010): $882 \, \text{kg m}^{-3}$ and $917 \, \text{kg m}^{-3}$, respectively. The two fixed snow density values used in investigation 3 (see above) are $240 \, \text{kg m}^{-3}$ and $340 \, \text{kg m}^{-3}$ and correspond to the mean wintertime minimum and maximum snow density, respectively (Warren et al., 1999).

4 Results

In the following we present the results of comparing the various data sets. We start with snow depth and (sea ice) freeboard and then continue with sea ice draft and thickness.

4.1 Snow depth

The results of the inter-comparison of collocated W99, OIB and AMSR-E are summarized for 2009 and 2010 in Table 2. OIB data from the Arctic Ocean, the Canadian Archipelago, and the Fram Strait region are used (see Fig. 1). Mean snow depth along the OIB tracks in the Arctic Ocean in 2009 is 0.36 m and 0.16 m over MYI and FYI, respectively. In 2010, OIB snow depth is 0.23 m and 0.13 m over MYI and FYI, respectively. Over MYI, OIB and W99 snow depths agree within 0.02 m in 2009 while in 2010 W99 overestimates OIB snow depth by 0.12 m. Over FYI, W99 overestimates OIB snow depths by 0.19 m and 0.21 m in 2009 and 2010, respectively. In April 2010, OIB flights tracks are located over FYI in the Arctic Ocean and in the Canadian Archipelago. For the latter region, we found a similar mean snow depth over FYI than in the Arctic Ocean. We did not compare OIB and W99 snow depths because in the Canadian Archipelago, W99 snow depth relies purely on extrapolation (Warren et al., 1999). In April 2010, OIB flight tracks covered the Fram Strait area (Fig. 1). These tracks also are north of 80° N and thus still in the region of valid W99 snow depth data. W99 overestimation of OIB snow depth is even larger than for

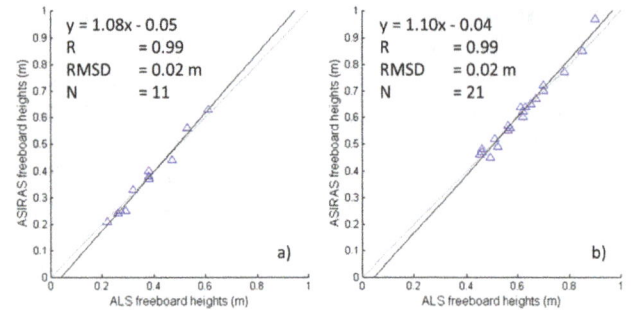

Figure 4. Scatterplot ASIRAS versus ALS total freeboard for the CryoVEx campaigns (see Fig. 1 for location) in 2008 (**a**) and 2011 (**b**).

the tracks in the Arctic Ocean. W99 snow depth is about 0.40 m while the mean snow depth along the OIB track is 0.17 m. In both years, 2009 and 2010, W99 snow depths are about twice as large as AMSR-E snow depth over FYI in the Arctic Ocean. The difference is 0.18 m (Table 2), which is of the same magnitude as the difference between OIB and W99 snow depth (see previous paragraph). AMSR-E and OIB snow depths agree on average by about 0.02 m for the flight tracks crossing the Arctic Ocean as well as those in the Canadian Archipelago. For the OIB flight in the Fram Strait region, none of the collocation regions contained enough FYI for a comparison between AMSR-E and OIB snow depths.

The results of our snow depth comparison agree with Kurtz and Farrell (2011) and Kurtz et al. (2013): over FYI AMSR-E data give a much better measure of the actual snow depth than W99. Snow depths from W99 are about twice as large as AMSR-E and OIB snow depths over FYI. Over MYI, OIB and W99 differ by only 0.02 m in 2009 but by 0.12 m in 2010. Only grid cells with at least 65 % MYI are used here. One possible explanation for the different degree of agreement could be inter-annual variation in snow depth over MYI. While in 2009 OIB snow depth was 0.36 m it was just 0.23 m in 2010. Mean W99 snow depth was 0.35 m and 0.34 m, respectively. Based on climatology, the W99 does not capture the inter-annual variability in snow depth. The W99 estimate for inter-annual variability for the snow depth in March is 0.06 m, explaining half of the observed difference in 2010.

4.2 Sea ice and total freeboard

During the CryoVEx campaigns in 2008 and 2011 in the Fram Strait, both the radar altimeter (ASIRAS) and the laser instrument (ALS) essentially sensed the snow surface as is illustrated in the scatterplots in Fig. 4. Radar penetration into the snow cover on sea ice in the Fram Strait during CryoVEx campaigns was close to zero although the radar is supposed to sense the ice–snow interface at the frequency used in Ku-band according to laboratory experiments (Beaven et al., 1995). There is growing evidence that this assumption is

Table 2. Summary of the comparison between OIB, W99 and AMSR-E snow depth in the Arctic Ocean. Absolute values are only given for OIB; all other values are differences. All values are given together with one standard deviation.

Data set	All	MYI (> 65 %)	FYI (> 95 %)	Can. Arch.
OIB 2009	(0.26 ± 0.11) m	(0.36 ± 0.04) m	(0.16 ± 0.02) m	–
OIB – W99	(-0.07 ± 0.11) m	(0.02 ± 0.04) m	(-0.19 ± 0.02) m	–
OIB – AMSR-E	–	–	(-0.01 ± 0.02) m	–
W99 – AMSR-E	–	–	(0.18 ± 0.03) m	–
OIB 2010	(0.21 ± 0.07) m	(0.23 ± 0.05) m	(0.13 ± 0.02) m	(0.13 ± 0.04) m
OIB – W99	(-0.13 ± 0.07) m	(-0.12 ± 0.05) m	(-0.21 ± 0.01) m	–
OIB – AMSR-E	–	–	(-0.03 ± 0.02) m	(-0.01 ± 0.03) m
W99 – AMSR-E	–	–	(0.18 ± 0.02) m	–

violated for more cases than previously thought (e.g. Ricker et al., 2014). Both freeboard measurements (ASIRAS and ALS) linearly agreed with a RMSD of 0.02 m, a bias of about 0.05 m, a slope close to 1 and a linear correlation coefficient of 0.99 for 2008 and 2011. Therefore from CryoVEx, only total freeboard is used in this study.

For 2011, CryoVEx ALS total freeboard underestimates RA-2 total freeboard computed using W99 snow depth by 0.06 m; for 2008, this underestimation is about 0.16 m. These values are larger than the uncertainties expected for transect lengths of 50 km for the ALS data. It has to be kept in mind that we look at only 11 and 21 data pairs in 2008 and 2011, respectively. During CryoVEx 2008, the sea ice in the measured area was primarily FYI, and by applying snow depth from AMSR-E (available for 9 out of 11 points) the comparison of total freeboards is improved. In addition both CryoVEx campaigns are south of 80° N, where W99 is solely based on extrapolation and is hence not very reliable.

OIB total freeboard observations of 2009 and 2010 are compared with RA-2 total freeboards computed from collocated RA-2 sea ice freeboard by adding the respective collocated OIB or W99 snow depth in the Arctic Ocean (Table 3, Fig. 5); observations in the Fram Strait and the Canadian Archipelago are excluded. Mean OIB total freeboard in the Arctic Ocean agrees overall within 0.02 m with RA-2 total freeboard when using collocated OIB snow depths. If instead W99 snow depth is used the agreement remains fine for 2009, but for 2010 RA-2 underestimates the overall mean OIB total freeboard by 0.11 m. This could be explained by the difference between OIB snow depth and W99 snow depth (see Sect. 4.1). However, it could also be explained by the different fraction of MYI in these data sets. For 2009 the selected OIB flight tracks were located over MYI only, while in 2010 about one-third of the OIB data of the selected OIB tracks were located over FYI. As shown in Sect. 4.1, OIB snow depth agrees much better with W99 snow depth over MYI than over FYI.

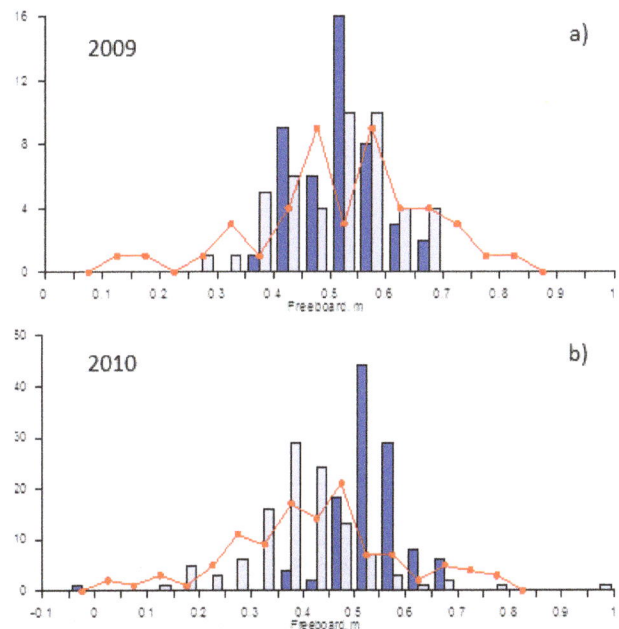

Figure 5. Histograms of OIB (red lines) and RA-2 (blue bars) freeboard for OIB data from the Arctic Ocean for 2009 (**a**) and 2010 (**b**). RA-2 freeboard is derived using OIB snow depth (light blue bars) and W99 snow depth (dark blue bars). Both MYI and FYI data are included. Note the different *y* axis ranges for the number of data per freeboard bin.

4.3 Sea ice draft

The results of the comparison of sea ice draft between ULS and radar altimeter is summarized in Tables 4 and 5. Sea ice draft observed by US submarine ULS in October 1996 is overestimated by ERS-1 RA by 0.13 m which is within the ULS uncertainty of 0.25 m to 0.3 m (Table 4). For April 1994, however, ERS-1 RA underestimates observed sea ice draft by 0.45 m which is outside the uncertainty range given for these ULS data. This discrepancy is illustrated in Fig. 6c and d: while both data sets show maximum probability in the same draft bin of 1.5 m to 2.0 m for 1996, the histograms are shifted relative to each other for April 1994 with largest

Table 3. Summary of overall mean observed (OIB) and computed (RA-2) snow freeboard using OIB or W99 snow depth; given are mean values plus/minus one standard deviation.

Data set	Snow freeboard (OIB)	Snow freeboard (RA-2 + OIB snow depth)	Snow freeboard (RA-2 + W99 snow depth)
OIB 2009	(0.52 ± 0.15) m	(0.51 ± 0.10) m	(0.52 ± 0.07) m
OIB 2010	(0.42 ± 0.16) m	(0.40 ± 0.12) m	(0.53 ± 0.08) m

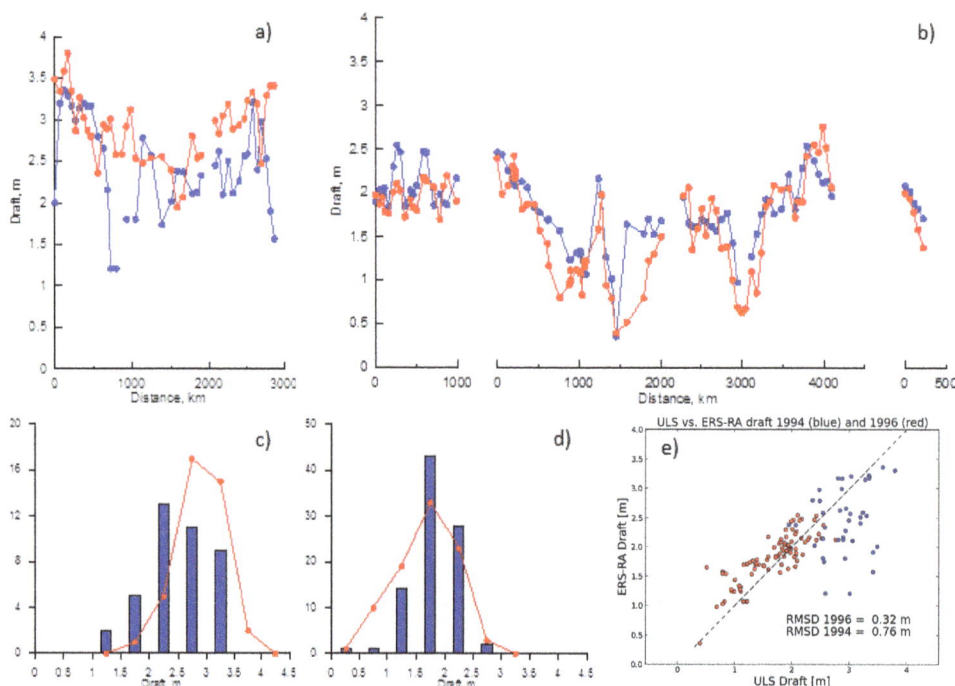

Figure 6. Comparison between sea ice draft observed from US submarine ULS (red) and computed from ERS-1 RA sea ice freeboard using W99 snow data (blue). Images (**a**) and (**b**) are profiles along submarine track for April 1994 and October 1996, respectively (see also Fig. 1); Images (**c**) and (**d**) show corresponding histograms; the y axis denotes the number of data per draft bin. Image (**e**) compares data from both cruises for 1994 (blue) and 1996 (red) together with the RMSD.

probability in bin 2.5 m to 3.0 m for the ULS data but 2.0 m to 2.5 m for RA data. The scatterplot in Fig. 6e underlines that the agreement is much better for October 1996 than for April 1994; in particular the RMSD for 1996 is less than half that for 1994.

Sea ice draft observed by UK submarine ULS in April 2007 is underestimated by RA-2 by 0.12 m (Table 4). However, the majority of this cruise took place north of 81.5° N (see also Fig. 1) and our comparison is therefore based on only 15 collocated data pairs, compared to about 90 and 40 data pairs for the US submarine cruises.

Mean winter sea ice draft observed by BGEP ULS agrees within 0.05 m with sea ice draft computed from RA-2 data using W99 snow depth and density, and standard sea ice and water density values. However, the seasonal range in sea ice draft is much lower for RA-2 than for BGEP ULS (Table 4, Fig. 7). Only for winters 2005/2006 and 2006/2007 does the seasonal range of sea ice draft agree in both data

sets. The area considered here was covered by almost 100% MYI from 2003 to 2007 (first four winters), whereas FYI entered the region in winter 2007/2008 (taken from AMSR-E snow depth data set, Cavalieri et al., 2004). Therefore, for the first four winters, one might need to use the MYI density instead of the value of $900 \, \text{kg m}^{-3}$ used. By doing so the RA-2 draft would decrease by between 0.1 m and 0.4 m, depending on season and year (Fig. 7, brown lines). This would result in a better agreement between BGEP ULS and RA-2 draft early in the winter season, but it would not improve the agreement in terms of the seasonal range. A possible explanation for our RA2 drafts not showing the same seasonal range as ULS drafts could be that during winter more new ice forms and thus the net ice density increases. Confirming this would however require direct ice density measurements. Note that usage of AMSR-E snow depth, possible for winter 2007/2008, results in RA-2 ice draft values that would be typical for 100 % MYI and a snow density of about

Table 4. Summary of observed and computed sea ice draft values using standard settings and W99 snow parameters; given are mean values plus/minus one standard deviation. The respective month the data set is valid for is given in the first column. See Table 1 for data set acronyms.

Data set	Observed draft (ULS)	Derived draft (RA, RA-2)
BS 1994 (April)	(2.92 ± 0.41) m	(2.47 ± 0.57) m
BS 1996 (October)	(1.68 ± 0.51) m	(1.81 ± 0.41) m
BSS 2007 (March)	(2.48 ± 0.46) m	(2.36 ± 0.54) m
BGEP 2003–2008 (October to March)	(1.59 ± 0.42) m	(1.64 ± 0.25) m

Table 5. Differences of mean and median observed minus computed sea ice draft from submarine and moored ULS (see Table 1) and algorithms A1 to A6 applied to radar altimeter data for the Arctic Ocean. Algorithms giving the smallest difference are highlighted in bold.

	Data set	A1	A2	A3	A4	A5	A6
	BS, 10/1996	0.13 **(0.03)**	−0.12 (−0.23)	0.06 (0.04)	0.13 **(0.03)**	0.49 (0.35)	**0.01** (−0.13)
Difference in mean (median) SID [m]	BGEP, 2002/03 to 2007/08	**−0.01** **(0.05)**	−0.22 (−0.19)	0.02 (0.09)	−0.04 **(0.05)**	0.16 (0.27)	−0.43 (−0.35)
	BSS, 03/2007	**0.00** **(0.01)**	−0.22 (−0.24)	0.08 (−0.15)	−0.36 (−0.33)	−0.46 (−0.40)	−0.69 (−0.70)

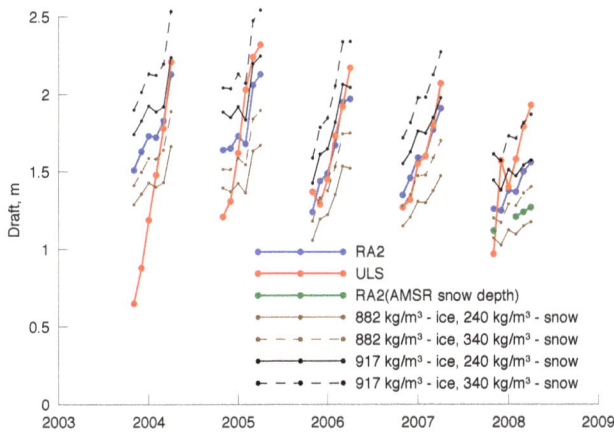

Figure 7. BGEP ULS draft data, averaged to monthly mean for the winter months October to March (red) compared to monthly mean draft computed from RA-2 sea ice freeboard using W99 snow depth and density and standard values: $\rho_i = 900 \, \text{kg m}^{-3}$ and $\rho_w = 1030 \, \text{kg m}^{-3}$ (blue); W99 snow depth but MYI density: $\rho_i = 882 \, \text{kg m}^{-3}$ (brown); W99 snow depth and FYI density: $\rho_i = 917 \, \text{kg m}^{-3}$ (black); and AMSR-E snow depth (green). Note that the latter is only possible for FYI areas. Snow density is set fixed to either $240 \, \text{kg m}^{-3}$ (solid lines) or $340 \, \text{kg m}^{-3}$ (broken lines) for the lines where sea ice density is varied (brown + black).

$290 \, \text{kg m}^{-3}$ (Fig. 7, green dots); these RA-2 ice drafts are much smaller than those observed by the ULS. However, as AMSR-E snow depth can only be obtained over FYI, the usage of MYI ice density and AMSR-E together may yield too

small draft estimates, and one might need to use the FYI density of $917 \, \text{kg m}^{-3}$ instead. This would shift the green dots by 0.3 m towards larger ice draft values (Fig. 7, compare blue and black lines) and would result in a slightly better agreement between ULS and RA-2 drafts. More investigations are needed to confirm this.

Furthermore, we compared ULS sea ice draft with sea ice draft computed from RA sea ice freeboard using six different realizations of the freeboard-to-draft conversion. Of the six realizations, one uses fixed ice density at $900 \, \text{kg m}^{-3}$, i.e. the average of typical FYI and MYI densities, and W99 snow depth (A1); one uses separate FYI and MYI densities and parameterizes W99 snow depth following (Laxon et al., 2013) (A2); one uses fixed FYI density at $910 \, \text{kg m}^{-3}$ combined with a freeboard dependent MYI density (Ackley et al., 1974) and W99 snow depth (A3); one uses fixed ice density at $900 \, \text{kg m}^{-3}$ (see A1) with full and half W99 snow depth over FYI and MYI, respectively (A4); one uses separate but fixed FYI and MYI snow depth and separate FYI and MYI densities (Alexandrov et al., 2010) (A5); one follows the empirical approach for thick MYI without including any snow depth information (Wadhams et al., 1992) (A6). All realizations use seasonally varying W99 snow density. Of these realizations only A1 is shown in Figs. 6 and 7. Table 5 summarizes the difference in the mean and median observed minus computed sea ice draft (SID) for the six realizations and the ULS data sets listed in Table 1. Methods A1, A3 and A4 agree equally well with the ULS sea ice draft data within their uncertainty bounds (about 0.3 m for BS and BSS and

0.05 m for BGEP), and A5 and A6 show the largest discrepancies.

4.4 Sea ice thickness

We computed sea ice thickness from RA-2 data collocated with the OIB tracks in the Arctic Ocean (see Fig. 1) using different snow depth data and compared the results to OIB (2009, 2010) sea ice thickness estimates using the thicknesses provided in the OIB data set (Kurtz et al., 2013). For the RA-2 freeboard-to-thickness conversion, we used the sea ice density of 900 kg m^{-3}. We omitted CryoVEx data from this comparison because of the ambiguous results reported in Sect. 4.2 and because W99 snow depth is less reliable in the area sensed during CryoVEx compared to the OIB track obtained in the Fram Strait in April 2010. Snow depth data sets used are W99 only, W99 over MYI and $0.5 \times$ W99 over FYI (Kurtz and Farrell (2011), henceforth abbreviated KF11), OIB only, and W99 over MYI, but AMSR-E over FYI. The results of this comparison are summarized in Table 6 for the OIB tracks from 2009 and 2010 in the Arctic Ocean and in Table 7 for the OIB track from 2010 in the Fram Strait.

For OIB 2009 data of the Arctic Ocean, none of the four snow data sets reveal a RA-2 sea ice thickness correlated with the OIB one better than 0.65. Using OIB snow depth gives highest correlation and smallest RMSD of 0.96 m. However, the RMSD is similar for the other three data sets. For OIB 2010 data of the Arctic Ocean, using OIB snow depth gives highest correlation, 0.38, but largest RMSD, 1.52 m (Table 6). Correlations and RMSD are smaller when using the other snow data sets. Using W99 data results in the lowest correlation but also the smallest RMSD (Table 6). This is illustrated by Fig. 8 which shows scatterplots of sea ice thickness computed using the mentioned snow depth data sets versus observed sea ice thickness during OIB for 2009 (images a to c) and 2010 (images d to f). Using W99 in combination with AMSR-E and KF11 results in a similar statistics because AMSR-E snow depth is found to be close to half the W99 snow depth and to agree with OIB snow depth within 0.02 m (see Table 2 and Kurtz and Farrell (2011)).

For the Fram Strait, OIB and RA-2 sea ice thickness agree well using either OIB 2010 or W99 snow depth data. The correlation between OIB and RA-2 are 0.84 (OIB 2010 snow) and 0.80 (W99 snow), see Table 7. Similar to the OIB tracks of 2010 in the Arctic Ocean (Table 6) the RMSD is smaller using W99 snow depth, 0.88 m, than using OIB snow depth, 1.03 m. The number of data points (only 13 data pairs; Fig. 8g, h) is, substantially smaller in this region than in the Arctic Ocean region, which limits the value of this comparison. Also the number of snow depth observations contributing to the W99 climatology is quite small in the Fram Strait area (see Warren et al., 1999), which might limit their usefulness for such a study in this area. However, the three boxes (5° latitude by 15° longitude) adjacent to the US and north-

ern Canadian coast contain a similarly small amount of snow depth observations in W99: 50, 43, and 9 compared to 20, 53, and 45 for the boxes north of Svalbard (Warren et al., 1999, Fig. 3).

5 Discussion

The present paper deals with an investigation of the quality and the usefulness of input parameters such as snow depth and densities of snow and sea ice for radar altimeter freeboard-to-thickness conversion. It further gives examples of inter-comparisons between independent estimates of sea ice parameters such as sea ice freeboard, total (sea ice + snow) freeboard, sea ice thickness and sea ice draft, and estimates of these parameters based on satellite radar altimetry. The evaluation of radar altimeter freeboard and the computation of a radar altimeter freeboard uncertainty are not aimed for in the present paper. We assume that the obtained sea ice freeboard is correct. For Envisat RA-2 data this is a fair assumption given the results of, e.g. Connor et al. (2009). An estimate of sea ice freeboard obtained by subtracting OIB snow depth from OIB total freeboard agrees within 0.02 m with colocated RA-2 sea ice freeboard. This is better than the accuracy of 0.05 m given for RA-2 and OIB freeboard data (Kurtz et al., 2013) and indicates that at least along OIB tracks in 2009 and 2010 in the Arctic Ocean, Envisat RA-2 sea ice freeboard is accurate.

Our main conclusion from the comparison of using different estimates for snow depth and ice density (see Table 5) is that methods A1, A3 and A4 agree equally well with the ULS sea ice draft data within their uncertainty bounds (about 0.3 m for BS and BSS and 0.05 m for BGEP), and that A5 and A6 show the largest discrepancies. Why is A2 (Laxon et al., 2013) biased low? Almost all ULS data are obtained under MYI. A2 uses a MY ice density of 882 kg m^{-3} while A1 and A4 use 900 kg m^{-3}. Such a difference in sea ice density can cause a negative bias in the obtained sea ice draft by 0.2 m (compare blue and brown lines in Fig. 7). However, the good agreement between A1 and A4 in mean and median sea ice draft (Table 5) does not mean these use the perfect combination of input parameters. As we can see in Fig. 7 for A1, agreement between observed and computed sea ice draft varies from month to month. As stated in Sect. 4.3, RA-2 sea ice draft does not very well capture the increase in ULS sea ice draft over winter. Generally the increase in RA-2 sea ice draft is smaller than the increase in ULS sea ice draft. There could be various reasons for this.

The area covered by the BGEP moorings (A, B, C and D) is approximately 4° in latitude by 10° in longitude while RA-2 sea ice draft is computed from an area of 12° in latitude by 30° in longitude to account for ice type changes due to drift during the freezing season and to ensure a large enough number of single RA-2 freeboard measurements (confer Fig. 2). Hence RA-2 SID is an average over

Table 6. Summary of comparison between RA-2 sea ice thickness computed using different snow depth data sets and OIB sea ice thickness for the Arctic Ocean. Total number of data pairs is $N = 43$ for 2009 and $N = 90$ for 2010.

month/year	04/2009				03+04/2010			
		AMSR-E				AMSR-E		
Snow data set	OIB	W99	+W99	KF11	OIB	W99	+W99	KF11
R	0.65	0.57	0.62	0.62	0.38	0.23	0.34	0.34
RMSD [m]	0.96	1.00	1.02	1.02	1.52	1.35	1.41	1.40

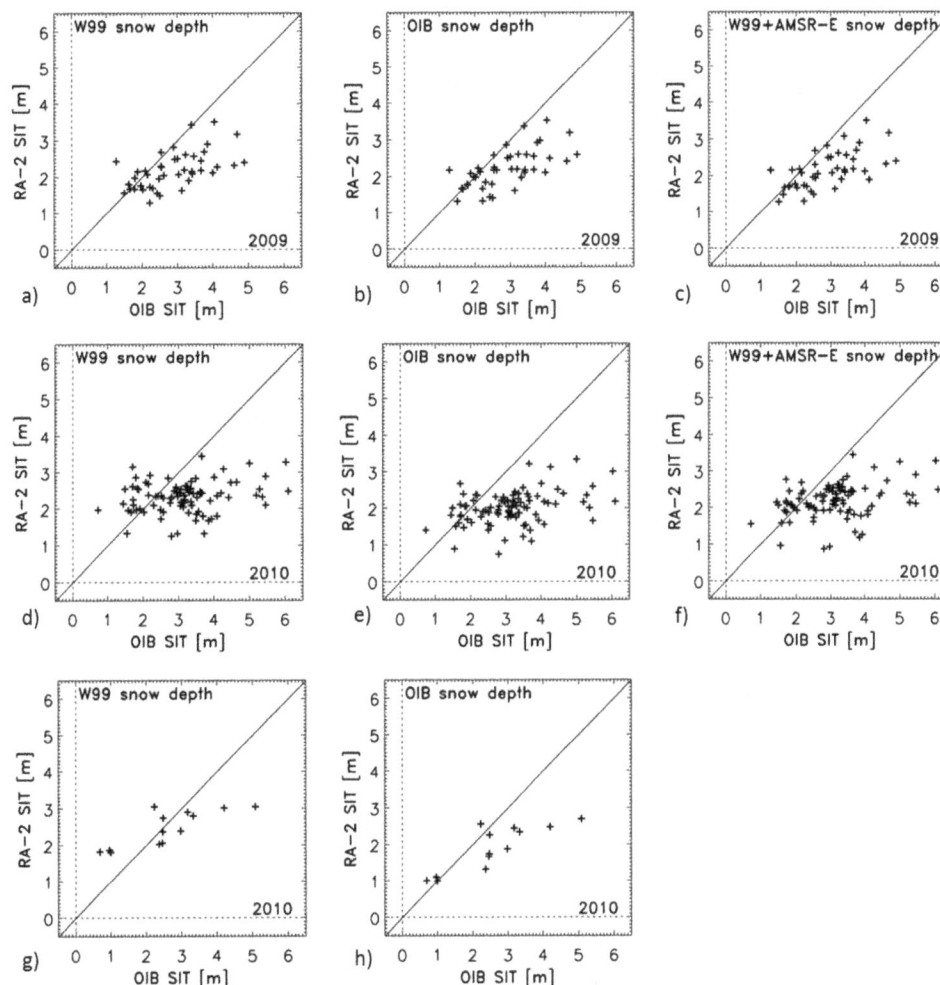

Figure 8. RA-2 sea ice thickness computed using different snow depth data sets versus OIB sea ice thickness for 2009 (**a** to **c**) and 2010 (**d** to **h**). Images (**a**) to (**f**) are for the Arctic Ocean, images (**g**) and (**h**) are for the Fram Strait area.

an almost 10-fold larger area which can explain the smaller seasonal amplitude.

Freeboard-to-thickness conversion is very sensitive to the correct choice of snow depth (see, e.g. Zygmuntowska et al. (2014) and Fig. 9b). We found that W99 snow depth is twice as large as OIB snow depth over FYI, as already reported by Kurtz and Farrell (2011) and Kurtz et al. (2013). AMSR-E snow depths over FYI agree with OIB snow depth

within 0.02 m. We find that even over MYI W99 might overestimate the actual snow depth, as is the case for April 2010. The climatological nature of W99 on the one hand and interannual variation of snow depth on the other hand explain part of the disagreement, but more snow depth inter-comparisons are required to further investigate this finding. It was shown recently that Soil Moisture and Ocean Salinity (SMOS) satellite data can be used to retrieve snow depth over thick Arctic

Table 7. Summary of comparison between RA-2 sea ice thickness computed using different snow depth data sets and OIB sea ice thickness for the Fram Strait area for April 2010. Total number of data pairs is $N = 13$.

Snow data set	R	RMSD [m]
W99	0.80	0.88
OIB	0.84	1.03

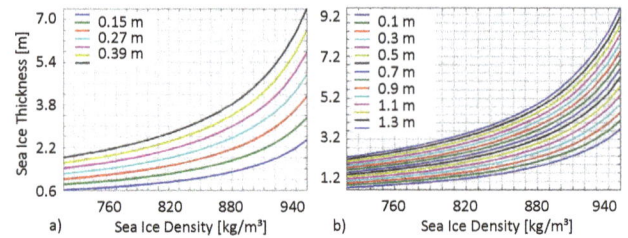

Figure 9. Sensitivity of sea ice thickness obtained from RA sea ice freeboard sea ice density and snow depth. (**a**) Sea ice thickness computed with Eq. (1) for different sea ice freeboard values (0.009 m to 0.45 m) and snow depth 0.3 m as function of sea ice density. (**b**) Similar to (**a**) but computed for different snow depths (0 m to 1.4 m) and sea ice freeboard 0.27 m as function of sea ice density.

sea ice, e.g. MYI (Maaß et al., 2013). Such data could be combined with snow depth from an AMSR-E sensor type of product. For this, however, a better quantification of the MYI fraction than is included in the AMSR-E snow depth product (Cavalieri et al., 2004) is mandatory. This would not only help to obtain a more realistic snow depth distribution but it would also help to choose correct sea ice densities (see below). For this purpose, we recommend carrying out an intercomparison of current sea ice type data sets in the Arctic as can be derived, for example, from satellite scatterometry e.g. QuikSCAT (Kwok, 2004; Swan and Long, 2012). For the Envisat RA-2 measurement period QuikSCAT products can be used. However, for the planned sea ice thickness data set for 1993 until the present, a harmonized sea ice type distribution data set needs to be developed, which is free of inconsistencies or biases due to changes between sensors, such as from ERS1/2 ESCAT to QuikSCAT to ASCAT.

We find that typical variations in sea ice density cause variations in sea ice thickness that are as large as those caused by snow depth variations. This is different to laser altimetry (Kwok and Cunningham, 2008). Under typical variations we understand the difference between MYI and FYI densities (Alexandrov et al., 2010) and the difference between snow depth on MYI compared to FYI (see Table 2). For typical sea ice freeboard values, the typical range in ice density induces variations in sea ice thickness between 0.4 m and 0.8 m (see Fig. 9a). Hence the freeboard-to-thickness conversion is quite sensitive to the choice of sea ice density. Consequently, CryoSat-2 sea ice thickness retrieval (Laxon et al., 2013) uses two different sea ice densities – one for FYI and one for MYI. The sensitivity due to sea ice density can be seen in Fig. 7, which shows differences of up to 0.7 m (March 2004 and March 2005) between RA-2 sea ice draft calculated using a typical FYI density (black lines) and a typical MYI density (brown lines).

We did not carry out a detailed investigation of the impact of snow density. According to the W99 climatology and other studies, e.g. Alexandrov et al. (2010), snow density varies seasonally between $< 100 \, \mathrm{kg \, m^{-3}}$ (fresh snow) to $> 400 \, \mathrm{kg \, m^{-3}}$ (old, compacted snow). Snow density can also vary on short spatial scales. However, in this study satellite RA data is used to obtain sea ice thickness at 100 km spatial scale and a temporal scale of a month. Therefore we feel confident in referring to Fig. 7 to illustrate the effect of

snow density. Snow densities range typically over values of $240 \, \mathrm{kg \, m^{-3}}$ to $340 \, \mathrm{kg \, m^{-3}}$. The change in mean sea ice draft associated with the snow density range applied is about 0.2 m to 0.3 m. This translates into a bias in sea ice thickness of a magnitude of 0.3 m and recommends using seasonally varying snow density when retrieving ice thickness from satellite RA data as is done in this paper.

It is important to bear in mind the different spatiotemporal scales which are involved. For instance, OIB data is obtained at fine spatiotemporal resolution along transects and is averaged over 50 km long segments for this study (see Sect. 2). RA-2 data, as are used here, comprise measurements from all overpasses within a month which fall into a disc of 100 km diameter centred at each 50 km OIB track segment. In addition the footprint of a single RA-2 measurement is 2 to 3 orders of magnitude larger than the footprint of a single OIB measurement. It is likely that RA-2 data provide an average ice thickness rather than the actual range of ice thickness values (see Fig. 8). This depends, however, on the degree by which different ice types and ice surface properties impact the radar backscatter and the waveform (Zygmuntowska et al., 2013, Ricker et al., 2014). More studies need to look into the different backscatter of sea ice of different types and roughness to quantify the impact of sea ice property variation on the radar altimeter signal and hence the sea ice freeboard.

OIB sea ice thickness is computed using a fixed sea ice density of $915 \, \mathrm{kg \, m^{-3}}$ (Kurtz et al., 2013). This density value represents FYI but results in a positive bias in draft and thickness for MYI because it is about $30 \, \mathrm{kg \, m^{-3}}$ higher than the average MYI density value suggested, e.g. by Alexandrov et al. (2010). This makes an assessment of the obtained sea ice thickness values a difficult task, in particular if the aim is to quantify the impact of different sea ice density values on the obtained sea ice thickness. Currently, OIB data are the only airborne data source for contemporary data of freeboard and snow depth.

Our interpretation of the CryoVEx data remains inconclusive because the ASIRAS instrument, which is supposed to sense the ice–snow interface and thus provide an independent sea ice freeboard measurement, failed to do so. Instead it provided the total freeboard as does the ALS sensor. By means of atmospheric re-analysis data, we identify snow cover property changes as a possible reason for CryoVEx 2011 but not for 2008. This suggests that even under freezing conditions sensors such as Envisat RA-2 or CryoSat-2 might not sense the sea ice surface. It is likely that vertical snow density gradients and/or volume scattering in the snow in general influence the radar signal, resulting in a less distinct signal from the ice–snow interface or in similarly strong returns from the snow surface or interior as was shown for Antarctic sea ice by Willatt et al. (2010).

We note that almost all sea ice draft data and many of our validation data are from MYI regions. A real assessment of approaches which includes ice-type dependent ice density and snow depth could therefore not be carried out in a systematic enough way. More work and more data are required here.

6 Summary and recommendations

Satellite radar altimetry (RA) has been providing surface elevation measurements of the Arctic Ocean for about 2 decades. With the assumption that these elevation measurements represent sea ice freeboard these are used to derive sea ice thickness (Laxon et al., 2013, 2003). Here we report on the results of an investigation of the sensitivity of satellite RA freeboard-to-thickness conversion to input parameters and assumptions carried out within the European Space Agency Climate Change Initiative Sea Ice Essential Climate Variable project using Envisat radar altimetry (RA-2). For RA sea ice freeboard uncertainty estimation, which is not part of the present paper, we refer to, e.g. Peacock and Laxon (2004); Zygmuntowska et al. (2013); Ricker et al. (2014); Kurtz et al. (2014) and Armitage and Davidson (2014).

We found the Warren snow depth climatology (W99, Warren et al., 1999) to be outdated, in agreement with earlier studies (Kwok et al., 2011; Kurtz and Farrell, 2011). Modal and mean sea ice draft computed from RA-2 sea ice freeboard using different realizations of the freeboard-to-draft conversion agree with upward looking sonar observations of the freezing season (October to March) sea ice draft in the Beaufort Sea within the uncertainty bounds – provided the realizations include spatiotemporally varying snow depth and density. However, none of the realizations are able to reproduce the seasonal range in sea ice draft. A change of sea ice densities and/or snow depths as a function of ice type can improve the agreement with observed sea ice draft values at the beginning or end of the freezing season but does not have an impact on the overall seasonal sea ice draft range obtained from RA-2 data. Sea ice thickness computed from RA-2 sea ice freeboard using different snow depth data sets overestimate (underestimate) small (large) OIB sea ice thickness. An improvement from using ice-type-dependent snow

depth is not evident in our results, but most likely this simply needs more data and a different inter-comparison strategy to be quantified.

Some of the independent data used in our study point towards a larger range in sea ice draft and thickness than observed by RA-2. This results from the impact of different ground resolutions of the compared sensors. Submarine and airborne sensors have a much finer sampling of the sea ice along their track; sampling by RA is coarser and in addition depends on floe size, lead concentration, waveform distortion and surface roughness. Averaging over a track length of 50 km or 100 km of a submarine or an airborne sensor can only be an approximation of the variability in sea ice freeboard obtained from RA-2 over a disc with diameter 100 km. Data from submarine and airborne campaigns cover a few days while RA-2 data are averages over a month. More emphasis needs to be put on the choice of the scales involved both for sea ice thickness computation and validation. Hence, for a better validation of both sea ice freeboard and thickness products at a spatiotemporal scale of 100 km and one month, more data from airborne campaigns are required. Data from airborne campaigns, which allow sea ice thickness retrieval, often suffer from (i) environmental conditions and their not yet fully known impact on snow and sea ice physical properties (see our results from CryoVEx 2008 and 2011); (ii) uncertainty sources are not yet well understood (Kurtz et al., 2013); (iii) assumptions and parameters, such as sea ice and snow densities, used for derivation of sea ice thickness or snow from airborne data may differ from campaign to campaign and to spaceborne data, and may not be state of the art in view of recent literature (e.g. Alexandrov et al., 2010; Laxon et al., 2013).

We formulate the following recommendations for freeboard-to-thickness conversion using radar altimetry for the Arctic Ocean:

1. The Warren climatology has to be used carefully. It is not valid over first-year ice and it is of limited use outside the central Arctic Ocean. The Warren climatology is still valuable when no other depth snow estimate is available but we recommend using the Warren climatology in combination with a second data set of snow depth over first-year ice. Furthermore we recommend that effort should be put into developing an inter-annually varying snow depth and density over sea ice product for the ice-covered oceans. Snow depth obtained from SMOS over thick sea ice might be an important contribution here (Maaß et al., 2013).

2. Using radar altimetry, the impact of sea ice density on sea ice thickness retrieval is as large as the impact of snow depth. The difference in sea ice densities of multiyear ice and first-year ice is large enough to explain a bias in sea ice thickness of the magnitude of 0.5 m or more. It is recommended to use an ice-type dependent set of sea ice densities. In addition it is important to

also consider the density difference between ridged and level ice. We need many more measurements of ice density and isostasy across first-year ice and multiyear ice ridges to derive area-averaged ice densities for ridged sea ice.

3. For a sophisticated inter-comparison and validation of the final sea ice thickness product from satellite altimetry it is mandatory to use independent and preferably non-altimetric validation data. The amount of such contemporary sea ice draft, snow depth and sea ice thickness data is clearly sub-optimal and needs to be improved.

4. Potential improvement from utilizing new sets of input parameters, e.g. densities, cannot be quantified without consistent input parameters for freeboard-to-thickness conversion. We call for a consistent internationally agreed-upon standard set of densities to be used for freeboard-to-thickness conversion to be applied to air- and spaceborne altimeter data.

Author contributions. S. Kern prepared the manuscript with contributions from all co-authors but mainly E. Rinne, prepared Figs. 1, 2, and 8, and supervised Z. S. Parsakhoo. K. Khvorostovsky contributed a large part of the data analysis of the ULS data, namely Figs. 5, 6, and 7. H. Skourup was responsible for the analysis and interpretation of CryoVEx data and prepared Fig. 4. E. Rinne was responsible for the Round Robin Data Package, contributed substantially with writing and prepared Figs. 3 and 6e. Z. S. Parsakhoo was responsible for the inter-comparison of the snow depth data. V. Djepa prepared Fig. 9. P. Wadhams contributed with the scientific supervision of V. Djepa. S. Sandven contributed with the overall scientific supervision, in particular of K. Khvorostovsky.

Acknowledgements. This work was funded by ESA/ESRIN (sea ice CCI). S. Kern acknowledges support from the Center of Excellence for Climate System Analysis and Prediction (CliSAP), University of Hamburg, Germany. This work is also supported by the Research Council of Norway under contract no. 207584 (ArcticSIV). We are grateful to numerous data providers for the present study, namely, National Snow and Ice Data Centre (NSIDC) for OIB data, AMSR-E snow depth, SSM/I and AMSR-E sea ice concentrations, and the US submarine ULS data; Woods Hole Oceanographic Institute for BGEP ULS data; ESA for re-processed ERS-1/2 and Envisat ASAR data. The authors are grateful to all the teams in the field, in the air and in the ship for providing all these valuable observations. S. Kern acknowledges support from the International Space Science Institute (ISSI), Bern, Switzerland, under project no. 245: Heil, P. and Kern, S. "Towards an Integrated Retrieval of Antarctic Sea Ice Volume". We thank three anonymous reviewers and our editor Julienne Stroeve for their efforts to improve the paper.

Edited by: J. Stroeve

References

Ackley, S. F., Hibler III, W. D., Kugzruk, F., Kovacs, A., and Weeks, W. F.: Thickness and roughness variations of Arctic multiyear sea ice, AIDJEX Bulletin, 25, 75–95, 1974.

Alexandrov, V., Sandven, S., Wahlin, J., and Johannessen, O. M.: The relation between sea ice thickness and freeboard in the Arctic, The Cryosphere, 4, 373–380, doi:10.5194/tc-4-373-2010, 2010.

Armitage, T. W. K. and Davidson, M. W. J.: Using the interferometric capabilities of the ESA Cryosat-2 mission to improve the accuracy of sea ice freeboard retrievals, Trans. Geosci. Rem. Sens., 51, 529–536, doi:10.1109/TGRS.2013.2242082, 2014.

Bröhan, D. and Kaleschke L.: A nine-year climatology of Arctic sea ice lead orientation and frequency from AMSR-E, Remote Sens., 6, 1451–1475, doi:10.3390/rs6021451, 2014.

Brucker, L. and Markus, T.: Arctic-scale assessment of satellite passive microwave derived snow depth on sea ice using operational icebridge airborne data, J. Geophys. Res.-Oceans, 118, 2892–2905, doi:10.1002/jgrc.20228, 2013.

Cavalieri, D. J., Markus, T., and Comiso, J. C.: AMSR-E/Aqua Daily L3 25 km Brightness Temperature & Sea Ice Concentration Polar Grids Version 2, Boulder, Colorado USA: NASA DAAC at the National Snow and Ice Data Center, 2004.

Cavalieri, D. J., Markus, T., Ivanoff, A., Miller, J. A., Brucker, L., Sturm, M., Maslanik, J., Heinrichs, J. F., Gasiewski, A. J., Leuschen, C., Krabill, W., and Sonntag, J.: A comparison of snow depth on sea ice retrievals using airborne altimeters and an AMSR-E Simulator, Trans. Geosci. Rem. Sens., 50, 3027–3040, 2012.

Comiso, J. C., Cavalieri, D. J., and Markus, T.: Sea ice concentration, ice temperature and snow depth using AMSR-E data, Trans. Geosci. Rem. Sens., 41, 243–252, 2003.

Connor, L. N., Laxon, S. W., Ridout, A. L., Krabill, W., and McAdoo, D.: Comparison of Envisat radar and airborne laser altimeter measurements over Arctic sea ice, Remote Sens. Environ., 113, 563–570, 2009.

ESA SICCI project consortium: D2.6: Algorithm Theoretical Basis Document (ATBDv1), ESA Sea Ice Climate Initiative Phase 1 Report SICCI-ATBDv1-04-13, version 1.1, 2013.

Farrell, S. L., Kurtz, N. T., Connor, L., Elder, B., Leuschen, C., Markus, T., McAdoo, D. C., Panzer, B., Richter-Menge, J., and Sonntag, J.: A first assessment of icebridge snow and ice thickness data over Arctic Sea Ice, Trans. Geosci. Rem. Sens., 50, 6, 2098–2111, 2012.

Giles, K. A. and Hvidegaard, S. M.: Comparison of space borne radar altimetry and airborne laser altimetry over sea ice in the Fram Strait, Int. J. Remote Sens., 27, 3105–3113, 2006.

Giles, K. A., Laxon, S. W., Wingham, D. J., Wallis, D. W., Krabill, W. B., Leuschen, C. J., McAdoo, D., Manizade, S. S., and Raney, R. K.: Combined airborne laser and radar altimeter measurements over the Fram Strait in May 2002, Remote Sens. Environ., 111, 182–194, 2007.

Giles, K. A., Laxon, S. W., and Ridout, A. L.: Circumpolar thinning of Arctic sea ice following the 2007 record ice extent minimum, Geophys. Res. Lett., 35, L22502, doi:10.1029/2008GL035710, 2008.

Haas, C., Pfaffling, A., Hendricks, S., Rabenstein, L., Etienne, J.-L., and Rigor, I.: Reduced ice thickness in Arctic Transpolar

Drift favours rapid ice retreat, Geophys. Res. Lett., 35, L17501, doi:10.1029/2008GL034457, 2008.

Haas, C., Hendricks, S., Eicken, H., and Herber, A.: Synoptic airborne thickness surveys reveal state of Arctic sea ice cover, Geophys. Res. Lett., 37, L09501, doi:10.1029/2010GL042652, 2010.

Hvidegaard, S. M. and Forsberg, R.: Sea ice thickness from laser altimetry over the Arctic Ocean north of Greenland, Geophys. Res. Lett., 29, 1952–1955, 2002.

Kaleschke, L., Tian-Kunze, X., Maaß, N., Mäkynen, M., and Drusch, M.: Sea ice thickness retrieval from SMOS brightness temperatures during the Arctic freeze-up period, Geophys. Res. Lett., 39, L05501, doi:10.1029/2012GL050916, 2012.

Kern, S., Ozsoy-Cicek, B., Willmes, S., Nicolaus, M., Haas, C., and Ackley, S. F.: An intercomparison between AMSR-E snow depth and satellite C- and Ku-Band radar backscatter data for Antarctic sea ice, Ann. Glaciol., 52, 279–290, 2011.

Krinner, G., Rinke, A., Dethloff, K., and Gorodetskaya, I. V.: Impact of prescribed Arctic sea ice thickness in simulations of the present and future climate, Clim. Dynam., 35, 619–633, doi:10.1007/s00382-009-0587-7, 2010.

Krishfield, R. and Proshutinsky, A.: BGOS ULS Data Processing Procedure Report Woods Hole Oceanographic Institute, available at: http://www.whoi.edu/fileserver.do?id=85684&pt=2&p=100409 (last access: 25 January 2014), 2006.

Kurtz, N. T. and Farrell, S. F.: Large-scale surveys of snow depth on Arctic sea ice from Operation IceBridge, Geophys. Res. Lett., 38, L20505, doi:10.1029/2011GL049216, 2011.

Kurtz, N. T., Farrell, S. L., Studinger, M., Galin, N., Harbeck, J. P., Lindsay, R., Onana, V. D., Panzer, B., and Sonntag, J. G.: Sea ice thickness, freeboard, and snow depth products from Operation IceBridge airborne data, The Cryosphere, 7, 1035–1056, doi:10.5194/tc-7-1035-2013, 2013.

Kurtz, N. T., Galin, N., and Studinger, M.: An improved CryoSat-2 sea ice freeboard retrieval algorithm through the use of waveform fitting, The Cryosphere, 8, 1217–1237, doi:10.5194/tc-8-1217-2014, 2014.

Kwok, R.: Annual cycles of multiyear sea ice coverage of the Arctic Ocean: 1999–2003, J. Geophys. Res., 109, C11004, doi:10.1029/2003JC002238, 2004.

Kwok, R. and Cunningham, G. F.: ICESat over Arctic sea ice: estimation of snow depth and ice thickness, J. Geophys. Res., 113, C08010, doi:10.1029/2008JC004753, 2008.

Kwok, R. and Maksym, T.: Snow depth of the Weddell and Bellingshausen sea ice covers from IceBridge surveys in 2010 and 2011: An examination, J. Geophys. Res.-Oceans, 119, doi:10.1002/2014JC009943, 2014.

Kwok, R., Nghiem, S. V., Yueh, S. H., and Huynh, D. D.: Retrieval of thin ice thickness from Multifrequency Polarimetric SAR data, Remote Sens. Environ., 51, 361–374, 1995.

Kwok, R., Cunningham, G. F., Wensnahan, M., Rigor, I., Zwally, H. J., and Yi, D.: Thinning and volume loss of the Arctic Ocean sea ice cover: 2003–2008, J. Geophys. Res., 114, C07005, doi:10.1029/2009JC005312, 2009.

Kwok, R., Panzer, B., Leuschen, C., Pang, S., Markus, T., Holt, B., and Gogineni, S. P.: Airborne surveys of snow depth over Arctic sea ice, J. Geophys. Res., 116, C11018, doi:10.1029/2011JC007371, 2011.

Kwok, R., Cunningham, G. F., Manizade, S. S., and Krabill, W. B.: Arctic sea ice freeboard from IceBridge acquisitions in 2009:

estimates and comparisons with ICESat, J. Geophys. Res., 117, C02018, doi:10.1029/2011JC007654, 2012.

Laxon, S., Peacock, N., and Smith, D.: High interannual variability of sea-ice thickness in the Arctic region, Nature, 425, 947–950, 2003.

Laxon, S. W., Giles, K. A., Ridout, A. L., Wingham, D. J., Willatt, R., Cullen, R., Kwok, R., Schweiger, A., Zhang, J., Haas, C., Hendricks, S., Krishfield, R., Kurtz, N., Farrell, S. L., and Davidson, M.: CryoSat-2 estimates of Arctic sea ice thickness and volume, Geophys. Res. Lett., 40, 1–6, 2013.

Lindsay, R.: New unified sea ice thickness climate data record, EOS, 91, 405–406, 2010.

Maaß, N., Kaleschke, L., Tian-Kunze, X., and Drusch, M.: Snow thickness retrieval over thick Arctic sea ice using SMOS satellite data, The Cryosphere, 7, 1971–1989, doi:10.5194/tc-7-1971-2013, 2013.

Maksym, T. and Markus, T.: Antarctic sea ice thickness and snow-to-ice conversion from atmospheric reanalysis and passive microwave snow depth, J. Geophys. Res., 113, C02S12, doi:10.1029/2006JC004085, 2008.

Markus, T. and Cavalieri, D. J.: Snow depth distribution over sea ice in the southern ocean from satellite passive microwave data, in: Antarctic Sea Ice: Physical Processes, Interactions, and Variability, edited by: Jeffries, M. O., AGU Antarctic Research Series, American Geophysical Union, Washington DC, 74, 19–39, 1998.

Martin, S., Drucker, R., Kwok, R., and Holt, B.: Estimation of the thin ice thickness and heat flux for the Chukchi Sea Alaskan coast polynya from Special Sensor Microwave/Imager data, 1990–2001, J. Geophys. Res., 109, C10012, doi:10.1029/2004JC002428, 2004.

Onana, V.-de-P., Kurtz, N. T., Farrell, S. L., Koenig, L. S., Studinger, M., and Harbeck, J. P.: A sea-ice lead detection algorithm for use with high-resolution airborne visible imagery, Trans. Geosci. Rem. Sens., 51, 38–56, 2013.

Ozsoy-Cicek, B., Kern, S., Ackley, S. F., Xie, H., and Tekeli, A. E.: Intercomparisons of Antarctic sea ice types from visual ship, RADARSAT-1 SAR, Envisat ASAR, QuikSCAT, and AMSR-E satellite observations in the Bellingshausen Sea, Deep-Sea Res. Pt. II, 58, 9–10, 1092–1111, doi:10.1016/j.dsr2.2010.10.031, 2011.

Panzer, B., Gomez-Garcia, D., Leuschen, C., Paden, J., Rodriguez-Morales, F., Patel, A., Markus, T., Holt, B., and Gogineni, S. P.: An ultra-wideband, microwave radar for measuring snow thickness on sea ice and mapping near-surface internal layers in polar firn, J. Glaciol., 59, 244–255, 2013.

Peacock, N. R. and Laxon, S. W.: Sea surface height determination in the Arctic Ocean from ERS altimetry, J. Geophys. Res., 109, C07001, doi:10.1029/2001JC001026, 2004.

Ricker, R., Hendricks, S., Helm, V., Gerdes, R., and Skourup, H.: Comparison of sea-ice freeboard distribution from aircraft data and CryoSat-2, Proceedings paper, 20 years of progress in radar altimetry, 24–29 September 2012, Venice, Italy, 2012.

Ricker, R., Hendricks, S., Helm, V., Skourup, H., and Davidson, M.: Sensitivity of CryoSat-2 Arctic sea-ice freeboard and thickness on radar-waveform interpretation, The Cryosphere, 8, 1607–1622, doi:10.5194/tc-8-1607-2014, 2014.

Rothrock, D. A. and Wensnahan, M.: The accuracy of sea-ice drafts measured from US Navy submarines, J. Atmos. Ocean Tech., 24, 1936–1949, doi:10.1175/JTECH2097.1, 2007.

Rothrock, D. A., Percival, D. B., and Wensnahan, M.: The decline in arctic sea-ice thickness: separating the spatial, annual, and interannual variability in a quarter century of submarine data, J. Geophys. Res., 113, C05003, doi:10.1029/2007JC004252, 2008.

Schweiger, A., Lindsay, R., Zhang, J., Steele, M., Stern, H., and Kwok, R.: Uncertainty in modeled Arctic sea ice volume, J. Geophys. Res., 116, C00D06, doi:10.1029/2011JC007084, 2011.

Spreen, G., Kern, S., Stammer, D., Forsberg, R., and Haarpaintner, J.: Satellite based estimation of sea ice volume flux through Fram Strait, Ann. Glaciol., 44, 321–328, 2006.

Stranne, C. and Björk, G.: On the Arctic Ocean ice thickness response to changes in external forcing, Clim. Dynam., 39, 3007–3018, doi:10.1007/s00382-011-1275-y, 2012.

Swan, A. M. and Long, D. G.: Multiyear Arctic sea ice classification using QuikSCAT, Trans. Geosci. Rem. Sens., 50, 9, 3317–3326, doi:10.1109/TGRS.2012.2184123, 2012.

Timco, G. W. and Frederking, R. M. W.: A review of sea ice density, Cold Reg. Sci. Technol., 24, 1–6, 1996.

Wadhams, P.: Arctic ice cover, ice thickness and tipping points, Ambio, 41, 23–33, 2012.

Wadhams, P., Tucker III, W. B., Krabill, W. B., Swift, R. N., Comiso, J. C., and Davis, N. R.: Relationship between sea ice freeboard and draft in the Arctic Basin, and implications for ice thickness monitoring, J. Geophys. Res., 97, 20325–20334, doi:10.1029/92JC02014, 1992.

Wadhams, P., Hughes, N., and Rodrigues, J.: Arctic sea ice thickness characteristics in winter 2004 and 2007 from submarine sonar transects, J. Geophys. Res., 116, C00E02, doi:10.1029/2011JC006982, 2011.

Warren, S. G., Rigor, I. G., Untersteiner, N., Radionov, V. F., Bryazgin, N. N., Aleksandrov, Y. I., and Colony, R.: Snow depth on Arctic sea ice, J. Climate, 12, 1814–1829, 1999.

Willatt, R. C., Giles, K. A., Laxon, S. W., Stone-Drake, L., and Worby, A. P.: Field investigations of Ku-Band radar penetration into snow cover on Antarctic sea ice, Trans. Geosci. Rem. Sens., 48, 365–372, doi:10.1109/TGRS.2009.2028237, 2010.

Worby, A. P., Markus, T., Steer, A. D., Lytle, V. I., and Massom, R. A.: Evaluation of AMSR-E snow depth product over East Antarctic sea ice using in situ measurements and aerial photography, J. Geophys. Res., 113, C05S94, doi:10.1029/2007JC004181, 2008.

Yu, Y. and Rothrock, D. A.: Thin ice thickness from satellite thermal imagery, J. Geophys. Res., 101, 25753–25766, 1996.

Zhang, J., Lindsay, R., Schweiger, A., and Rigor, I. G.: Recent changes in the dynamic properties of declining Arctic sea ice: a model study, Geophys. Res. Lett., 39, L20503, doi:10.1029/2012GL053545, 2012.

Zygmuntowska, M., Khvorostovsky, K., Helm, V., and Sandven, S.: Waveform classification of airborne synthetic aperture radar altimeter over Arctic sea ice, The Cryosphere, 7, 1315–1324, doi:10.5194/tc-7-1315-2013, 2013.

Zygmuntowska, M., Rampal, P., Ivanova, N., and Smedsrud, L. H.: Uncertainties in Arctic sea ice thickness and volume: new estimates and implications for trends, The Cryosphere, 8, 705–720, doi:10.5194/tc-8-705-2014, 2014.

5

Large-area land surface simulations in heterogeneous terrain driven by global data sets: application to mountain permafrost

J. Fiddes[1], **S. Endrizzi**[1], **and S. Gruber**[2]

[1]Department of Geography, University of Zurich, Zurich, Switzerland
[2]Department of Geography and Environmental Studies, Carleton University, Ottawa, Canada

Correspondence to: J. Fiddes (joel.fiddes@geo.uzh.ch)

Abstract. Numerical simulations of land surface processes are important in order to perform landscape-scale assessments of earth systems. This task is problematic in complex terrain due to (i) high-resolution grids required to capture strong lateral variability, and (ii) lack of meteorological forcing data where they are required. In this study we test a topography and climate processor, which is designed for use with large-area land surface simulation, in complex and remote terrain. The scheme is driven entirely by globally available data sets. We simulate air temperature, ground surface temperature and snow depth and test the model with a large network of measurements in the Swiss Alps. We obtain root-mean-squared error (RMSE) values of 0.64 °C for air temperature, 0.67–1.34 °C for non-bedrock ground surface temperature, and 44.5 mm for snow depth, which is likely affected by poor input precipitation field. Due to this we trial a simple winter precipitation correction method based on melt dates of the snowpack. We present a test application of the scheme in the context of simulating mountain permafrost. The scheme produces a permafrost estimate of $2000\,\text{km}^2$, which compares well to published estimates. We suggest that this scheme represents a useful step in application of numerical models over large areas in heterogeneous terrain.

1 Introduction

Numerical simulation is an increasingly important tool for assessment of the energy and mass balance at the earth's surface for many fields of research and application (e.g. Wood et al., 2011; Barnett et al., 2005; Gruber, 2012). In addition, numerical methods allow for transient assessment of past and future states, an essential step for change detection of (near-)surface conditions (Etzelmüller, 2013). Numerical approaches may also provide the means to simulate land surface variables where there are insufficient data for statistical method,s e.g. remote areas or future periods.

Landscapes that are heterogeneous in terms of e.g. topography, vegetation or redistribution of snow (e.g. Smith and Riseborough, 2002; Liston and Haehnel, 2007) provide a great challenge in this respect, as surface and subsurface conditions may vary on various, and often short, length scales, creating highly spatially differentiated surface–atmosphere interactions. This poses, in particular, a challenge to large-area simulations, which can be summarised as follows: (1) high-resolution grids are required to capture surface heterogeneity, which is often numerically prohibitive over large areas, and efficient methods are therefore required to make this task scalable; (2) there is often a lack of a representative forcing at the site or scale that it is required, particularly in remote regions.

Recent efforts in this respect include spatially explicit or distributed simulations; e.g. Jafarov et al. (2012) and Westermann et al. (2013) produced a transient run of the ground thermal state in Alaska to assess permafrost dynamics under IPCC change scenarios. Another meso-scale modelling effort, that of Gisnå s et al. (2013), provides an equilibrium model of permafrost distribution in Norway at a spatial resolution of $1\,\text{km}^2$. While representing major steps in application of numerical models over large areas, the grid resolution of 1–2 km is too coarse to simulate relevant spatial differentiation on fine scales, particularly under heterogeneous terrain. At site scales several studies have applied numerical models to investigate the ground thermal regime at specific permafrost sites (Scherler et al., 2010, 2014); however, in a few studies downscaled climate data have been used to force such a model (e.g. Marmy et al., 2013).

In global climate models, a spatially detailed representation of the sub-grid land surface still remains somewhat behind the implementation of the atmosphere, yet is recognised to be key in accurately simulating feedbacks to the atmosphere (Pitman, 2003), e.g. surface albedo–atmosphere exchanges in the energy balance (Betts, 2009). For example, land surface heterogeneity is often represented in tiled approaches (Koster and Suarez, 1992), where surface types are represented by a limited number of surface types (or even a single one). Energy and mass balance is then computed independently, and finally aggregated at grid level. Here too, methods capable of representing fine-scale land surface heterogeneity efficiently could be useful. Finally, methods exist (e.g. SAFRAN-Crocus scheme; Durand et al., 1993, 1999) which classify topography according to fixed classes based on terrain parameters and enable application of numerical models over large areas in a semi-distributed fashion.

Gubler et al. (2011) have shown that fine-scale variability of surface processes can be high in complex terrain – e.g. variation in soil moisture, ground cover and local shading – can cause differences of as much as 3 °C mean annual ground surface temperature (MAGST) within a 10 m × 10 m grid. This underscores the importance of scale-appropriate evaluation of models. There are many studies in the literature where models operating on grids of 10s–100s or, in extreme cases, 1000s of metres are evaluated by point-scale measurements, and this is known to pose a serious challenges to model evaluation (Randall et al., 2003; Li, 2005). However, methods that provide simulation results over large areas capable of exploiting distributed site-scale ground truth are rare.

In previous papers (Fiddes and Gruber, 2012, 2014), methods have been developed and tested which enable (i) physically based land surface models (LSMs) to be applied over large areas using a sub-grid scheme that samples land surface heterogeneity and (ii) a method that scales gridded climate data necessary to drive an LSM to the sub-grid using atmospheric profiles. The philosophy behind these approaches is to develop methods that depend only on globally available data sets to derive high-quality local results in heterogeneous and/or remote regions.

The main aim of this study is to establish this combined method as a proof of concept and perform an initial evaluation of its performance in the context of the ground thermal regime and specifically permafrost occurrence in the European Alps, as a test case. That said, the aim is not to provide a best-possible result for e.g. permafrost (as the example subject of this study) but to provide a demonstration of this method using simple data sets. It is well known that precipitation bias is a common problem when using climate model or reanalysis data (e.g.Dai, 2006; Boberg et al., 2008) and a key driver of the energy and mass balance at the land surface.

Therefore, an additional aim is to explore a simple method that may be useful in addressing precipitation bias using the parameter melt date (MD) of the snowpack. Specifically this paper will

1. conduct a test application of the combined schemes together with the LSM GEOtop (Endrizzi et al., 2014) to derive land surface/near-surface variables air temperature (TAIR), ground surface temperature (GST) and snow depth (SD) over a large area of the European Alps at a resolution of 30 m, and additionally a derived permafrost estimate;

2. evaluate the performance of the combined schemes against a large network of TAIR, GST and SD measurements in the Swiss Alps;

3. demonstrate a simple bias correction method for the precipitation field;

4. interpret results together with uncertainties in the model chain.

2 Methods

The model chain used in this study uses two previously described methods, (i) TopoSUB (Fiddes and Gruber, 2012, hereafter FG2012) and (ii) TopoSCALE (Fiddes and Gruber, 2014, hereafter FG2013), together with a numerical LSM, GEOtop (Endrizzi et al., 2014). A brief synopsis of the tools used is given here to enable full understanding of the current study, but for further details and results of testing of these tools please see the respective publications.

2.1 TopoSUB

TopoSUB is a scheme which samples land surface heterogeneity at high resolution (here, 30 m). Input predictors describing relevant dimensions of variability are clustered with a K-means algorithm to reduce computational units in a given simulation domain (here, $0.75° \times 0.75°$). A 1-D LSM is then applied to each sample. For example, in FG2012 we show that reduction of a domain from 10^6 pixels to 258 samples yields comparable results to a full distributed 2-D baseline simulation. The main outcome is that the computational load is effectively reduced by a factor of 10^4, with an acceptable reduction in the quality of results. The scheme transfers model results to high-resolution pixels by membership functions (crisp or fuzzy) for a spatialised mapping of simulation results or statistical descriptions of the sub-grid domain. Additionally, we have an optional informed-scaling training routine, which regresses model results against input predictors after a training run in order to adjust the weighting of each input according to its significance; in doing so, it improves the quality of the final result. Limitations to this fundamentally 1-D approach include the fact that lateral mass and energy transfers can only be parameterised, not modelled explicitly. This approach allows for (1) modelling of processes at fine grid resolutions, (2) efficient statistical descriptions of sub-grid behaviour, (3) efficient aggregation of simulated variables to coarse grids and (4) comparing results and ground truth derived from similar scales.

2.2 TopoSCALE

TopoSCALE is a scheme which provides forcing to the LSM at fine scale using gridded climate data sets. It works on the assumption that vertical gradients are often more important than horizontal gradients in complex topography. Climate data sets are employed as they provide consistent fields required for LSM simulation in 3-D, therefore providing a detailed description of the atmospheric profile. In addition, they provide data with global coverage and so enable simulation in remote, data-poor regions. Finally, they provide the possibility of simulating future conditions. The basic principles of the scheme are as follows: (1) interpolate data available on pressure levels air temperature (TAIR), relative humidity (RH), wind speed (Ws) and wind direction (Wd) vertically above and below the target site to provide a scaling according to atmospheric conditions at each model timestep; (2) downwelling longwave radiation (LWin) is scaled according to TAIR, RH and sky emissivity; (3) topographic correction is made to downwelling radiation fields (SWin/LWin); and (4) lapse rate with elevation is applied to precipitation, P (optional disaggregation scheme based on climatology for site simulation only as this is spatially explicit). The final output is the time series of meteorological variables required to drive a numerical model at 3 h timestep. It is a flexible scheme that can be used to supply inputs to models in 1-D/2-D or lumped configurations. The scheme has been shown in FG2013 to improve the scaling of driving daily fields compared to reference methods such as fixed lapse rates.

2.3 Land surface model

The LSM used in this study, GEOtop, is a physically based model originally developed for hydrological research. It should be noted that this model is not an LSM in the conventional sense (e.g. Mosaic, CLM, NOAH; Koster and Suarez, 1992; Dai et al., 2003), as it has not been designed to feed back to the atmosphere. In addition this model has not been designed for global or hemispheric application. However, it couples the ground heat and water budgets, represents the energy exchange with the atmosphere, has a multilayer snowpack and represents the water and energy budget of the snow cover. GEOtop simulates the temporal evolution of the snow depth and its effect on ground temperature. It solves the heat conduction equation in one dimension and the Richards equation for water transport in one or three dimensions, describing water infiltration in the ground as well as freezing and thawing processes in the ground. We have used the ther-

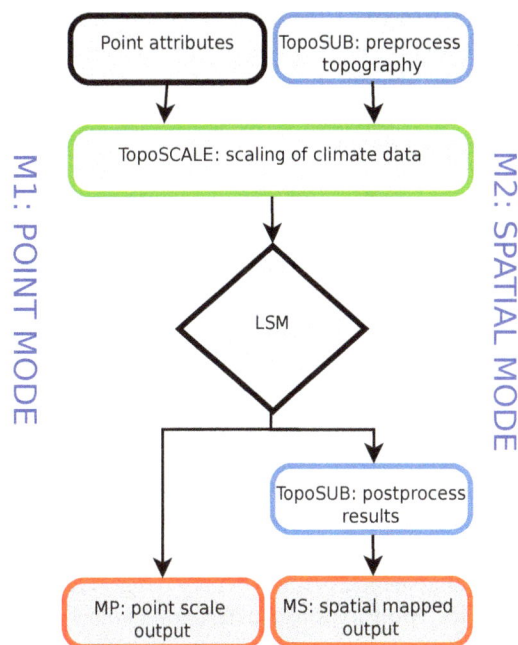

Figure 1. Overview of how the model chain of TopoSUB, TopoSCALE and LSM operate together. Two main modes of operation, (MP) point and (MS) spatial, are shown.

mal conductivity parameterisation given by Cosenza et al. (2003). It provides higher conductivities than Calonne et al. (2011), who show their parameterisation to be preferable to that of Sturm and Benson (1997), which provides rather low estimates. For densities below about $300 \, \text{kg m}^{-3}$, the results of Löwe et al. (2013) are also higher than Calonne et al. (2011). We did not investigate in detail the effects of different parameterisations on soil surface temperature. Given the other uncertainties discussed in this paper, we do not expect it to be significant. GEOtop is therefore a suitable tool to model permafrost relevant variables such as snow and ground temperatures both at the surface and at depth. It can be applied in high mountain regions and allows accounting for topographic and other environmental variability. Further information is given by Bertoldi et al. (2006), Rigon et al. (2006), Endrizzi (2007) and Dall'Amico et al. (2011). Further details specifically relevant to this study are given in Sect. 3. A full description of the model is given in Endrizzi et al. (2014), and a description of model uncertainty and sensitivity is given by Gubler et al. (2013).

2.4 Model chain and modes

The model chain can be employed in two main configurations: point mode (MP) and spatial mode (MS) (Fig. 1). MS requires TopoSUB and TopoSCALE, while MP requires only TopoSCALE. In terms of output, MP simulates point-scale results, whilst MS simulates a spatially explicit mapped result from samples. The basic model chain employs

TopoSCALE to derive a forcing at simulation points or samples, depending upon the mode employed. The LSM simulates target variables at the computed points or sample centroids. TopoSUB is used in MS to pre-process topography and post-process results.

2.5 Snow correction method

Precipitation is highly variable in time and space, and fields computed by climate models often do not capture the frequency and/or intensity distribution of events correctly (Piani et al., 2009; Manders et al., 2012; Dai, 2006). Additionally, sub-grid topographic features may place large controls on the distribution of precipitation events (Leung and Ghan, 1998). Therefore, a method is required to correct magnitudes of precipitation inputs due to the important influence of this field on land surface processes. The method we test in this study relies on detection of the MD of the snowpack, a parameter which summarises both energy and mass inputs to the snowpack. We vary a parameter in the model which applies a multiplicative correction on precipitation inputs called snow correction factor (SCF). We vary this parameter over the range 0.5–3 in steps of 0.25 and run a simulation for each correction factor. MDs are computed according Schmid et al. (2012) for each simulation and observation site using GST (which avoids circularity). We fit the simulation MDs to observed MDs to obtain a correction factor for precipitation input. This method is based on cumulative winter precipitation and assumes summer and winter distributions of precipitation biases are similar, which is likely not the case. However, our primary aim is to address the thermal influence of the winter snowpack. The approach shown here could potentially be used together with satellite imagery in order to estimate snowfall bias based on MD. However, this paper evaluates the point-based performance of the new method without bias correction.

3 Data

3.1 Input data

All input data used in this experiment are available free of charge, globally. This does not imply, however, that data quality is consistent globally.

3.1.1 Driving climate

Driving climate data are obtained from the ERA-Interim (ERA-I) data set, which is an atmospheric reanalysis produced by the ECMWF (Dee et al., 2011). ERA-I provides meteorological data from 1 January 1979 and continues to be extended in near-real time. Gridded products include a large variety of 3-hourly (00:00, 03:00, 06:00, 09:00, 12:00, 15:00, 18:00 and 21:00 UTC) grid-surface fields (GRID) and 6-hourly (00:00, 06:00, 12:00, 18:00 UTC) upper-atmosphere

Table 1. Description of surface and sub-surface parameters used in this study. These are generic values of natural materials obtained from the literature.

Parameter	Unit	Bedrock	Debris	Vegetation
Residual water content	–	0	0.055	0.056
Saturated water content	–	0.05	0.374	0.431
Van Genuchten parameter, α	–	0.001	0.1	0.002
Van Genuchten parameter, n	–	1.2	2	2.4
Hydraulic conductivity	$mm\,s^{-1}$	10^{-6}	1	0.044

products available on 60 pressure levels (PLs) with the top of the atmosphere located at 1 mb. ERA-I relies on a 4-D-Var assimiliation scheme which uses observations within the windows of 15:00 to 03:00 UTC and 03:00 to 15:00 UTC (on the next day) to initialise forecast simulations starting at 00:00 and 12:00 UTC, respectively. In order to allow sufficient spin-up, the first 9 h of the forecast simulations are not used. All fields used in this study were extracted on the ECMWF reduced Gaussian N128 grid ($0.75° \times 0.75°$). Six PLs are used in this study covering the range of 1000–500 mb (1000, 925, 850, 775, 650, 500), corresponding to approximately an elevation range of 150–5500 m a.s.l.

3.1.2 Surface data

The DEM used in this study is the Advanced Spaceborne Thermal Emission and Reflection Radiometer (ASTER) Global Digital Elevation Model Version 2 (GDEM V2) (Tachikawa, 2011) available at approximately 30 m. Landcover was derived by a combined bedrock–debris classification which relies primarily on slope angle and a vegetation mask from a soil-adjusted vegetation index (SAVI) derived from Landsat Thermal Mapper/Enhanced Thermal Mapper (TM/ETM+) sensors. Full details together with description of uncertainty are given in Boeckli et al. (2012a). This resulted in three landcover classes, which also define sub-surface properties according to Gubler et al. (2013): (i) bedrock, (ii) coarse blocks and (iii) vegetation. Table 1 gives a description of sub-surface properties associated with each class.

3.2 Validation data sets

The validation data set covers a broad elevation range of 1560–3750 m a.s.l.; full range of slopes from flat to vertical rock walls; full range of aspects; and main Alpine surface cover types: Alpine meadows, coarse debris and bedrock (Fig. 3). The entire Alpine space of Switzerland is well sampled by the data sets (Fig. 2). Table 2 gives an overview of each data set.

3.2.1 SLF IMIS stations

The WSL-Institut für Schnee- und Lawinenforschung (SLF) Intercantonal Measurement and Information System (IMIS)

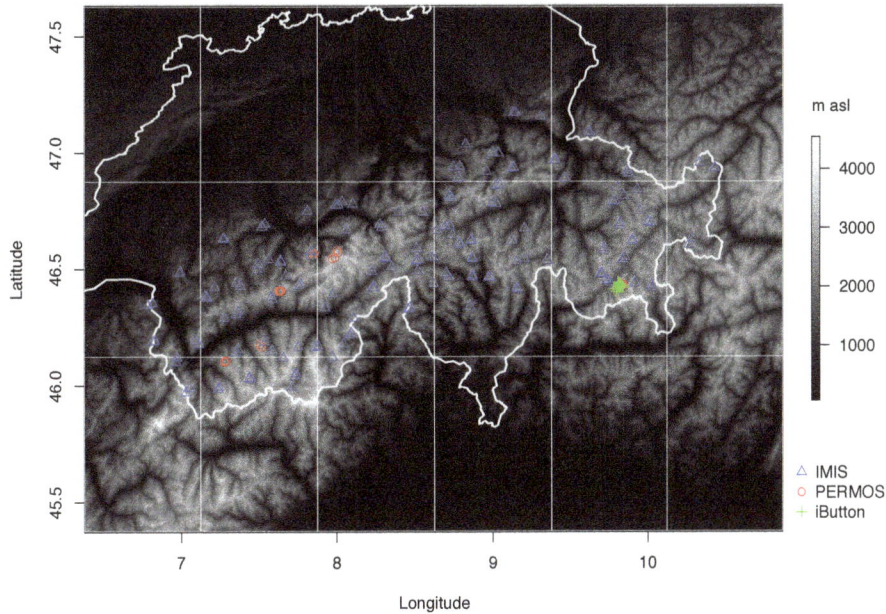

Figure 2. Experiment domain centred on the Swiss Alps together with evaluation data set locations. The ERA-I grid is overlaid in white.

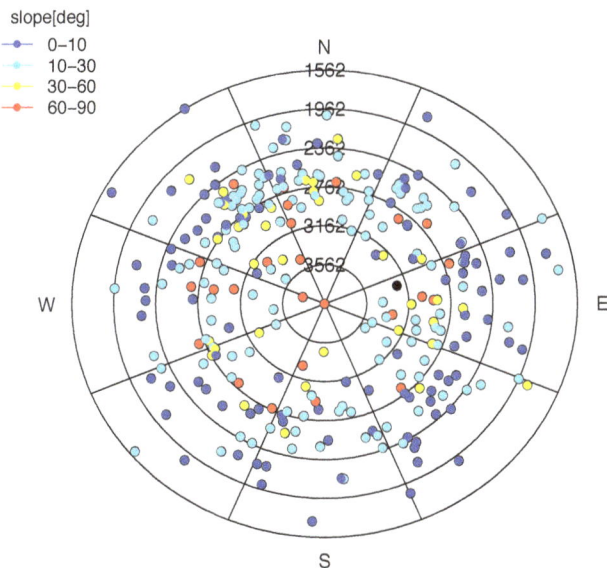

Figure 3. Polar plots describing topographic distribution of validation sites. Elevation range 1560–3750 m a.s.l.; slopes 0–> 90°; and all aspects are represented.

stations are used to evaluate TAIR, GST and SD. This network is biased towards high Alpine locations (there are few valley stations) but represents strong topographical heterogeneity, in terms of elevation, slope and aspect. The network elevation range is 1562–3341 m a.s.l. This data set is quite well behaved in that generally sites represent mainly elevation gradients. The data set used covers years 1996–2011;

GST is measured with a white temperature probe resting on the ground surface. It does not contain MAGST below 0 °.

3.2.2 Data loggers

The data logger data set comprises two individual data sets and is used to evaluate GST only. Sensors measure GST a few centimetres below the terrain surface to avoid radiation effects. The dataset PERMOS data set (Swiss Permafrost Monitoring Network, http://www.permos.ch) contains data loggers of various types covering years 1995–2012 (ongoing) distributed throughout the Swiss Alps and managed by a number of institutions in Switzerland. The data set is not homogeneous but has been compiled using consistent methods. This data set covers a great diversity of locations but is biased towards permafrost monitoring sites and therefore clustered around MAGST of 0 °C. In the analysis, PERMOS data are split into two groups: (a) PERMOS1: predominantly coarse debris; (b) PERMOS2: bedrock (mainly steep rock walls). The second logger data set, iBUTTONS, originates from a single study (Gubler et al., 2011; Schmid et al., 2012). It covers years 2010–2011 in a single region in the Engadin. While broader climatic heterogeneity is not represented by this data set, it does cover strong topographic variability. The data set is arranged as 10 m × 10 m "footprints" each containing 10 data loggers. In this study footprint mean values are used.

3.2.3 Data quality control

Observations outside acceptable limits were removed automatically by applying physically plausible thresholds to all

Table 2. Description of evaluation data sets used in this study.

Data set	Stations/ sites	Type	Variables	Period	Coverage (ERA boxes)
IMIS	81	Station	GST/TAIR/SD	1996–2011	12
PERMOS	77	Logger	GST	1995–2012 (variable)	4
iButtons	40	Logger	GST	2010–2011	1

data sets. Non-changing values beyond prescribed time limits were screened from wind direction data. These checks follow the methods of Meek and Hatfield (1994). Thresholds of a maximum of 10 % missing data in any given year qualified that year as a valid MAGST value. As data sets and sites within data sets differ in number of valid MAGST years (as defined above), validation values are computed as the mean of all available MAGST years and compared to the mean of the same modelled years.

4 Simulation experiments

The simulation domain covers an area of approximately 500 km × 250 km, centred over the Swiss Alps (Fig. 2). The domain contains 18 coarse-grid ERA-I boxes which supply the driving climate data. We simulate results for both MP and MS modes. TopoSUB is run at 200 sample resolution on each coarse-grid unit. The simulation period is 1984–2011. Spin-up is performed over 50 years (10 times, 1979–1983 period). This is necessary to obtain soil temperatures at depth that reflect atmospheric conditions and are independent of their initial value. The LSM runs on an hourly timestep. LSM model parameters are fixed in all simulations as a mean value of prior distributions defined in Gubler et al. (2013). We compute mean annual air temperature (MAAT), MAGST and mean annual snow depth (MASD). Focus is placed on mean annual values as we are primarily interested in analysing the performance of the spatial prediction of the scheme. In computing a permafrost estimate, we define a permafrost pixel as one in which the maximum daily ground temperature at 10 m depth (GT_{10}) over the entire observation period time is $\leq 0\,°C$. Results are analysed statistically using the root-mean-squared error (RMSE), correlation coefficient (CORR) and mean bias (BIAS), defined as

$$\mathrm{BIAS} = \overline{\mathrm{mod} - \mathrm{obs}}. \tag{1}$$

5 Results

5.1 Evaluation: simulated variables

Figure 4 gives MP and MS simulated results validated against measurement sites for MAGST, MAAT and MASD.

MAGST results are validated against IMIS, PERMOS and iBUTTON data sets. The scheme most successfully simulates IMIS sites with low error and bias; however, there is cold (warm) bias at cold (warm) sites. The iBUTTON data cover the largest range of MAGST and demonstrate good performance of the scheme in cold (i.e. MAGST < 0) locations. Both iBUTTON and PERMOS site validation shows the ability of the scheme to capture results influenced by the fine-scale variability of the topography (Fig. 3). PERMOS1 sites (non-bedrock) are reproduced with greater success than PERMOS2 sites (steep bedrock). MP gives improved results for IMIS and PERMOS2, whereas the converse is true for IBUTTON and PERMOS1. Over all data sets an RMSE of 1.29 is obtained for MS and 1.21 for MP. However, these figures should be interpreted with caution as there is an implicit weighting based on available data points, which are unlikely to be representative of the distribution of underlying surfaces in the simulation domain.

All MAGST results display some degree of positive bias, with the exception of PERMOS2 data. This is likely due to a negatively biased snowpack. This fits with PERMOS2 being the exception as locations of steep rockwalls and therefore no snowpack. Underestimated snowpacks may have various and opposing effects on the ground thermal regime, e.g. greater cooling in winter or greater warming in late spring with earlier melt. The balance of these effects depends on their relative magnitude. This issue is further explored in Sect. 5.2.

MAAT is well modelled at IMIS stations with low error and a bias of only 0.18 °C, although a counteracting slight cold (warm) bias at warm (cold) sites is visible in the data. MS and MP give statistically identical results. MASD is not captured well due to large biases in driving precipitation fields. The bias becomes more pronounced at higher values of SD. A slight improvement is seen in MP over MS. Overall, MP generally shows an improvement over MS in reproducing observations, which would be expected as the spatial uncertainty introduced by TopoSUB is removed. However the difference is generally quite small (most significant difference is MAGST IMIS), and this is encouraging in that it seems MS does not introduce significant uncertainty over MP simulation. Figure 5 gives a visual impression of MS simulated MAGST results as a transect through the experiment domain.

A simple classification of results according to permafrost and no permafrost based on MAGST > 0 °C and

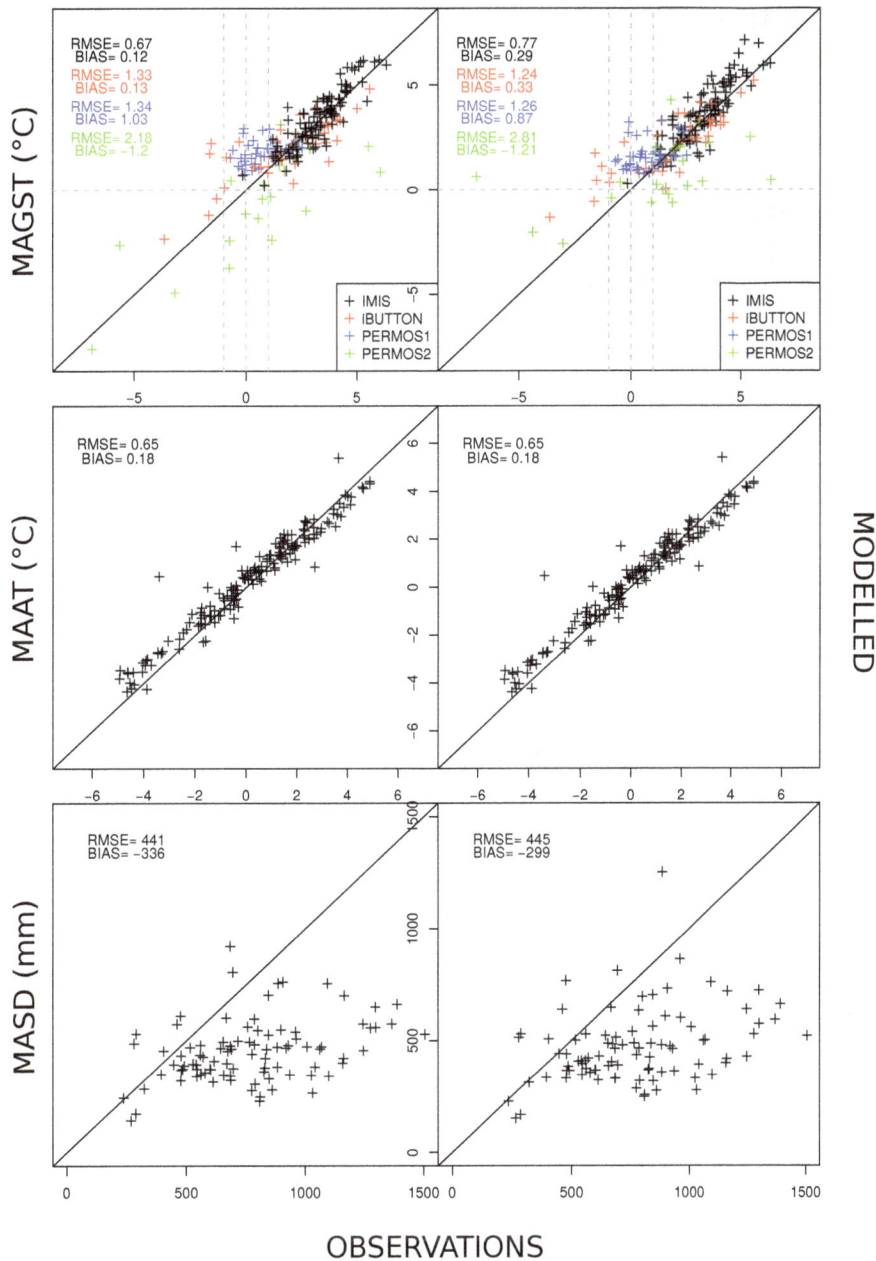

Figure 4. A comparison of MP and MS results. Modelled MAGST, MAAT and MASD evaluated against IMIS sites (2006–2011) together with statistics. Sensitive thresholds of 0 and ±1 °C are added to aid interpretation.

MAGST < 0 °C, respectively, shows that the model does not perform very well for PERMOS1 data. However, this data set is strongly clustered around 0 °C, and therefore small uncertainties lead to misclassification. In addition, strong bias due to likely underestimated snowpack skews the simulated results positively.

5.2   Snow bias correction method

Figure 6 illustrates the snow bias correction (SBC) method with an example from a high-snowfall region (Val Bedretto)

where underestimated snow depth drives warm bias in summer (cold bias in winter due to inadequate insulation of simulated snowpack). The SBC successfully corrects simulated snow depths and as a consequence the ground thermal regime. A broader evaluation is given by Fig. 7, which shows ground surface temperature (a) without and (b) with SBC at all available IMIS stations (64). The plots give data by season, with winter mean ground surface temperature (MGST, blue), summer MGST (red) and MAGST (green) shown. Negative bias is seen in winter, whereas a positive bias is

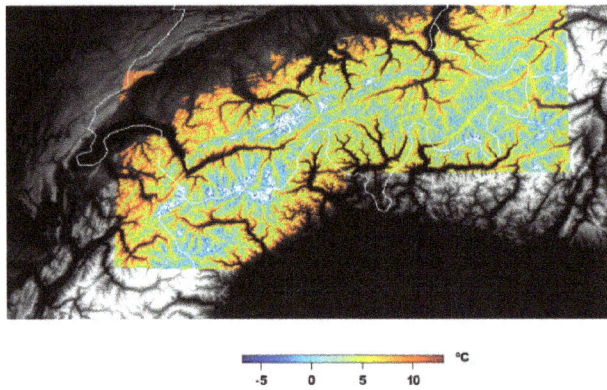

Figure 5. A visual impression of MS spatialised results: a section of a large-area simulation of MAGST with glacier mask for areas above 1000 m a.s.l. (UTM zone 32° N). Switzerland's southwestern border is overlaid for orientation.

Table 3. Comparison of PE ($km^2 \times 10^3$) obtained by this study compared to other methods in the literature.

Author	Value	Method	Relevance
This study	1.97	numerical	global
Gruber (2012)	0.7–2.5	statistical	global
Boeckli et al. (2012a)	2.2	statistical	regional
Keller et al. (1998)	1.7–2.5	statistical	regional

visible in summer. Annually the positive bias dominates due to the greater magnitude of values. The effect of correcting snow depth via the SBC method has more effect on summer MGST than winter MGST. This is explained by ensuring the correct end of snowpack date therefore averting strongly positive GST too early in summer. This agrees with the findings of Marmy et al. (2013), who found that snow duration rather than maximum snow height was the most important factor controlling simulated ground temperatures at a high Alpine permafrost site. MASD (c) without and (d) with SBC evaluated at the same stations. This shows the ability of the method to correct MASD, albeit while retaining a systematic bias. This could be due to the fact that SWE is reproduced accurately but parameters governing density of the snowpack or wind erosion are not correct. However, without SWE evaluation data at these stations, this is difficult to confirm.

Additionally, in permafrost areas basal temperatures of the snowpack may vary much more than in non-permafrost areas (such as those given by the IMIS stations). Therefore underestimation of the snowpack may have even stronger effect on model bias.

5.3 Test application: permafrost estimate

In this study we produced an estimate of 1974.9 km^2 permafrost within Switzerland based on the stated definition. Figure 8 gives a visual comparison of permafrost extent computed by this study with a state-of-the-art statistical model (Boeckli et al., 2012b) derived from Alpine specific data sets. The current method compares well in terms of spatial patterns with results of Boeckli et al. (2012b). Comparison of model results, despite differences in the definition of permafrost area and in observation periods, is intended to demonstrate the similarity of patterns resulting from both approaches (cf. face validity, Rykiel, 1996). Boeckli et al. (2012b) is based on climate normals 1961–1990 whereas the current estimate covers the period 1984–2011. The method

we have shown benefits from the simplicity of definition in actually simulating permafrost (i.e. ground $< 0\,°C$ for more than two years), although depth of simulation remains an arguable point. In addition, Table 3 shows that the current study produces an estimate that fits a range of key estimates from the literature, well.

5.4 Macroclimatic distribution of error

Figure 9 shows the distribution of bias for all IMIS stations for TAIR, GST and SD at the macroclimatic scale. The purpose was to investigate whether there are any significant biases or sign changes of bias at the mountain range scale. Such biases would largely be a result of how well the driving climate is simulated in different topo-climatic settings, e.g. north or south slope of the Alps, inner-Alpine regions or west to east. TAIR bias is well distributed in sign and generally small in magnitude. There is no clear pattern in error distribution, although the north slope seems most well modelled. GST bias is well distributed in magnitude but positively biased (as shown in Fig. 8). Again north slope seems to be modelled most successfully. SD bias is very negative and error magnitude seems to fit magnitudes of precipitation i.e. greater north and south of the main Alpine ridge and less in inner-Alpine regions (Frei and Schaer, 1998). However, stations on the north-slope of the main Alpine chain appear to be modelled well. Overall, there is no clear evidence of topo-climatic gradients in error patterns at the mountain range scale.

However, Fig. 9 is supporting evidence that generally negatively biased snowpack (too shallow) is most likely driving a positively biased MAGST (too warm) at least at the IMIS stations given in this figure. In addition, air temperature seems to be excluded as a driver of bias in MAGST as it displays no obvious bias pattern.

6   Discussion

6.1   Model chain uncertainty

In order to place these results in context we provide a semi-quantitative analysis of uncertainty through the model chain. The main sources of uncertainty we identify are: (1) bias in driving fields, (2) error in scaling approach, (3) uncertainty in

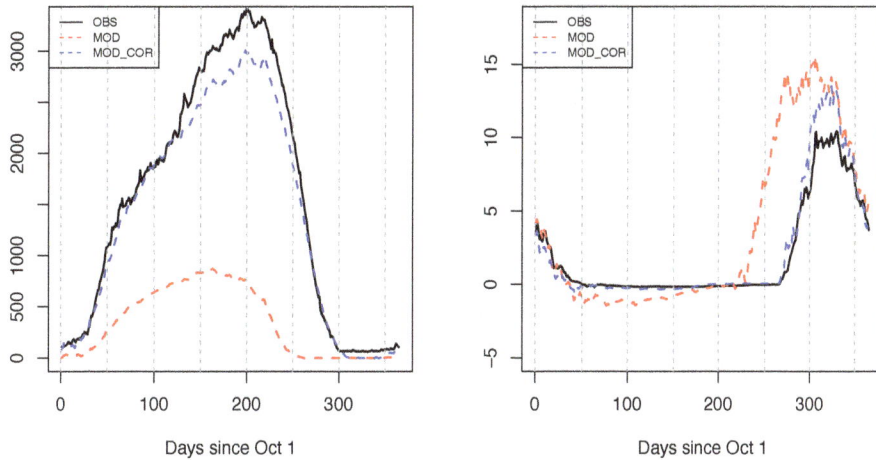

Figure 6. Evolution of SD and GST at a high-snowfall IMIS site (Val Bedretto); linkage between GST and SD is very clear. Correction of SD using the SBC method improves GST estimate.

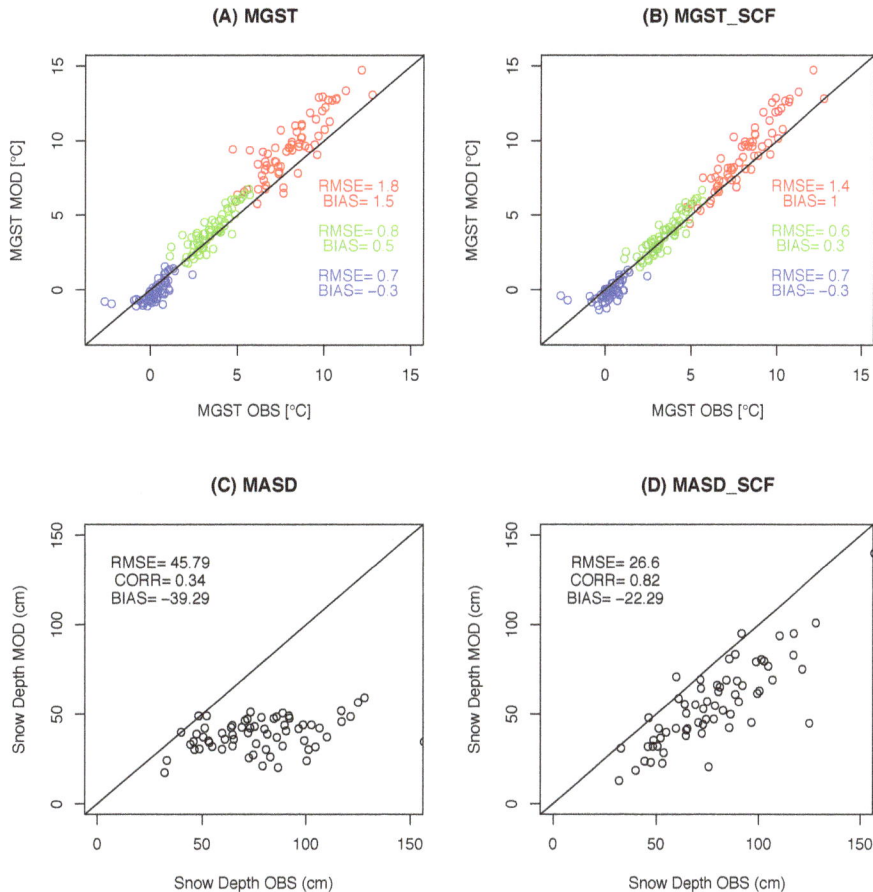

Figure 7. Ground surface temperature (**a**) without and (**b**) with SBC method evaluated at all available IMIS stations (64). Winter (blue) and summer (red) MGST, and MAGST (green) are shown. Negative bias is seen in winter and positive bias in summer. Annually the positive bias dominates due to greater magnitude of values. MASD (**c**) without and (**d**) with SBC method evaluated at the same stations. This shows the ability of the method to correct MASD, albeit while retaining a systematic bias.

Figure 8. Visual comparison of permafrost extent computed by **(a)** this study (MS setup) and **(b)** a state-of-the-art statistical model (Boeckli et al., 2012b). Comparison of model results, despite differences in the definition of permafrost area and in observation periods, is intended to demonstrate the similarity of patterns resulting from both approaches.

the lumped scheme, (4) LSM uncertainties (parameters and processes) and, (5) surface data based uncertainties (scale discussed separately).

1. Uncertainty in the driving fields can be due to bias, spatial/temporal issues or model physics and parameterisations. This issue was explored in FG2013 and reasonable to good results were reported for the variables tested. The exception being precipitation. Additionally, reanalysis data sets are expected to vary spatially and temporally with density of observations assimilated. Bias in driving precipitation is a commonly reported problem in atmospheric models (e.g. Dai, 2006; Boberg et al., 2008), and we have attempted to address this issue with the correction method detailed in this study. Two notes of caution are worth mentioning with respect to this method: (a) this method is only valid currently at site scale, and (b) it relies on GST measurements. However, the approach shown here could potentially be used together with satellite imagery in order to estimate snowpack bias based on MD to enable scalability of the method. However this is beyond the scope of this paper. Other uncertainties in modelling snow precipitation lie in the definition of the snow/rain threshold; the fact that errors are cumulative over a season; and also that significant inputs are relatively infrequent, discrete events, which means that missing an event can have a large impact on season totals. Bias associated with snow-based precipitation may have strong impacts on the ground thermal regime due to the thermal properties of the snowpack, duration of snowpack or even cooling effects of very shallow snowpacks where the albedo effect may dominate. In this study the snowpack tends to be negatively biased (too shallow), leading to often negatively biased winter temperatures (too cold, possibly exacerbated by choice of snow thermal conductivity parameterisation) due to lack of adequate insulation, and positively biased (too warm) summer temperatures, due

to early melt of spring snowpack (Fig. 7. Out of these seasonal effects the summer warm bias dominates on average due to higher magnitude of values.

2. In discussing uncertainty in the scaling approach (TopoSCALE), we focus on TAIR as this is the only driving variable evaluated in this study due to its importance in driving the ground thermal regime. Other driving fields (including TAIR) were previously evaluated in FG2013. Frei (2014) reports a TAIR RMSE of 1.5 °C in the Alps using a sophisticated interpolation technique of station data. While these are daily values and cannot be compared directly to an RMSE of 0.64 as obtained in this study, in FG2013 we show that TopoSCALE is able to achieve an RMSE of 1.93 on daily TAIR values. To place this in context, the method of Frei (2014) interpolates station data to model non-linearities in the vertical thermal profile together with a distance-weighting scheme to account for terrain effects. Given this, TopoSCALE compares quite favourably given that only vertical profiles are modelled explicitly. In addition there are possibly advantages in the gridded ERA data set over interpolated station data in terms of representing larger-scale, synoptic conditions. It should be noted that TAIR at the majority of stations is modelled at considerably lower RMSE but the overall value is affected by four key outliers (RMSE is sensitive to outliers), which degrade the overall result (Figs. 4 and 9).

3. Uncertainty of the sub-grid scheme (TopoSUB) has two main sources: (1) the resolution of the base DEM and (2) the description of surface cover. The resolution of the DEM defines the range of parameter space, irrespective of number of samples computed. For example, the base DEM of 30 m in this study produces in several cases a steepest sample of under 60°, whereas in reality vertical slopes exist. This has an effect on both mass and energy balance computed at such sites. In this study,

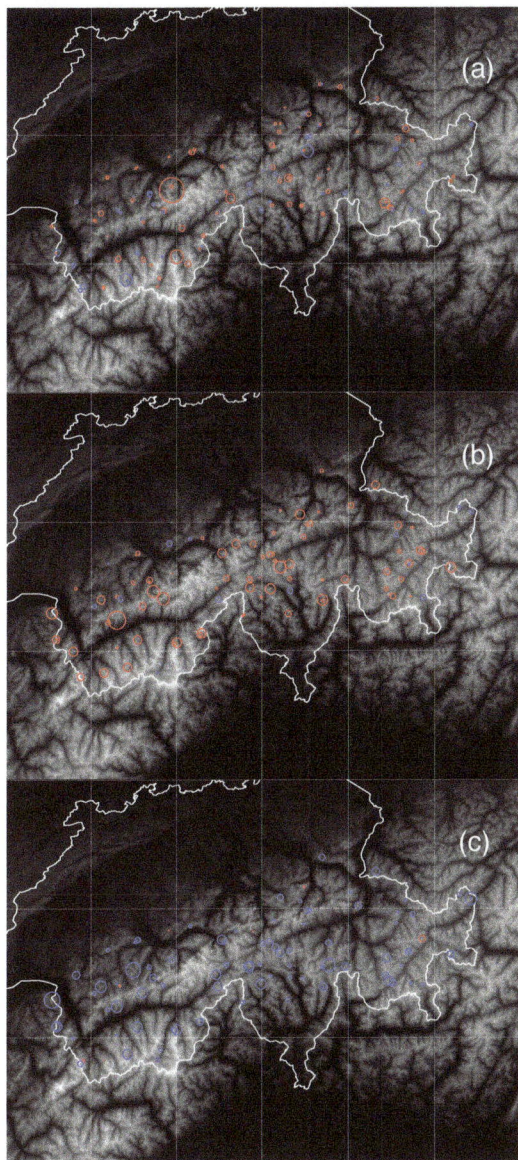

Figure 9. Distribution of bias in **(a)** TAIR, **(b)** GST and **(c)** SD at the macroclimatic scale. Blue indicates negative bias (model colder/less); red indicates positive bias (model warmer/more). Size of circle indicates the relative magnitude. Sites correspond to all IMIS stations included in the analysis.

surface cover is prescribed as an average value of surface characteristics within a TopoSUB sample, which are derived from a simple landcover data set. Landcover could however be used as a predictor in sample formation if this was deemed to be important, e.g. vegetation mosaics that significantly affect soil moisture, wind drift or energy balance at the surface. While surface cover is often a function of topographic predictors in the study domain, samples with complex surface characteristics – e.g. vegetation, boulder and bedrock matrix – will exhibit a degree of uncertainty due to the fact that all

members are modelled as the modal surface type. Due to the significance of surface (and prescribed sub-surface) characteristics in simulating surface (sub-surface) processes, this may constitute an important source of uncertainty.

4. Uncertainties due to the LSM have three main source: (i) process description (or omission), (ii) parameterisation of processes not explicitly modelled and (iii) values given to sensitive parameters. This topic has been well discussed in the literature (e.g. Gupta et al., 2005; Beven, 1995), and so here we focus on parameter values that are sensitive and therefore have a large influence on the final result. Parameter values were fixed and taken from a distribution described by Gubler et al. (2013). The exception to this is sub-surface properties (Table 1) which vary as a function of surface type. Gubler et al. (2013) provide a thorough analysis of sensitivities and uncertainties related to parameter values used in the model GEOtop, and their analysis is likely applicable to many other LSMs that have similar process description to GEOtop. In this study the authors found a total parametric uncertainty based on intensive Monte Carlo simulation of 0.1–0.5 for clay silt and rock and 0.1–0.8 for peat sand and gravel – the higher values being related to higher hydraulic conductivity of these surface classes. Therefore a portion of the error statistics given in Fig. 4 could be explained by LSM uncertainty alone.

While addressing all these sources of uncertainty within the analysis is beyond the scope of the paper, an important outcome of this work is that through the improved efficiency of simulation by several orders of magnitude, intensive simulation-based uncertainty analysis starts to become feasible.

6.2 Scale issues

While scale issues are a central topic of this work in scaling between atmospheric forcing and surface simulation, another important aspect of scale mismatch arises in validation. Evaluation exercises are often carried out in the literature where model results representing cells with side lengths of 10s–100s or, in extreme cases, 1000s of metres are compared to point-scale measurements. In this study the PERMOS data set is point scale in both measurements and topographic properties upon which modelled results are based, as these properties have been measured locally and not extracted from the DEM. The IMIS data set is comprised of point-scale measurements. The iButton measurements are aggregated to a footprint mean, representing a 10 m pixel. Modelled results of both data sets are based on properties derived from a 10 m DEM. While this seems a reasonable comparison, Gubler et al. (2011) demonstrated large differences in surface conditions within such a scale domain. Smoothing of slope angles by DEM resolution, localised shading or snow drifting at a measurement point may cause large differences in measured and modelled conditions. Additionally, as stated above, there is a limitation based on resolution of base DEM (30 m) which under-represents steep slopes in Topo-SUB sampling.

6.3 Important limitations

Key limitations are discussed in terms of (a) TopoSCALE and (b) TopoSUB. TopoSCALE based limitations primarily originate from the horizontal resolution of driving fields. In this study ERA-I fields at 0.75° × 0.75° are used. This resolution is far too coarse to represent sub-grid effects such as valley temperature inversions. In addition, topographic precipitation barriers are unresolved in regions like the Mattertal (SW Switzerland) which produce important rain shadows. Finally, spatial patterns of sub-grid effects such as shallow (mainly cumulus) convective precipitation, which is important in simulating correct precipitation intensities, only start to become resolved at resolutions of around 1 km (e.g. Kendon et al., 2012). In the case of ERA-I this is because shallow convection is parameterised by a bulk mass flux scheme, as described by Tiedtke (1989), which cannot resolve the level of spatial differentiation that is present in the measurements. This process is particularly significant in spring and summer months, as surface heating occurs during a typical diurnal cycle, driving convective mass fluxes. An outlook in this respect is that the presented scheme is readily scalable to higher-resolution driving climate data that will likely come into the public domain in the next few years.

Another key limitation of the TopoSCALE scheme is that boundary effects are not included in the scaling of atmospheric profiles that represent the free atmosphere. It was shown in FG2013 that the diurnal amplitude of fields such as TAIR and shortwave (SW) radiation was not as great as surface-affected measurements within the boundary layer. This may have important implications for processes which are driven by strong daily amplitudes, such as spring melt of the snowpack.

TopoSUB-based limitations are largely derived from the scale of the base DEM on which sampling is based (as previously discussed), together with the description of surface cover. However, an important limitation comes from the inherent 1-D structure of samples as simulation units. This means that all lateral processes can only be parameterised and not modelled explicitly. For example, in computing horizon angles that are important for cast-shadow calculations (Dubayah and Rich, 1995), we apply a mean horizon angle derived from the sky-view factor and local slope. This has obvious problems when the horizon is highly asymmetric; e.g. consider a steep, south-facing mountain slope overlooking a plain. In this situation the horizon angle would be artificially raised in the southerly direction to a mean level, therefore reducing radiation inputs and consequently introducing a cold bias. This effect may also give biases in

terms of reduced radiation in westerly directions under convective systems (e.g. Marty et al., 2002) or under strongly anisotropic local horizons. Another example of neglected 2-D effects is illustrated well by the PERMOS2 results (steep rock walls, Fig. 4). The results here are negatively biased, indicating that part of the energy balance is missing or poorly described. Missing or inadequately described physical processes is a well-known and common characteristic of most physical models (Arneth et al., 2012; Beven, 1995); however, as testing of the physical model GEOtop is not the focus of this study, we provide only a limited discussion on this topic. GEOtop computes the emissivity (LW) and albedo (SW) of its hypothetical surroundings as identical to that of the point itself – in this case a steep rock wall. This will reduce the SW radiation reflected from surrounding terrain, which can be a significant energy input to steep rock walls when a winter snowpack is present. From a mass balance perspective we do not model redistribution of snow by wind or avalanche. This has an important effect on the surface energy balance where melt dates can be several weeks later due to heavy accumulations at bases of avalanche slopes (Harris et al., 2009) or earlier on wind-eroded slopes (Bernhardt et al., 2010). This sub-grid effect can be parameterised by computing multiple cases for increased/decreased snow cover, but corresponding results will be difficult to spatialise. We model the loss of snow on steep slopes as a function of slope angle; however this is not a mass-conservation method as the removed mass is not redistributed.

6.4 Snowpack issues

The winter snowpack is extremely important in controlling the ground thermal regime (cf. Smith, 1975; Goodrich, 1982; Ling and Zhang, 2003; Zhang et al., 1996; Zhang, 2005b). Therefore, here we summarise the main issues with respect to this paper together with a brief assessment of the modelling approach used.

Precipitation is often the most difficult model driver to estimate both in quantity and timing, whether originating from a model or extrapolated from a nearby station. This challenge is increased in the case of winter precipitation, which is often harder to measure (gauge undercatch, satellite insensitivity) or model (solid/liquid thresholds, densities etc.). In our study the use of output from an atmospheric model (even reanalysed) has been shown to have inherent biases. This can be clearly seen in how precipitation is distributed (cf. Fiddes and Gruber (2014), Figs. 7 and 8), for example in the absence of high-intensity precipitation events which contribute strongly to build-up of the winter snowpack. Additionally, errors accumulate over the winter season, which can be a challenging characteristic of the modelled seasonal snowpack.

All our results display some degree of positive bias in MAGST with the exception of steep rock wall sites, indicating the importance of the snowpack (and driving precipitation) as a controlling variable. We have shown that correction of the snowpack can be extremely important in successfully modelling the ground thermal regime. Underestimated snowpacks may have various and opposing effects on ground temperatures, e.g. greater cooling in winter due to albedo effects or greater warming in late spring with earlier melt of a shallow snowpack. The balance of these effects over an annual cycle depends upon their relative magnitude. In our study region we have shown that a shallow snowpack is likely driving positively biased MAGST. However, at a seasonal scale we see negatively biased winter temperatures due to radiative cooling and higher thermal conductivities (generally cooling atmosphere) and positively biased spring/summer temperatures due to earlier melt of snowpack and exposure to atmosphere (generally warming). Negative biases in winter are not necessarily insignificant and may have process-relevant effects, but they are often masked in this study by positive summer biases. In addition these opposing biases can cause a cancelling effect of errors at annual timescales. How these results scale out to other geographical regions very much depends upon the topo-climatic conditions. However, some generally applicable conclusions may be drawn: (1) biased snowpacks can be a common feature of model simulations due to bias in driving precipitation field, (2) correction of snowpack bias is often important to address bias in the underlying ground thermal regime and (3) determining the sign of the effect of a biased snowpack upon ground temperatures is complex and likely varies strongly with location and season.

The approach taken in this study offers a straightforward correction, which can compliment the downscaling strategy in its simplicity. In this sense, the primary aim is to correct major biases in the water balance and therefore the magnitude of resulting biases in the ground's energy balance. However, as this correction is used only on input precipitation, it assumes that this is the dominating bias. It does not address possible biases in temperature (or other variables influencing snowpack evolution, e.g. wind) or biases due to snowpack-related model parameters, such as thermal conductivities.

6.5 Applications and outlook

In this study we provide a large-scale permafrost model estimate as a test case. However, the scheme is generic in that it is able to generate surface fields of any variable the LSM is able to simulate. The scheme can be used to simulate high-resolution maps of current conditions as well as recent dynamics which can be used to generate estimates of near-future trajectories of change. In longer-term planning applications the scheme, when driven by suitable climate model data, can be used to produce scenarios of site-specific future conditions. A core strength of the scheme is computation reduction, which means that multiple repeat simulations are more likely to be possible. This can be utilised by producing a range of outcomes that consider significant uncertainties in the model chain and therefore a range of scenarios

that should be considered in a given study. Such scenarios could be interpreted together with site-specific knowledge to provide an improved quality of result, or a range of outcomes to be planned for in terms of uncertainty related to future conditions or other unknowns. In terms of model evaluation the scheme has two important contributions: it provides model data at an appropriate scale for validation measurements (e.g. site of measurement); secondly, by utilising an LSM, the scheme can generate a wide range of variables, in order to maximise use of all available evaluation data and therefore provide more robust evaluations.

With respect to this permafrost application (but also relevant to other land surface variables), it is important to remember that we currently have a strong environmental change and we assume permafrost thaw in many areas; therefore it is not so much only the classification of permafrost vs. non-permafrost that we are interested in, i.e. a classic distribution map, but also the ability to describe the evolution over time or, in other words, map a process. Much of the inherent uncertainty (including whether there is permafrost in the first place) will have to be accepted and estimated. For this, our scheme provides a way to reduce the computational effort of a thorough uncertainty analysis.

7 Conclusions

This study has shown that the presented scheme is able to simulate GST and TAIR reasonably well over large areas in heterogeneous terrain, using global data sets. We have presented a simple method that enables correction of winter precipitation inputs and thus greatly improves simulation of SD, where data are available. As a test application, an estimate of permafrost distribution in Switzerland has been computed with the scheme, which is comparable to published statistical model results. However, the scheme described in this study is additionally capable of producing transient simulations; results in remote areas; and many more useful variables besides the simple variable of distribution, such as changing subsurface properties (e.g. ground ice loss). This underscores a key strength of the scheme: through efficiency gains, it allows for application of LSMs at high resolutions over large areas with transient simulation possible. This opens a number of new possibilities in the field of land surface change assessments in heterogeneous environments. In addition it allows model results to be validated at an appropriate scale by a wide range of measurement types due to the comprehensive set of physically consistent outputs that are generated. We summarise the main contributions and insights of this work as the following:

- the presented scheme works well in large-area simulation of the tested variables due to an efficient sub-grid sampling of surface heterogeneity and scaling of driving climate;

- simple bias correction of winter precipitation may be possible based on the melt date of the snowpack;

- the scheme produces an estimate of permafrost area in the Swiss Alps that is comparable to statistical methods;

- all inputs are derived from global data sets, suggesting that consistent application globally in heterogeneous and/or remote terrain is possible.

Acknowledgements. We would like to thank the SLF for the IMIS data set together with M. Dall'Amico and P. Pogliotti for compiling the IMIS stations meta-data. We would like to thank ECMWF for availability of the reanalysis data set ERA-Interim. We thank PERMOS for the GST logger data set and work done within the TEMPS project to make these data accessible. This work was also supported by GC3: Grid Computing Competence Centre (www.gc3.uzh.ch) with customised libraries (gtsub_control and GC3Pie) and user support. This project was funded by the Swiss National Science Foundation projects CRYOSUB and X-Sense. Finally we thank three anonymous reviewers and T. Zhang, all of whom contributed to significantly improving this paper.

Edited by: T. Zhang

References

Arneth, A., Mercado, L., Kattge, J., and Booth, B. B. B.: Future challenges of representing land-processes in studies on land-atmosphere interactions, Biogeosciences, 9, 3587–3599, doi:10.5194/bg-9-3587-2012, 2012.

Barnett, T. P., Adam, J. C., and Lettenmaier, D. P.: Potential impacts of a warming climate on water availability in snow-dominated regions, Nature, 438, 303–309, 2005.

Bernhardt, M., Liston, G. E., Strasser, U., Zängl, G., and Schulz, K.: High resolution modelling of snow transport in complex terrain using downscaled MM5 wind fields, The Cryosphere, 4, 99–113, doi:10.5194/tc-4-99-2010, 2010.

Bertoldi, G., Rigon, R., and Over, T. M.: Impact of watershed geomorphic characteristics on the energy and water budgets, J. Hydrometeorol., 7, 389–403, 2006.

Betts, A. K.: Land–surface–atmosphere coupling in observations and models, J. Adv. Model. Earth Syst., 2, 4, doi:10.3894/JAMES.2009.1.4, 2009.

Beven, K.: Linking parameters across scales: subgrid parameterizations and scale dependent hydrological models, Hydrol. Process., 9, 507–525, 1995.

Boberg, F., Berg, P., Thejll, P., Gutowski, W. J., and Christensen, J. H.: Improved confidence in climate change projections of precipitation evaluated using daily statistics from the PRUDENCE ensemble, Clim. Dynam., 32, 1097–1106, doi:10.1007/s00382-008-0446-y, 2008.

Boeckli, L., Brenning, A., Gruber, S., and Noetzli, J.: A statistical approach to modelling permafrost distribution in the European Alps or similar mountain ranges, The Cryosphere, 6, 125–140, doi:10.5194/tc-6-125-2012, 2012a.

Boeckli, L., Brenning, A., Gruber, S., and Noetzli, J.: Permafrost distribution in the European Alps: calculation and evaluation of an index map and summary statistics, The Cryosphere, 6, 807–820, doi:10.5194/tc-6-807-2012, 2012b.

Calonne, N., Flin, F., Morin, S., Lesaffre, B., du Roscoat, S. R., and Geindreau, C.: Numerical and experimental investigations of the effective thermal conductivity of snow, Geophys. Res. Lett., 38, L23501, doi:10.1029/2011GL049234, 2011.

Cosenza, P., Guerin, R., and Tabbagh, A.: Relationship between thermal conductivity and water content of soils using numerical modelling, Eur. J. Soil Sci., 54, 581–587, 2003.

Dai, A.: Precipitation characteristics in eighteen coupled climate models, J. Climate, 19, 4605–4630, doi:10.1175/JCLI3884.1, 2006.

Dai, Y., Zeng, X., Dickinson, R. E., Baker, I., Bonan, G. B., Bosilovich, M. G., Denning, A. S., Dirmeyer, P. A., Houser, P. R., Niu, G., Oleson, K. W., Schlosser, C. A., and Yang, Z.-L.: The common land model, B. Am. Meteorol. Soc., 84, 1013–1023, doi:10.1175/BAMS-84-8-1013, 2003.

Dall'Amico, M., Endrizzi, S., Gruber, S., and Rigon, R.: A robust and energy-conserving model of freezing variably-saturated soil, The Cryosphere, 5, 469–484, doi:10.5194/tc-5-469-2011, 2011.

Dee, D. P., Uppala, S. M., Simmons, A. J., Berrisford, P., Poli, P., Kobayashi, S., Andrae, U., Balmaseda, M. A., Balsamo, G., Bauer, P., Bechtold, P., Beljaars, A. C. M., van de Berg, L., Bidlot, J., Bormann, N., Delsol, C., Dragani, R., Fuentes, M., Geer, A. J., Haimberger, L., Healy, S. B., Hersbach, H., Hólm, E. V., Isaksen, L., Kållberg, P., Köhler, M., Matricardi, M., McNally, A. P., Monge-Sanz, B. M., Morcrette, J.-J., Park, B.-K., Peubey, C., de Rosnay, P., Tavolato, C., Thépaut, J.-N., and Vitart, F.: The ERA-Interim reanalysis: configuration and performance of the data assimilation system, Q. J. Roy. Meteorol. Soc., 137, 553–597, doi:10.1002/qj.828, 2011. TS8: No refence given in reference list you mean, I think: Rykiel, J. E. J.: Testing ecological models: the meaning of validation, Ecol.

Dubayah, R. and Rich, P.: Topographic solar radiation models for GIS, Int. J. Geogr. Inf. Sci., 9, 405–419, 1995.

Durand, Y., Brun, E., Merindol, L., Guyomarc'h, G., Lesaffre, B., and Martin, E.: A meteorological estimation of relevant parameters for snow models, Ann. Glaciol., 18, 65–71, 1993.

Durand, Y., Giraud, G., Brun, E., Merindol, L., and Martin, E.: A computer-based system simulating snowpack structures as a tool for regional avalanche forecasting, J. Glaciol., 45, 469–484, 1999.

Endrizzi, S.: Snow cover modelling at a local and distributed scale over complex terrain, Diss. PhD thesis, PhD dissertation, Dept. of Civil and Environmental Engineering, University of Trento, Trento, Italy, 2007.

Endrizzi, S., Gruber, S., Dall'Amico, M., and Rigon, R.: GEOtop 2.0: simulating the combined energy and water balance at and below the land surface accounting for soil freezing, snow cover and terrain effects, Geosci. Model Dev., 7, 2831–2857, doi:10.5194/gmd-7-2831-2014, 2014.

Etzelmüller, B.: Recent advances in mountain permafrost research, Permafrost Periglac., 24, 99–107, doi:10.1002/ppp.1772, 2013.

Fiddes, J. and Gruber, S.: TopoSUB: a tool for efficient large area numerical modelling in complex topography at sub-grid scales, Geosci. Model Dev., 5, 1245–1257, doi:10.5194/gmd-5-1245-2012, 2012.

Fiddes, J. and Gruber, S.: TopoSCALE v.1.0: downscaling gridded climate data in complex terrain, Geosci. Model Dev., 7, 387–405, doi:10.5194/gmd-7-387-2014, 2014.

Frei, C.: Interpolation of temperature in a mountainous region using nonlinear profiles and non-Euclidean distances, Int. J. Climatol., 34, 1585–1605, doi:10.1002/joc.3786, 2014.

Frei, C. and Schaer, C.: A precipitation climatology of the Alps from, Int. J. Climatol., 900, 873–900, 1998

Gisnås, K., Etzelmüller, B., Farbrot, H., Schuler, T. V., and Westermann, S.: CryoGRID 1.0: permafrost distribution in Norway estimated by a spatial numerical model, Permafrost Periglac., 24, 2–19, doi:10.1002/ppp.1765, 2013.

Goodrich, L.: The influence of snow cover on the ground thermal regime, Can. Geotech. J., 19, 421–432, 1982.

Gruber, S.: Derivation and analysis of a high-resolution estimate of global permafrost zonation, The Cryosphere, 6, 221–233, doi:10.5194/tc-6-221-2012, 2012.

Gubler, S., Fiddes, J., Keller, M., and Gruber, S.: Scale-dependent measurement and analysis of ground surface temperature variability in alpine terrain, The Cryosphere, 5, 431–443, doi:10.5194/tc-5-431-2011, 2011.

Gubler, S., Endrizzi, S., Gruber, S., and Purves, R. S.: Sensitivities and uncertainties of modeled ground temperatures in mountain environments, Geosci. Model Dev., 6, 1319–1336, doi:10.5194/gmd-6-1319-2013, 2013.

Gupta, H. V., Beven, K. J., and Wagener, T.: Model Calibration and Uncertainty Assessment, John Wiley & Sons, Ltd, New York, 2005.

Harris, C., Arenson, L. U., Christiansen, H. H., Etzelmüller, B., Frauenfelder, R., Gruber, S., Haeberli, W., Hauck, C., Hölzle, M., Humlum, O., Isaksen, K., Kääb, A., Kern-Lütschg, M. A., Lehning, M., Matsuoka, N., Murton, J. B., Nötzli, J., Phillips, M., Ross, N., Seppälä, M., Springman, S. M., and Vonder Mühll, D.: Permafrost and climate in Europe: monitoring and modelling thermal, geomorphological and geotechnical responses, Earth-Sci. Rev., 92, 117–171, doi:10.1016/j.earscirev.2008.12.002, 2009.

Jafarov, E. E., Marchenko, S. S., and Romanovsky, V. E.: Numerical modeling of permafrost dynamics in Alaska using a high spatial resolution dataset, The Cryosphere, 6, 613–624, doi:10.5194/tc-6-613-2012, 2012.

Keller, F., Frauenfelder, R., Gardaz, J. M., Hölzle, M., Kneisel, C., Lugon, R., Philips, M., Reynard, E., and Wenker, L.: Permafrost map of Switzerland, in: Proceedings, 7th International Conference on Permafrost, Collection Nordicana, vol. 57, edited by: Lewkowicz, A. G. and Allard, M., Université Laval, Quebec, 557–562, 1998.

Kendon, E. J., Roberts, N. M., Senior, C. A., and Roberts, M. J.: Realism of rainfall in a very high-resolution regional climate model, J. Climate, 25, 5791–5806, doi:10.1175/JCLI-D-11-00562.1, 2012.

Koster, R. D. and Suarez, M. J.: Modeling the land surface boundary in climate models as a composite of independent vegetation stands, J. Geophys. Res., 97, 2697, doi:10.1029/91JD01696, 1992.

Leung, L. R. and Ghan, S. J.: Parameterizing subgrid orographic precipitation and surface cover in climate models, Mon. Weather Rev., 126, 3271–3291, 1998.

Li, Z.: Natural variability and sampling errors in solar radiation measurements for model validation over the Atmospheric Radiation Measurement Southern Great Plains region, J. Geophys. Res., 110, D15S19, doi:10.1029/2004JD005028, 2005.

Ling, F. and Zhang, T.: Impact of the timing and duration of seasonal snow cover on the active layer and permafrost in the Alaskan Arctic, Permafrost Periglac. Process., 14, 141–150, doi:10.1002/ppp.445, 2003.

Liston, G. and Haehnel, R.: Instruments and methods simulating complex snow distributions in windy environments using SnowTran-3D, J. Glaciol., 53, 241–256, 2007.

Löwe, H., Riche, F., and Schneebeli, M.: A general treatment of snow microstructure exemplified by an improved relation for thermal conductivity, The Cryosphere, 7, 1473–1480, doi:10.5194/tc-7-1473-2013, 2013.

Manders, A. M. M., van Meijgaard, E., Mues, A. C., Kranenburg, R., van Ulft, L. H., and Schaap, M.: The impact of differences in large-scale circulation output from climate models on the regional modeling of ozone and PM, Atmos. Chem. Phys., 12, 9441–9458, doi:10.5194/acp-12-9441-2012, 2012.

Marmy, a., Salzmann, N., Scherler, M., and Hauck, C.: Permafrost model sensitivity to seasonal climatic changes and extreme events in mountainous regions, Environ. Res. Lett., 8, 035048, doi:10.1088/1748-9326/8/3/035048, 2013.

Marty, C., Philipona, R., Fr, C., and Ohmura, A.: Altitude dependence of surface radiation fluxes and cloud forcing in the alps: results from the alpine surface radiation budget network, Theor. Appl. Climatol., 72, 137–155, 2002.

Meek, D. and Hatfield, J.: Data quality checking for single station meteorological databases, Agr. Forest Meteorol., 69, 85–109, 1994.

Piani, C., Haerter, J. O., and Coppola, E.: Statistical bias correction for daily precipitation in regional climate models over Europe, Theor. Appl. Climatol., 99, 187–192, doi:10.1007/s00704-009-0134-9, 2009.

Pitman, A. J.: Review: the evolution of, and revolution in, land surface schemes, Int. J. Climatol., 510, 479–510, doi:10.1002/joc.893, 2003.

Randall, D., Krueger, S., Bretherton, C., Curry, J., Duynkerke, P., Moncrieff, M., Ryan, B., Starr, D., Miller, M., Rossow, W., Tselioudis, G., and Wielicki, B.: Confronting Models with Data: The GEWEX Cloud Systems Study, B. Am. Meteorol. Soc., 84, 455–469, doi:10.1175/BAMS-84-4-455, 2003.

Rigon, R., Bertoldi, G., and Over, T. M.: GEOtop: A Distributed Hydrological Model with Coupled Water and Energy Budgets, J. Hydrometeorol., 7, 371–388, doi:10.1175/JHM497.1, 2006.

Rykiel, J. E. J.: Testing ecological models: the meaning of validation, Ecol. Model., 90, 299–244, 1996.

Scherler, M., Hauck, C., Hoelzle, M., Stähli, M., and Völksch, I.: Meltwater infiltration into the frozen active layer at an alpine permafrost site, Permafrost Periglac. Process., 21, 325–334, doi:10.1002/ppp.694, 2010.

Scherler, M., Schneider, S., Hoelzle, M., and Hauck, C.: A two-sided approach to estimate heat transfer processes within the active layer of the Murtèl–Corvatsch rock glacier, Earth Surf. Dynam., 2, 141–154, doi:10.5194/esurf-2-141-2014, 2014.

Schmid, M.-O., Gubler, S., Fiddes, J., and Gruber, S.: Inferring snowpack ripening and melt-out from distributed measurements of near-surface ground temperatures, The Cryosphere, 6, 1127–1139, doi:10.5194/tc-6-1127-2012, 2012.

Smith, M. W.: Microclimatic Influences on Ground Temperatures and Permafrost Distribution, Mackenzie Delta, Northwest Territories, Can. J. Earth Sci., 12, 1421–1438, doi:10.1139/e75-129, 1975.

Smith, M. W. and Riseborough, D. W.: Climate and the limits of permafrost: a zonal analysis, Permafrost Periglac., 15, 1–15, doi:10.1002/ppp.410, 2002.

Sturm, M. and Benson, C. S.: Vapor transport, grain growth and depth-hoar development in the subarctic snow, J. Glaciol., 43, 42–59, 1997.

Tachikawa T., Hato, M., Kaku, M., and Iwasaki, A.: The characteristics of ASTER GDEM version 2, Proc. IGARSS 2011 Symposium, 24–29 July 2011, Vancouver, Canada, 3657–3660, 2011.

Tiedtke, M.: A comprehensive mass flux scheme for cumulus parameterization in large-scale models, Mon. Weather Rev., 117, 1779–1800, 1989.

Westermann, S., Schuler, T. V., Gisnås, K., and Etzelmüller, B.: Transient thermal modeling of permafrost conditions in Southern Norway, The Cryosphere, 7, 719–739, doi:10.5194/tc-7-719-2013, 2013.

Wood, E. F., Roundy, J. K., Troy, T. J., van Beek, L. P. H., Bierkens, M. F. P., Blyth, E., de Roo, A., Döll, P., Ek, M., Famiglietti, J., Gochis, D., van de Giesen, N., Houser, P., Jaffé, P. R., Kollet, S., Lehner, B., Lettenmaier, D. P., Peters-Lidard, C., Sivapalan, M., Sheffield, J., Wade, A., and Whitehead, P.: Hyperresolution global land surface modeling: meeting a grand challenge for monitoring Earth's terrestrial water, Water Resour. Res., 47, 1–10, doi:10.1029/2010WR010090, 2011.

Zhang, T.: Influence of the seasonal snow cover on the ground thermal regime: An overview, Rev. Geophys., 43, 1–23, doi:10.1029/2004RG000157, 2005b.

Zhang, T., Osterkamp, T. E., and Stamnes, K.: Influence of the depth hoar layer of the seasonal snow cover on the ground thermal regime, Water Resour. Res., 32, 2075–2086, 1996.

Organic carbon pools in permafrost regions on the Qinghai–Xizang (Tibetan) Plateau

C. Mu[1], T. Zhang[1], Q. Wu[2], X. Peng[1], B. Cao[1], X. Zhang[1], B. Cao[1], and G. Cheng[2]

[1]College of Earth and Environmental Sciences, Lanzhou University, Lanzhou Gansu 730000, China
[2]State Key Laboratory of Frozen Soil Engineering, Cold and Arid Regions Environmental and Engineering Research Institute, CAS, Lanzhou Gansu 730000, China

Correspondence to: T. Zhang (tjzhang@lzu.edu.cn)

Abstract. The current Northern Circumpolar Soil Carbon Database did not include organic carbon storage in permafrost regions on the Qinghai–Xizang (Tibetan) Plateau (QXP). In this study, we reported a new estimation of soil organic carbon (SOC) pools in the permafrost regions on the QXP up to 25 m depth using a total of 190 soil profiles. The SOC pools were estimated to be 17.3 ± 5.3 Pg for the 0–1 m depth, 10.6 ± 2.7 Pg for the 1–2 m depth, 5.1 ± 1.4 Pg for the 2–3 m depth and 127.2 ± 37.3 Pg for the layer of 3–25 m depth. The percentage of SOC storage in deep layers (3–25 m) on the QXP (80 %) was higher than that (39 %) in the yedoma and thermokarst deposits in arctic regions. In total, permafrost regions on the QXP contain approximately 160 ± 87 Pg SOC, of which approximately 132 ± 77 Pg (83 %) stores in perennially frozen soils and deposits. Total organic carbon pools in permafrost regions on the QXP was approximately 8.7 % of that in northern circumpolar permafrost region. The present study demonstrates that the total organic carbon storage is about 1832 Pg in permafrost regions on northern hemisphere.

1 Introduction

Soil organic carbon (SOC) storage in permafrost regions has received worldwide attention due to its direct contribution to the atmospheric greenhouse gas contents (Ping et al., 2008a; Tarnocai et al., 2009; Zimov et al., 2009). Climate warming will thaw permafrost, which can cause previously frozen SOC become available for mineralization (Zimov et al., 2006). Permafrost has potentially the most significant carbon-climate feedbacks not only due to the intensity of climate forcing, but also the size of carbon pools in permafrost regions (Schuur et al., 2008; Mackelprang et al., 2012; Schneider von Deimling et al., 2012).

Recently, carbon stored in permafrost regions has created many concerns because of the implication on global carbon cycling (Ping et al., 2008; Burke et al., 2012; Zimov et al., 2006; Michaelson et al., 2013; Hugelius et al., 2013). It has been estimated that permafrost regions of circum-Arctic areas contain approximately 1672 Pg of organic carbon, which include 495.8 Pg for the 0–1 m depth, 1024 Pg for the 0–3 m depth and 648 Pg for 3–25 m depth. Based on newly available regional soil maps, the estimated storage of SOC in 0–3 m depth is estimated to 1035 ± 150 Pg (Hugelius et al., 2014), about 1 % higher than the previous estimate by Tarnocai et al. (2009). The thawing of permafrost would expose the frozen organic carbon to microbial decomposition, and thus may initiate a positive permafrost carbon feedback on climate (Schuur et al., 2008). The strength and timing of permafrost carbon feedback greatly depend on the distribution of SOC in permafrost regions. Therefore, understanding soil carbon storage in permafrost regions is critical for better predicting future climate change. However, the present knowledge of SOC pool in permafrost regions only limited to the circum-Arctic areas. Little is known about the SOC pools in the low-altitude permafrost regions.

The Qinghai–Xizang (Tibetan) Plateau (QXP) in China has the largest extent of permafrost in the low-middle latitudes of the world, with permafrost regions of about 1.35×10^6 km^2 and underlying ~ 67 % of the QXP area (Ran et al., 2012). It has been suggested that SOC in permafrost regions on the QXP was very sensitive to global warming, due to the permafrost characteristics of high temperature ($< \sim 2.0°$), thin thickness (< 100 m) and unstable thermal states (Cheng and Wu, 2007; Li et al., 2008; Wu and Zhang, 2010). Mean annual permafrost temperatures at 6.0 m depth increased by a range of $0.12°$ to $0.67°$ from 1996 to 2006 (Wu and Zhang, 2008), and increased $\sim 0.13°$ from 2002 to 2012 (Wu et al., 2015). Active layer thickness increased, on average, approximately ~ 4.26 cm yr^{-1} along the Qinghai–Tibetan Highway from 2002 to 2012 (Wu et al., 2015). In addition, the carbon stored in permafrost area was labile and a great part of the carbon was mineralizable (Mu et al., 2014; Wu, et al., 2014).

Some studies have been conducted on SOC pools in 0–1 m depth on the QXP (Wang et al., 2002, 2008; Yang et al., 2008, 2010; Liu et al., 2012; Wu et al., 2012). It was estimated that total SOC for the top 0.7 m was about 30–40 Pg in the grassland of the plateau. The disagreement among the studies on the SOC pools was attributed to the limited sampling points and the quality of the SOC data gathered to date. Despite the importance of SOC in permafrost areas, there are still few reports to the SOC storage in permafrost regions of the QXP. So far, the current Northern Circumpolar Soil Carbon Database does not include the SOC in permafrost regions on the QXP (Tarnocai et al., 2009).

Perennially frozen soils are important earth system carbon pools because of their vulnerability to climate change (Koven et al., 2011). Some of the movement of SOC from surface to few meter depth is accomplished through cryoturbation (Bockheim et al., 1998), which is caused by cracking due to soil freeze-thaw cycles and by soil hydrothermal gradients (Ping et al., 2008b). It was reported that the total yedoma region contains $211 + 160/-153$ Pg C in deep soil deposits (Strauss et al., 2013). Current studies have shown the importance of deep organic carbon in permafrost regions and its feedback with climate change (Hobbie et al., 2000; Davidson and Janssens, 2006; Schuur et al., 2009). Deep organic carbon can be more sensitive to temperature increasing compared with that in the active layer (Waldrop et al., 2010). Therefore, it is essential to study the distribution of organic carbon content in deep layers of permafrost regions.

For the top layer, important factors controlling SOC pools are vegetation type and climate (Jobbagy and Jackson, 2000). The vegetation type and climate conditions related closely to each other on the QXP (Wang et al., 2002). Thus it is possible to calculate the SOC pools at 0–2 m depth according to the area of vegetation type (Chinese Academy of Sciences, 2001) in the permafrost regions (LIGG/CAS, 1988). For deep layers, the geomorphology and lithological conditions play an important role in the distribution of SOC pools (Hugelius et al., 2013). Thus it is reasonable to estimate the SOC pools at 2–25 m depth according to the area of Quaternary geological stratigraphy in permafrost regions on the QXP.

The objective of this study is to assess the SOC pools in permafrost regions on the QXP, based on the published data

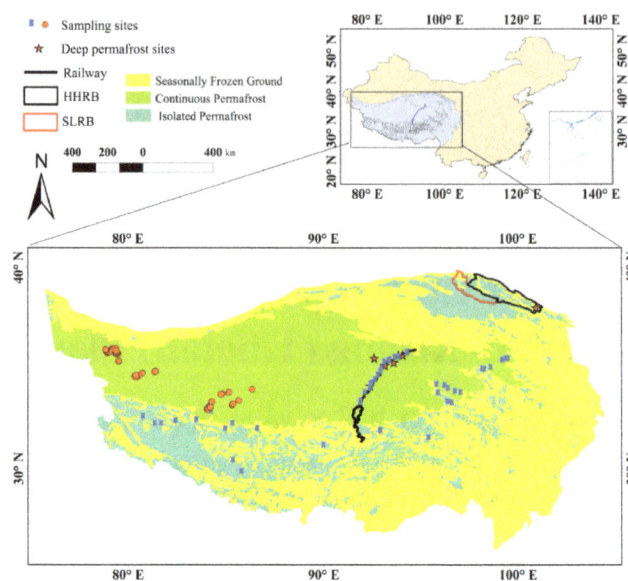

Figure 1. Location of sampling sites on the QXP, shown on the background of QXP permafrost distribution (blue points were sampling sites in Yang et al., 2010; orange points were in Wu et al., 2012; red box was Shule River basin (SLRB) in Liu et al., 2012; black box was Heihe River basin (HHRB) in Mu et al., 2013).

and new field sampling through deep drilling from this study. The new estimation focuses on the permafrost regions and includes deeper layers, down to 25 m. SOC storages of the plateau were estimated using the published data of 190 soil profiles and 11 deep sampling sites from this study in combination with the vegetation map, permafrost map and geological stratigraphy map of the QXP (Figs. 1–3). The result would update current estimation of surface organic carbon pools and deep organic carbon storage in permafrost regions of the QXP, which can provide new insights in permafrost carbon on the global scale.

2 Materials and methods

2.1 Soil carbon database in previous reports

The soil carbon databases in 0–1 m depth were retrieved from the previous reports (Yang et al., 2010; Liu et al., 2012; Wu et al., 2012; Dorfer et al.,2013; Mu et al., 2013) (Table 1). We integrated the databases from Yang et al. (2010), Dorfer et al. (2013) and Ohtsuka et al. (2008) because these studies were all performed in the middle and eastern parts of the QXP. The data of Wu et al. (2012), Liu et al. (2012) and Mu et al. (2013) in the soil carbon database in 0–1 m depth were calculated separately, since their study regions of western QXP, Shule River basin (SLRB) and Heihe river basin (HHRB) belonged to the isolated permafrost zone and the climate conditions differed greatly with the continuous permafrost zones of the QXP. The total organic carbon pools in

Table 1. Organic carbon pools in the 0–1 m depth with different vegetation type on the QXP.

Vegetation types	References	Analytical methods	Study area	Site data (n)	Area ($\times 10^6$ km^2)	SOC stock (kg m^{-2})	SOC storage (Pg)
Alpine meadow	Yang et al., (2010)	Wet oxidation	QXP	22	0.224	9.3 ± 3.9	10.7 ± 3.8
	Ohtsuka et al. (2008)	Heat combustion	QXP	1		13.7	
	Dorfer et al. (2013)	Heat combustion	QXP	2		10.4	
	Mu et al. (2013)	Heat combustion	HHRB	11	0.0065	39.0 ± 17.5	0.3 ± 0.1
	Liu et al. (2012)	Wet oxidation	SLRB	-42	0.013	8.7 ± 1.2	0.1 ± 0.02
Alpine steppe	Yang et al. (2010)	Wet oxidation	QXP	33	0.772	3.7 ± 2.0	5.3 ± 2.8
	Wu et al. (2012)	Wet oxidation	western QXP	52		7.7 ± 3.2	
	Liu et al. (2012)	Wet oxidation	SLRB	-42		9.2 ± 1.1	
Alpine desert	Wu et al. (2012)	Wet oxidation	western QXP	25	0.175	3.3 ± 1.5	0.7 ± 0.3
	Liu et al. (2012)	Wet oxidation	SLRB	~ 42		4.4 ± 0.7	

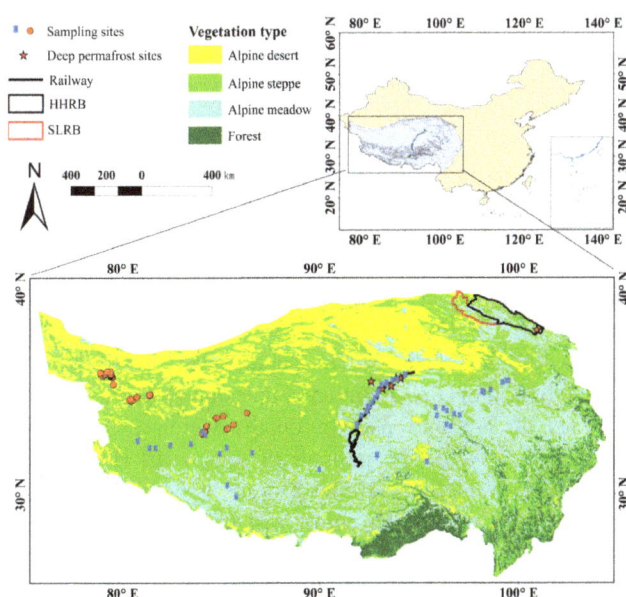

Figure 2. Location of sampling sites on the QXP, shown on the background of QXP vegetation atlas at a scale of 1 : 400 000 (Chinese Academy of Sciences, 2001). (Sampling sites were the same as those shown on the background of permafrost distribution.)

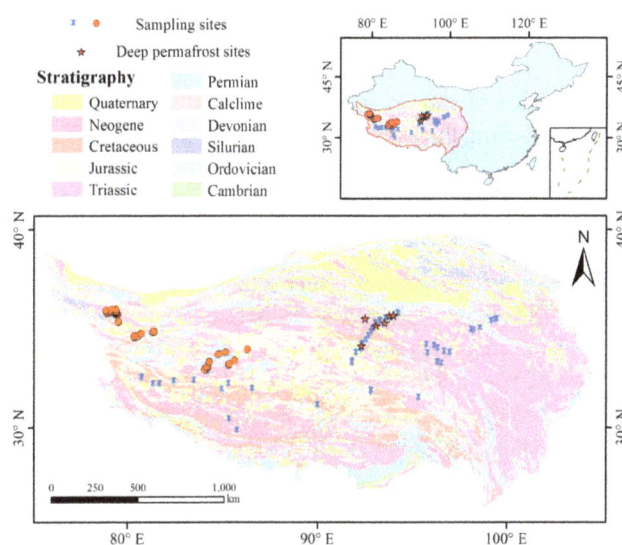

Figure 3. Location of sampling sites on the QXP, shown on the background of the QXP Quaternary geological map. (Sampling sites were the same as those shown on the background of permafrost distribution.)

0–1 m depth in permafrost regions on the QXP were calculated using 190 profile sites from published sources.

2.2 Field sampling

To calculate the deep carbon pools (2–25 m) in permafrost regions, 11 boreholes on the QXP were drilled from 2009 to 2013 (Fig. 1). Geographic location for the 11 boreholes, together with the active layer depth, sampling depth, vegetation type, geological stratigraphies, SOC contents, bulk density, water contents and soil texture are provided in the supplement materials.

The deep sampling sites were mainly located in three vegetation types of alpine meadow, alpine stepper and alpine desert (Fig. 2). Three sampling sites (KXL: *KaiXin Ling*, HLH-1: *HongLiang He-1*, HLH-2: *HongLiang He-2*) were located in the vegetation type of alpine steppe. Another site was near *ZhuoEr Hu* (ZEH) in *Kekexili*, with soil formed from lacustrine deposits. It was typical alpine desert and perennially frozen, containing less amounts of organic carbon. Five sampling sites (KL150: *KunLun150*, KL300: *KunLun300*, KL450: *KunLun450*, WDL: *WuDao Liang*, XSH: *XiuShui He*) were located in the vegetation type of alpine meadow. In addition, two sites in permafrost regions of the Heihe river basin (HHRB: Heihe-1, Heihe-2) with vegetation type of alpine meadow were rich in organic carbon with high soil water contents (Mu et al., 2013).

The deep sampling sites were mainly distributed in three geological stratigraphies: ZEH, WDL, XSH, Heihe-1 and Heihe-2 were in the Quaternary stratigraphy, KL150, KL300, KL450, HLH-1 and HLH-2 were in the Triassic stratigraphy, and KXL was in the Permian stratigraphy (Fig. 3).

2.3 Analytical methods

For SOC analyses, the homogenized samples were quantified by dry combustion on a vario EL elemental analyzer (Elemental, Hanau, Germany). During measurement, 0.5 g dry soil samples were pretreated by HCl (10 mL 1 mol L^{-1}) for 24 h to remove carbonate (Sheldrick, 1984). Bulk density was determined by measuring the volume (length, width, height) of a section of frozen core, and then drying the segment at 105° (for 48 h) and determining its mass.

2.4 Calculation of soil carbon pools

For the stock of soil organic carbon (SSOC, kg m^{-2}), it was calculated using the Eg. (1) (Dorfer et al., 2013):

$$SSOC = C \times BD \times T \times (1 - CF), \quad (1)$$

where C was the organic carbon content (wt %), BD was the bulk density (g cm^{-3}), T was the soil layer thickness and CF was the coarse fragments (wt %). Using this information, the SSOC was calculated for the 0–1, 1–2, 2–3 and 3–25 m depths, respectively. Then, SOC storage (Pg) was estimated by multiplying the SSOC at different depth by the distribution area.

For the organic carbon storage in 0–1 m depth, the reported SOC densities data of 190 sampling sites were collected through their distribution in permafrost regions (Fig. 1). The area of alpine meadow, alpine steppe and alpine desert in permafrost regions was calculated through overlaying the vegetation map over the QXP permafrost regions (Fig. 2). For the organic carbon storage in 1–2 m depth, the organic carbon densities of 11 boreholes were extrapolated to the located vegetation-type area.

For the organic carbon storage in 2–3 and 3–25 m depths, the area of permafrost regions in the Quaternary, Triassic and Permian stratigraphies on the QXP was calculated through overlaying the distribution of geological stratigraphies over the permafrost map (Fig. 3). The organic carbon pools of 2–3 and 3–25 m depth were estimated through deep organic carbon densities multiplied by the area of geological stratigraphies. The three geological stratigraphies had thick sediments of about 25 m (Fang et al., 2002, 2003; Qiang et al., 2001). As for other geological stratigraphies, the poor soil development was reported and soil thickness was usually less than 3 m (Wu et al., 2012; Yang et al., 2008; Hu et al., 2014). Thus other stratigraphies were not considered in the estimation of deep organic carbon pools in the permafrost regions.

3 Results

3.1 Organic carbon pools in the 0–1 m depth

Based on the vegetation data on the QXP (Figs. 1, 2), the area of permafrost regions in the alpine meadow, alpine steppe and alpine desert are 0.302×10^6 km^2, 0.772×10^6 km^2 and 0.175×10^6 km^2, respectively, with a total area of approximately 1.249×10^6 km^2.

Organic carbon storage of the permafrost regions in the 0–1 m depth on the QXP was approximately 17.3 ± 5.3 Pg, of which approximately 11.3 ± 4.0 Pg (65 %) in the alpine meadow, 5.3 ± 2.8 Pg (31 %) in the alpine steppe, and 0.7 ± 0.3 Pg (4 %) in the alpine desert, respectively (Table 1). There were great variations in SOC contents among the sites under alpine meadow area. SOC store in the HHRB (39.0 ± 17.5 kg m^{-2}) was much higher than that of most sites in the predominately continuous permafrost zone on the QXP. In contrast, the SOC stores showed little variation over the sites in the alpine steppe and alpine desert areas, with the ranges of 6.9 ± 3.6 and 3.9 ± 1.5 kg m^{-2}, respectively.

3.2 Distribution of deep organic carbon

According to the distribution of sampling sites at the geological stratigraphies, for the Quaternary stratigraphy, average SOC contents at 2–3 and 3–25 m depths were 0.8 ± 0.6 and 0.8 ± 0.7 %. For the Triassic stratigraphy, average SOC contents at 2–3 and 3–25 m depths were 1.1 ± 0.3 and 1.2 ± 0.6 %. For the Permian stratigraphy, average SOC contents at 2–3 and 3–25 m depths were 1.5 ± 0.4 and 1.1 ± 0.3 %. As for the permafrost regions in HHRB, the SOC contents (Heihe-1, Heihe-2) were higher than those of predominately continuous permafrost zone on the QXP, with a range of 5.1 ± 3.7 and 2.7 ± 2.4 % to depth of 19 m. SOC contents decreased with depth in most deep boreholes, while SOC contents in deeper layers were higher than those in the top layer at the XSH, KL150 and KL300 (Fig. 4).

With the deep soil data, a relationship between SOC contents (SOC %) and soil depth (h) in deep soils of permafrost regions can be characterized by a power Eq. (2) (Fig. 4):

$$SOC\% = 14.11h^{-1.20} (R^2 = 0.68, p < 0.01, n = 362). \quad (2)$$

3.3 Deep organic carbon pools

Based on the Quaternary stratigraphies data in permafrost regions of the QXP (Fig. 3), the area of permafrost regions in the Quaternary, Triassic and Permian stratigraphies are 0.194×10^6, 0.238×10^6 and 0.135×10^6 km^2 respectively, with a total area of approximately 0.567×10^6 km^2, about 45 % of permafrost regions on the QXP.

Organic carbon storages in permafrost regions on the QXP were approximately 10.6 ± 2.7 Pg in the 1–2 m depth, 5.1 ± 1.4 Pg in the 2–3 m depth and 127.2 ± 37.3 Pg in deep depth of 3–25 m (Table 2). In total, it contains approximately

Table 2. Permafrost organic carbon storage to the depth of 25 m on the QXP.

Vegetation types	Alpine meadow			Alpine steppe		Alpine desert		
Soil depth (m)	SOC (kg m^{-2})	SOC storage (Pg) QTP	HHRB	SOC (kg m^{-2})	SOC storage (Pg)	SOC (kg m^{-2})	SOC storage (Pg)	Total (Pg)
0–1 m	–	11.0 ± 3.9	0.3 ± 0.1	6.9 ± 3.6	5.3 ± 2.8	3.8 ± 1.5	0.7 ± 0.3	17.3 ± 5.3
1–2 m	16.7 ± 4.7	4.9 ± 1.4	0.2 ± 0.1	6.5 ± 2.2	5.0 ± 1.7	3.0 ± 1.3	0.5 ± 0.2	10.6 ± 2.7
Total (Pg)		16.4 ± 5.2			10.3 ± 2.7		1.2 ± 0.3	27.9 ± 6.2
Geological stratigraphies	Quaternary			Triassic		Permian		
Soil depth (m)	SOC (kg m^{-2})	SOC storage (Pg) QTP	HHRB	SOC (kg m^{-2})	SOC storage (Pg)	SOC (kg m^{-2})	SOC storage (Pg)	Total (Pg)
2–3 m	9.8 ± 8.4	1.9 ± 1.6	0.1 ± 0.06	9.6 ± 4.5	2.3 ± 1.1	5.6 ± 0.9	0.8 ± 0.1	5.1 ± 1.4
3–25 m	134.9 ± 115.3	26.2 ± 22.4	2.3 ± 1.4	281.9 ± 191.7	67.1 ± 45.6	234.2 ± 86.0	31.6 ± 11.6	127.2 ± 37.3
Total (Pg)		30.5 ± 16.6			69.4 ± 52.8		32.4 ± 20.2	132.3 ± 76.8

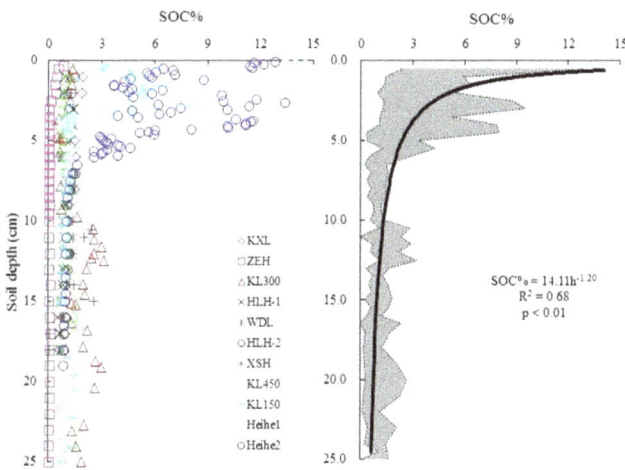

Figure 4. Distributions of soil organic carbon contents in deep soils in permafrost regions on the QXP.

160 ± 87 Pg of organic carbon at depth of 25 m in permafrost regions on the QXP.

Active layer thickness on the QXP varies from 0.8 to 4.6 m, and in most regions, active layer thickness was about 2 m (Cheng and Wu, 2007; Wu and Zhang, 2008; Zhao et al., 2010; Wu et al., 2012). Thus we consider the upper 2 m as the active layer. According to this depth, the organic carbon storage in permafrost layers of 132 ± 77 Pg was approximately five times of that (28 ± 6 Pg) in the active layer.

SOC storages in Quaternary, Triassic and Permian stratigraphies were 31 ± 17, 69 ± 53 and 32 ± 20 Pg at depth of 2–25 m, respectively. More than a half of organic carbon is stored in permafrost layers which belonged to the Triassic stratigraphy.

4 Discussions

Our estimates indicate that organic carbon storage in permafrost regions in the 0–1 m depth on the QXP was approximately 17.3 ± 5.3 Pg. However, previous soil carbon pools on the alpine grasslands of the whole QXP were estimated to be 33.5 Pg of 0–0.75 m (Wang et al., 2002), and 10.5 Pg of 0–0.30 m (Yang et al., 2010). The difference, in large part, between our new estimate and previous reports can be explained as follows: (i) area of vegetation types in permafrost regions was recalculated. The area of permafrost regions of about 1.249×10^6 km^2 was smaller than that of Wang et al. (2002) (1.63×10^6 km^2) and Yang et al. (2010) (1.26×10^6 km^2). (ii) Carbon density data of sampling sites located in permafrost regions was collected. The integration of carbon data from the results of recent publications (Ohtsuka et al., 2008; Dorfer et al., 2013; Wu et al., 2012) and our field data resulted in a higher carbon density than those of previous reports (Wang et al., 2002; Yang et al., 2010). (iii) The regions of SLRB and HHRB were not considered in previous SOC pool estimate. The organic carbon storages of 0.43 ± 0.11 Pg in SLRB and 0.25 ± 0.11 Pg in HHRB were added in the present study.

It is worth to mention that there were wide variations in organic carbon contents in permafrost regions on the QXP in previous reports (Wang et al., 2002; Yang et al., 2010; Liu et al., 2012; Wu et al., 2012; Dorfer et al., 2013; Ohtsuka et al., 2008; Mu et al., 2013). A possible explanation is the spatial heterogeneity of SOC contents in permafrost regions of the QXP. In addition, the different analytical methods may also contribute to the differences of carbon contents (Table 1). It has been demonstrated that if taking the dry combustion method as standard, the recovery of organic carbon was 99 % for wet combustion and 77 % for the Walkley–Black procedure (Kalembasa and Jenkinson, 1973; Nelson and Sommers, 1996).

The SOC stocks at 0–1 m depth (17.3 kg m^{-2}) in the alpine meadow on the QXP is higher than that in subarctic alpine permafrost ($0.9 \, \text{kg m}^{-2}$) (Fuchs et al., 2014), and similar to that of the lowland and hilly upland soils in the North American Arctic region (55.1, $40.6 \, \text{kg m}^{-2}$) (Ping et al., 2008a). It implies that SOC of the alpine meadow in permafrost regions has a large proportion in permafrost carbon pools. The SOC contents at 0–1 m depth ($3.9 \pm 1.5 \, \text{kg m}^{-2}$) in the alpine desert on the QXP was similar to that (3.4, $3.8 \, \text{kg m}^{-2}$) in rubble-land and mountain soils in the North American Arctic region (Ping et al., 2008a). These results suggest that the SOC stocks are closely related to the vegetation type in the permafrost regions.

SOC decreases with the depth on the QXP (Fig. 4), which is in good agreement with those reported in circum-Arctic regions (Strauss et al., 2013; Zimov et al., 2006). This could be explained by the dynamics of Quaternary deposit and SOC formation in permafrost regions (Strauss et al., 2013). However, the organic carbon contents of deep layers in some sites (XSH, KL150 and KL300) were higher than those in the top layers (Fig. 4), which may be caused by the cryoturbation and sediment burying process (Ping et al., 2010), and Quaternary deposits following the uplift of Tibetan Plateau (Li et al., 1994, 2014). Overall, SOC decreases exponentially with depth (Eq. 1) in permafrost regions on the QXP, which is in agreement with results from other regions (Don et al., 2007). Certainly, more efforts are still needed in studying the distribution of deep organic carbon density in permafrost regions.

In the present study, it is the first time to study the deep organic carbon in permafrost regions, and quantify the carbon storage below 1.0 m depth on the QXP. The mean SOC content of 11 boreholes in permafrost regions on the QXP (2.5 wt %) was similar to that in the yedoma deposits (3.0 wt %) (Strauss et al., 2013), and that of lowland steppe-tundra soils in Siberia and Alaska (2.6 wt %) (Zimov et al., 2006). Since it has been pointed out that yedoma deposits contain a large amount of organic carbon, it would be reasonable to infer that deep soil carbon in permafrost regions on the QXP may also have a great contribution to carbon pools. Our estimations indicate that the soils on the QXP contains $33.0 \pm 13.2 \, \text{Pg}$ of organic carbon in the top 3.0 m of soils, with an additional $127.2 \pm 37.3 \, \text{Pg}$ C distributed in deep layers (3–25 m) of the Quaternary, Triassic and Permian stratigraphies in permafrost regions. In northern circumpolar permafrost region, 1024 Pg of organic carbon was in the 0–3 m depth and 648 Pg (39 %) of carbon was stored in deep layers of yedoma and deltaic deposits (Tarnocai et al., 2009). The percentage of SOC storage in deep layers (3–25 m) on the QXP (80 %) is much higher than that (39 %) in the yedoma and thermokarst deposits in Siberia and Alaska. This could be explained as that the paleoenvironment of the QXP was wet and warm, or lacustrine sediment in most regions (Zhang et al., 2003; Lu et al., 2014), which always links to the well formation of soil organic matter (Kato et al., 2004; Piao et al., 2006; Chen et al., 1990).

In total, there is approximately $160 \pm 87 \, \text{Pg}$ of organic carbon stored at 0–25 m depth in permafrost regions on the QXP,

which would update the total carbon pools to 1832 Pg in permafrost regions of northern hemisphere. The total carbon pools on the QXP permafrost regions account for approximately 8.7 % of the total carbon pools in permafrost regions in northern hemisphere. Since the permafrost region on the QXP was about 6 % of the northern permafrost area (Ran et al., 2012), it could be seen that SOC in permafrost regions on the QXP should be paid more attention in the future studies.

5   Conclusions

1. According to the organic carbon data in previous analysis and field exploration of deep boreholes in permafrost regions, the organic carbon storages in permafrost regions on the QXP were estimated to approximately $17.3 \pm 5.3 \, \text{Pg}$ in the 0–1 m, $10.6 \pm 2.7 \, \text{Pg}$ in the 1–2 m, $5.1 \pm 1.4 \, \text{Pg}$ in the 2–3 m and $127.2 \pm 37.3 \, \text{Pg}$ in deep depth of 3–25 m.

2. The percentage of SOC storage in deep layers (3–25 m) of permafrost regions on the QXP was 80 %, which was higher than that in the yedoma and thermokarst deposits in Siberia and Alaska.

3. In total, organic carbon pools in permafrost regions on the QXP are approximately $160 \pm 87 \, \text{Pg}$, of which $132 \pm 76 \, \text{Pg}$ occurs in permafrost layers. The total carbon pools in permafrost regions in northern hemisphere are now updated to 1832 Pg.

Acknowledgements. This work was supported by the National Key Scientific Research Project (Grant 2013CBA01802), National Natural Science Foundation of China (Grants 91325202, 41330634), and the Open Foundations of State Key Laboratory of Cryospheric Sciences (Grant SKLCS-OP-2014-08) and State Key Laboratory of Frozen Soil Engineering (Grant SKLFSE201408). The authors gratefully acknowledge the reviewers, Gustaf Hugelius and Chien-Lu Ping, as well as the editor, Steffen M. Noe, for their constructive comments and suggestions.

Edited by: S. M. Noe

References

Bockheim, J. B., Walker, D. A., Everett, L. R., Nelson, F. E., and Shikolmanov, N. I.: Soils and cryoturbation in moist nonacidic and acidic tundra in the Kuparuk river basin arctic Alaska, USA, Arct. Alp. Res., 30, 166–174, 1998.

Burke, E. J., Hartley, I. P., and Jones, C. D.: Uncertainties in the global temperature change caused by carbon release from permafrost thawing, The Cryosphere, 6, 1063–1076, doi:10.5194/tc-6-1063-2012, 2012.

Chen, K. Z., Bowler, J. M., and Kelts, K.: Palaeoclimate evolution within the Qinghai-Xizang (Tibet) plateau in the last 40 000 years, Quaternary Sci., 1, 22–30, 1990.

Cheng, G. D. and Wu, T. H.: Responses of permafrost to climate change and their environmental significance, Qinghai-Tibet Plateau, J. Geophys. Res., 112, F02S03, doi:10.1029/2006JF000631, 2007.

Davidson, E. A. and Janssens, I. A.: Temperature sensitivity of soil carbon decomposition and feedbacks to climate change, Nature, 440, 165–173, 2006.

Don, A., Schumacher J., Scherer-Lorenzen, M., Scholten, T., and Schulze, E.D.: Spatial and vertical variation of soil carbon at two grassland sites – Implications for measuring soil carbon stocks, Geoderma, 141, 272–282, 2007.

Dorfer, C., Kuhn, P., Baumann, F., He, J. S., and Scholten, T.: Soil Organic Carbon Pools and Stocks in Permafrost-Affected Soils on the Tibetan Plateau, PLoS ONE, 8, e57024, doi:10.1371/journal.pone.0057024, 2013.

Editorial Board of Vegetation Map of China, Chinese Academy of Sciences, Vegetation Atlas of China (1 : 1 000 000), Beijing, Science Press, 2001.

Fang, X. M., Lu, L. Q., Yang, S. L., Li, J. J., An, Z. S., Jiang, P. A., and Chen, X. L.: Loess in Kunlun Mountains and its implications on desert development and Tibetan Plateau uplift in west China, Science China, 45, 291–298, 2002.

Fang, X. M., Lu, L. Q., Mason, J. A., Yang, S. L., An, Z. S., Li, J. J., and Guo, Z. L.: Pedogenic response to millennial summer monsoon enhancements on the Tibetan Plateau, Quaternary Internat., 106–107, 79–88, 2003.

Fuchs, M., Kuhry, P., and Hugelius, G.: Low soil organic carbon storage in a subarctic alpine permafrost environment, The Cryosphere Discuss., 8, 3493–3524, doi:10.5194/tcd-8-3493-2014, 2014.

Hobbie, S. E., Schimel, J. P., Trumbore, S. E., and Randerson, I. R.: Controls over carbon storage and turnover in high-latitude soils, Glob. Change Biol., 6, 196–210, 2000.

Hu, G. L., Fang, H. B., Liu, G. M., Zhao, L., Wu, T. H., Li, R., and Wu, X. D.: Soil carbon and nitrogen in the active layers of the permafrost regions in the Three Rivers' Headstream, Environ. Earth. Sci, 72, 5113–5122, 2014.

Hugelius, G., Tarnocai, C., Broll, G., Canadell, J. G., Kuhry, P., and Swanson, D. K.: The Northern Circumpolar Soil Carbon Database: spatially distributed datasets of soil coverage and soil carbon storage in the northern permafrost regions, Earth Syst. Sci. Data, 5, 3–13, doi:10.5194/essd-5-3-2013, 2013.

Hugelius, G., Strauss, J., Zubrzycki, S., Harden, J. W., Schuur, E. A. G., Ping, C.-L., Schirrmeister, L., Grosse, G., Michaelson, G. J., Koven, C. D., O'Donnell, J. A., Elberling, B., Mishra, U., Camill, P., Yu, Z., Palmtag, J., and Kuhry, P.: Estimated stocks of circumpolar permafrost carbon with quantified uncertainty ranges and identified data gaps, Biogeosciences, 11, 6573–6593, doi:10.5194/bg-11-6573-2014, 2014.

Jobbagy, E. G. and Jackson, R. B.: The vertical distribution of soil organic carbon and its relation to climate and vegetation, Ecol. Appl., 10, 423–436, 2000.

Kalembasa, S. J. and Jenkinson, D. D.: A comparative study of titrimetric and gravimetric methods for the determination of organic carbon in soil, J. Sci. Food Agricul., 24, 1085–1090, 1973.

Kato, T., Tang, Y., and Gu, S.: Carbon dioxide exchange between the atmosphere and an alpine meadow ecosystem on the Qinghai–Tibetan Plateau, China, Agric. For. Meteorol., 124, 121–34, 2004.

Koven, C. D., Ringeval, B., Friedlingstein, P., Ciais, P., Cadule, P., Khvorostyanov, D., Krinner, G., and Tarnocai, C.: Permafrost carbon-climate feedbacks accelerate global warming, P. Natl. Acad. Sci. USA, 108, 14769–14774, 2011.

Lanzhou Institute of Glaciology and Geocryology, Chinese Academy of Sciences, Map of Snow, Ice and Frozen Ground in China (1 : 4 000 000). Cartographic Publishing House, Beijing, China, 1988 (in Chinese).

Li, J. J., Zhang, Q. S., and Li, B. Y.: Main processes of geomorphology in China in the past fifteen years, Ac. Geogr. Sin., 49, 642–648, 1997.

Li, X., Cheng, G. D., Jin, H. J., Kang, E. S., Che, T., Jin, R., Wu, L. Z., Nan, Z. T., Wang, J., and Shen, Y. P.: Cryospheric Change in China, Glob. Planet. Change, 62, 210–218, 2008.

Liu, W. J., Chen S. Y., Qin, X., Baumann, F., Scholten, T., Zhou, Z. Y., Sun, W. J., Zhang, T. Z., Ren, J. W., and Qin, D. H.: Storage, patterns, and control of soil organic carbon and nitrogen in the northeastern margin of the Qinghai–Tibetan Plateau, Environ. Res. Lett., 7, 1–12, 2012.

Li, J. J., Fang, X. M., Song, C. H., Pan, B. T., Ma, Y. Z., and Yan, M. D.: Late Miocene–Quaternary rapid stepwise uplift of the NE Tibetan Plateau and its effects on climatic and environmental changes, Quaternary Res., 81, 400–423, 2014.

Lu, H. Y. and Guo, Z. T.: Evolution of the monsoon and dry climate in East Asia during late Cenozoic: A review, Sci. China Earth Sci., 57, 70–79, 2014.

MacDougall, A. H., Avis, C. A., and Weaver, A. J.: Significant contribution to climate warming from the permafrost carbon feedback, Nat. Geosci., 5, 719–721, 2012.

Michaelson, G. J., Ping, C. L., and Clark, M.: Soil pedon carbon and nitrogen data for Alaska: An analysis and update, Open J. Soil Sci., 3, 132–142, 2013.

Mu, C. C., Zhang, T. J., Cao, B., Wan, X. D., Peng, X. Q., and Cheng, G. D.: Study of the organic carbon storage in the active layer of the permafrost over the Eboling Mountain in the upper reaches of the Heihe River in the Eastern Qilian Mountains, J. Glaciol. Geocryol., 35, 1–9, 2013.

Mu, C. C., Zhang, T. J., Wu, Q. B., Zhang, X. K., Cao, B., Wang, Q. F., Peng, X. Q., and Cheng, G. D.: Stable carbon isotopes as indicators for permafrost carbon vulnerability in upper reach of Heihe River basin, northwestern China, Quaternary Internat., 321, 71–77, 2014.

Nelson, D. E. and Sommers, L. E.: Total carbon, organic carbon, and organic matter, Methods of soil analysis, Part 3 – chemical methods, 961–1010, 1996.

Ohtsuka, T., Hirota, M., Zhang, X., Shimono, A., Senga, Y., Du, M., Yonemura, S., Kawashima, S., and Tang, Y.: Soil organic carbon pools in alpine to nival zones along an altitudinal gradient (4400–5300 m) on the Tibetan Plateau, Polar Sci., 2, 277–285, 2008.

Piao, S. L., Fang, J. Y., and He, J. S.: Variations in vegetation net primary production in the Qinghai–Xizang Plateau, China, from 1982 to 1999, Clim. Chang, 74, 253–67, 2006.

Ping, C. L., Michaelson, G. J., Jorgenson, T., Kimble, J. M., Epstein, H., Romanovsky, V. E., and Walker, D. A.: High stocks of soil organic carbon in the North American arctic region, Nat. Geosci., 1, 615–619, 2008a.

Ping, C. L., Michaelson, G. J., Kimble, J. M., Romanovsky, V. E., Shur, Y. L., Swanson, D. K., and Walker, D. A.: Cryogenesis and soil formation along a bioclimate gradient in Arctic North America, J. Geophys. Res., 113, G03S12, doi:10.1029/2008JG000744, 2008b.

Qiang, X. K., Li, Z. X., Powell, C. McA., and Zheng, H. B.: Magnetostratigraphic record of the Late Miocene onset of the East Asian monsoon, and Pliocene uplift of northern Tibet, Earth Planet. Sci. Lett., 187, 83–93, 2001.

Ran, Y. H., Li, X., Cheng, G. D., Zhang, T. J., Wu, Q. B., Jin, H. J., and Jin, R.: Distribution of Permafrost in China: An Overview of Existing Permafrost Maps, Permafr. Perigl. Proc., 23, 322–333, 2012.

Schneider von Deimling, T., Meinshausen, M., Levermann, A., Huber, V., Frieler, K., Lawrence, D. M., and Brovkin, V.: Estimating the near-surface permafrost-carbon feedback on global warming, Biogeosciences, 9, 649–665, doi:10.5194/bg-9-649-2012, 2012.

Schuur, E. A. G., Vogel, J. G., Crummer, K. G., Lee, H., Sickman, J. O., and Osterkamp, T. E.: The effect of permafrost thaw on old carbon release and net carbon exchange from tundra, Nature, 459, 556–559, 2009.

Sheldrick, B. H.: Analytical Methods Manual, Land Resour. Res. Inst., Res. Branch, Agric. Can., Ottawa, 212 pp., 1984.

Strauss, J., Schirrmeister, L., Grosse, G., Wetterich, S., Ulrich, M., Herzschuh, U., and Hubberten, H. W.: The deep permafrost carbon pool of the yedoma region in Siberia and Alaska, Geophys. Res. Lett., 40, 6165–6170, doi:10.1002/2013GL058088, 2013.

Tarnocai, C., Canadell, J. G., Schuur, E. A. G., Kuhry, P., Mazhitova, G., and Zimov, S.: Soil organic carbon pools in the northern circumpolar permafrost region, Global Biogeochem. Cy., 23, GB2023, doi:10.1029/2008GB003327, 2009.

Waldrop, M. P., Wickland, K. P., White, R., Berhe, A. A., Harden, J. W., and Romanovsky, V. E.: Molecular investigations into a globally important carbon pool: permafrost-protected carbon in Alaskan soils, Glob. Change Biol., 16, 2543–2554, 2010.

Wang, G. X., Qian, J., Cheng, G. D., and Lai, Y. M.: Soil organic carbon pool of grassland soils on the Qinghai-Tibetan Plateau and its global implication, Sci. Total Environ., 291, 207–217, 2002.

Wang, G., Li, Y., Wang, Y., and Wu, Q.: Effects of permafrost thawing on vegetation and soil carbon losses on the Qinghai-Tibet Plateau, China, Geoderma, 143, 143–152, 2008.

Wu, X. D., Zhao, L., Fang, H. B., Yue, G. Y., Chen, J., Pang, Q. Q., Wang, Z. W., and Ding, Y. J.: Soil Organic Carbon and Its Relationship to Vegetation Communities and Soil Properties in Permafrost of Middle-western Qinghai-Tibet Plateau, Permafr. Perigl. Proc., 23, 162–169, 2012.

Wu, X. D., Fang, H. B., Zhao, L., Wu, T. H., Li, R., Ren, Z. W., Pang, Q. Q., and Ding, Y. J.: Mineralization and Fractions Changes in Soil Organic Matter in Soils of Permafrost Region in Qinghai-Tibet Plateau, Permafr. Perigl. Proc., 25, 35–44, 2014.

Wu, Q., Hou, Y., Yun, H., and Liu, Y.: Changes in active-layer thickness and near-surface permafrost between 2002 and 2012 in alpine ecosystems, Qinghai–Xizang (Tibet) Plateau, China, Glob. Planet. Change, 124, 149–155, 2015.

Wu, Q. B. and Zhang, T. J.: Recent permafrost warming on the Qinghai-Tibetan Plateau, J. Geophys. Res., 113, D13108, doi:10.1029/2007JD009539, 2008.

Wu, Q. B. and Zhang, T. J.: Changes in active layer thickness over the Qinghai-Tibetan Plateau from 1995 to 2007, J. Geophys. Res., 115, D09107, doi:10.1029/2009JD012974, 2010.

Yang, M., Nelson, F. E., Shiklomanov, N. I., Guo, D., and Wan, G.: Permafrost degradation and its environmental effects on the Tibetan Plateau: A review of recent research, Earth-Sci. Rev., 103, 31–44, 2010.

Yang, Y. H., Fang, J. Y., Tang, Y. H., Ji, C. J., Zheng, C. Y., He, J. S., and Zhu, B.: Storage, patterns and controls of soil organic carbon in the Tibetan grasslands, Glob. Change Biol., 14, 1592–1599, 2008.

Zimov, N. S., Zimov, S. A., Zimova, A. E., Zimova, G. M., Chuprynin, V. I., and Chapin III, F. S.: Carbon storage in permafrost and soils of the mammoth tundra-steppe biome: Role in the global carbon budget, Geophys. Res. Lett., 36, L02502, doi:10.1029/2008GL036332, 2009.

Zimov, S. A., Schuur, E. A. G., and Chapin, F. S.: Permafrost and the global carbon budget, Science, 312, 1612–3, 2006.

Zhang, Q. B., Cheng, G. D., Yao, T. D., Kang, X. C., and Huang, J. G.: A 2,326-year tree-ring record of climate variability on the northeastern Qinghai-Tibetan Plateau, Geophys. Res. Lett., 30, 1739, doi:10.1029/2003GL017425, 2003.

Zhao, L., Wu, Q. B., Marchenko, S. S., and Sharkhuu, N.: Thermal state of permafrost and active layer in Central Asia during the international polar year, Permafr. Perigl. Proc., 21, 198–207, 2010.

7

A 1-D modelling study of Arctic sea-ice salinity

P. J. Griewank and D. Notz

Max Planck Institute for Meteorology, Bundesstr. 53, 20146 Hamburg, Germany

Correspondence to: P. J. Griewank (philipp.griewank@mpimet.mpg.de)

Abstract. We use a 1-D model to study how salinity evolves in Arctic sea ice. To do so, we first explore how sea-ice surface melt and flooding can be incorporated into the 1-D thermodynamic Semi-Adaptive Multi-phase Sea-Ice Model (SAMSIM) presented by Griewank and Notz (2013). We introduce flooding and a flushing parametrization which treats sea ice as a hydraulic network of horizontal and vertical fluxes. Forcing SAMSIM with 36 years of ERA-interim atmospheric reanalysis data, we obtain a modelled Arctic sea-ice salinity that agrees well with ice-core measurements. The simulations thus allow us to identify the main drivers of the observed mean salinity profile in Arctic sea ice. Our results show a 1.5–4 g kg^{-1} decrease of bulk salinity via gravity drainage after ice growth has ceased and before flushing sets in, which hinders approximating bulk salinity from ice thickness beyond the first growth season. In our simulations, salinity interannual variability of first-year ice is mostly restricted to the top 20 cm. We find that ice thickness, thermal resistivity, freshwater column, and stored energy change by less than 5 % on average when the full salinity parametrization is replaced with a prescribed salinity profile.

1 Introduction

Sea ice is a multiphase material consisting of salty brine, fresh ice, and gas bubbles and is far from static. Brine moves through the ice and across the ice–ocean interface, transporting dissolved tracers such as salt. The thermal properties of sea ice change along with the phase composition; bubbles form, dissolve, and escape into the atmosphere while chemical and biologic processes occur in the brine. Salt is a core component of sea ice as it, along with temperature, dictates the phase composition of sea ice through the liquidus relationship. It also influences the brine density, the chemical properties, the small-scale sea-ice structure, and the vertical

stratification of the underlying ocean via salt transport to the mixed layer. Unfortunately, the salinity of sea-ice is an elusive quantity that is difficult to observe. Many open questions related to the salinity evolution can not be answered due to the limited amount and the isolated nature of ice-core measurements, such as to what extent gravity drainage occurs during ice melt, what causes interannual salinity variability, how first-year ice transforms to multiyear ice, and how bulk salinity is linked to ice thickness. To fill these gaps in our understanding, we here study the salinity evolution of Arctic sea ice and quantify the impact of the salinity evolution on various sea-ice properties using an expanded version of the Semi-Adaptive Multi-phase Sea-Ice Model (SAMSIM) introduced in Griewank and Notz (2013).

To do so, SAMSIM needed to be expanded to model sea-ice surface melt. The surface of melting sea ice is complex and highly heterogeneous. Meltwater flows horizontally through snow and ice into melt ponds and cracks or percolates vertically through the ice. The properties of melting wet snow differ strongly from those of dry fresh snow, and the ice surface also deteriorates during melt and can form a layer of white deteriorated ice which is visually similar to snow (Eicken et al., 2002). All these processes influence albedo. Due to the large influence ice albedo has on sea-ice evolution, the sea-ice modelling community has produced many albedo and melt pond parametrizations (e.g. Flocco and Feltham, 2007; Pedersen et al., 2009), but otherwise surface melt has received very little attention. All 1-D thermodynamic models since Maykut and Untersteiner (1971) have disregarded the physical structure and high gas fraction of the surface during melt and have treated melting sea ice as freshwater ice with modified thermal properties.

Over the last decade, researchers have begun to parametrize the sea-ice salinity evolution (e.g. Vancoppenolle et al., 2006, 2007, 2009; Wells et al., 2011; Hunke et al., 2011; Rees Jones and Worster, 2013; Turner et al.,

2013) to study the biogeochemical and physical processes in and below sea ice (e.g. Vancoppenolle et al., 2010; Tedesco et al., 2010, 2012; Jeffery et al., 2011; Saenz and Arrigo, 2012; Jardon et al., 2013). Despite these developments, the only published sea-ice model with a fully parameterized salinity evolution is the Louvain-la-Neuve (LIM) 1-D model of Vancoppenolle et al. (2007) based on the 1-D thermodynamic model of Bitz and Lipscomb (1999). Accordingly, many possible approaches for modelling surface melt and parametrizing salinity remain unexplored in 1-D sea-ice models. We introduce new schemes to parametrize surface melt, flooding, and flushing within our 1-D sea-ice model SAMSIM, making it capable of simulating the full growth and melt cycle of sea ice, including the salinity evolution.

We force SAMSIM with Arctic reanalysis data to study the desalination processes and the resulting salinity evolution in the Arctic. This is the first general multiyear model study of sea-ice salinity throughout the Arctic. The only previous model study of sea-ice salinity is the study by Vancoppenolle et al. (2007), which focuses on two ice-core sites of land-fast ice from 1999 to 2001. Model studies are necessary because measurement campaigns can only provide brief glimpses of the full salinity evolution, whereas we can easily explore a far greater diversity of conditions over a longer time frame. The simulated salinity profiles are compared to ice-core measurements to evaluate the model performance.

We have decided to limit the study to the Arctic because flooding and the corresponding snow-ice formation play a large role in the Antarctic. As explained in detail in Sect. 2.4.5, we treat the flooding parametrizations currently implemented in SAMSIM as ad hoc solutions only suitable for dealing with isolated and sporadic flooding events. Accordingly, we will refrain from studying Antarctic ice until flooding is better understood.

The final topic we address is how parametrizing the salinity affects various sea-ice properties important to climate models. As sea-ice components of climate models are slowly becoming more sophisticated and modellers have begun to treat sea-ice salinity as a variable instead of a prescribed value or profile (e.g. Vancoppenolle et al., 2009; Turner et al., 2013), it remains unclear how much model performance can be improved by fully parametrizing the temporal salinity evolution and how sophisticated the parametrizations should be to balance the improvements against the increase in computational cost and code complexity.

This paper is organized as follows. In Sect. 2 we detail how surface melt, flooding, and flushing are implemented in SAMSIM. The section ends with a description of the three separate salinity approaches used to parametrize salinity in SAMSIM. In Sect. 3 we conduct an idealized melting experiment to study flushing and to determine how sensitive SAMSIM responds to changes of key parameters. In Sect. 4 we study the salinity evolution of 36 years of simulated sea ice forced with ERA-interim reanalysis data taken from throughout the Arctic. The simulations are split into first-

year and multiyear ice, which are analyzed separately and compared to ice-core data. Readers who are primarily interested in the geophysical insights gained by our simulations can understand most of this section without reading Sects. 2 and 3. The final section uses the same atmospheric forcing as Sect. 4 to quantify the impact of the various salinity approaches on quantities relevant to climate models in order to evaluate whether climate models would benefit from a fully parametrized temporal salinity evolution in their sea-ice sub models.

2   Model description

For the purpose of this paper we expand the SAMSIM model which we first described in Griewank and Notz (2013). SAMSIM is a 1-D column model which employs a semi-adaptive grid. In this section we will introduce how SAMSIM treats surface ablation and processes related to surface melting as well as flooding.

We provide a very brief description of the fundamentals of SAMSIM in Sect. 2.1; a detailed description including the underlying equations and numerics can be found in Griewank and Notz (2013). Following the brief description of SAMSIM we address a small modification of the gravity drainage parametrizations originally presented in Griewank and Notz (2013) in Sect. 2.2. Section 2.3 addresses how sea ice melts in reality and in SAMSIM. The final additions to SAMSIM are the parametrizations of flushing and flooding introduced in Sect. 2.4. In Sect. 2.5 we describe the three salinity set-ups used in SAMSIM.

All parametrizations introduced in this section were designed for SAMSIM. As SAMSIM has some unique characteristics, such as a gas volume fraction (see Griewank and Notz, 2013) none of the proposed parametrizations can be applied in precisely the same way to the commonly used models of Semtner (1976), Bitz and Lipscomb (1999), and Winton (2000). The differences between the models are mostly related to specific definitions of the ice–ocean interface, snow–ice interactions, meltwater formation, and tracer advection. We have made sure to include all assumptions from which the various parametrizations were derived so that corresponding parametrizations for other models can be derived. An evaluation of our new parametrizations is given in Sect. 4.3.

2.1   SAMSIM

Each layer of SAMSIM is defined by the four fundamental variables mass m, absolute salinity S_{abs}, absolute enthalpy H_{abs}, and thickness $\triangle z$. Absolute values are simply the integral over the mass-weighted bulk salinity S_{bu} and enthalpy H. The solid and liquid mass fractions ψ_s and ψ_l, as well as the solid, liquid, and gas volume fraction ϕ_s, ϕ_l, and ϕ_g, are derived from the fundamental variables. A salt-free snow

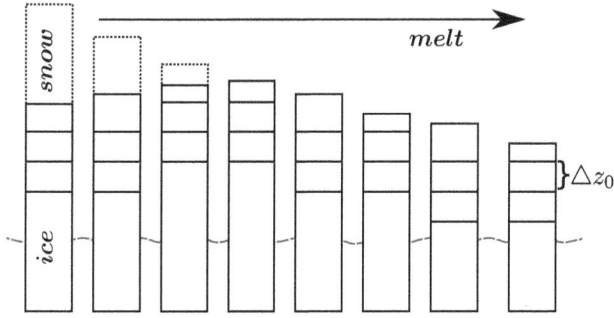

Figure 1. Sketch of SAMSIM grid evolution for three top ice layers during snow melt and following surface ablation as explained in Sect. 2.3.

Table 1. Default model settings and free parameter values of salinity parametrizations.

$\triangle z_0$	1 cm
dt	10 s
N_{top}	20
N_{mid}	60
N_{bot}	20
$\phi_{s,\,\text{min}}$	0.05
$\phi_{s,\,\text{melt}}$	0.4
$\phi_{g,\,\text{melt}}$	0.2
alb	0.75
pen	0.3
κ	2 1 L min^{-1}
α	$5.84 \times 10^{-4}\,\text{kg m}^{-3}\,\text{s}^{-1}$
R_{crit}	4.89
γ	0.99
β	1
δ	0.5
ϵ	0.1
ζ_{max}	5 cm

layer can exist on the ice, which has a variable density that affects the snow thermal conductivity. However, the only process currently implemented in SAMSIM which affects the snow density is rainfall into snow. This occurs when rain falls while snow is present, during which the snow thickness remains unchanged while the rain displaces some of the previous gas fraction and increases the mass of the snow layer. A full description of the snow and ice thermodynamics is included in Griewank and Notz (2013).

In this paper, we refer to a specific layer by an upper right index counting from top to bottom, with the exception of the snow layer which is marked with "snow". For example, m^6 is the mass of the sixth layer from the surface, m^1 is the mass of the top ice layer, and m^{snow} is the mass of the snow layer.

SAMSIM is the only sea-ice model to employ a semi-adaptive grid which grows and shrinks in discrete steps of $\triangle z_0$ at the ice–ocean interface (Griewank and Notz, 2013). However, at the ice–atmosphere boundary it is necessary to have a freely adjustable boundary to deal with incremental surface ablation and snow-to-ice conversion. This is addressed by letting the top ice layer thickness vary freely between $1/2\,\triangle z_0$ and $3/2\,\triangle z_0$. Once the top ice layer grows thicker than $3/2\,\triangle z_0$ it is split into two layers, the lower layer of the two with a thickness of $\triangle z_0$. Similarly, when the top ice layer shrinks below $1/2\,\triangle z_0$ it is merged together with the second layer. A sketch of how a grid with three top ice layers evolves during melt is shown in Fig. 1.

This semi-adaptive grid differs in a few crucial aspects from those used in other models such as the one introduced by Bitz and Lipscomb (1999). Firstly, the number of layers of the semi-adaptive grid is not constant and changes with ice thickness (Fig. 1 in Griewank and Notz, 2013), while other models use a fixed amount of layers which grow and shrink with ice thickness. Secondly, the ice–ocean boundary is not defined in the SAMSIM grid, as discussed in Griewank and Notz (2013). Thirdly, the thickness of the upper layers remains constant throughout the run, with the exception of the top layer. Fourthly, as the upper layer boundaries only move in steps of $\triangle z_0$, there is no numerical diffusion in the upper

layers which results from the constant thickness adjustments used in other models.

The short-wave radiation properties of the ice are set with a number of parameters which determine how much radiation is absorbed at the ice surface and how much of the radiation penetrates into the ice and is absorbed in the lower layers. These parameters are the albedo, "alb", the fraction of penetrating short-wave radiation, "pen", and the optical thickness of the ice, κ. Various parametrizations have been proposed which define the optical properties based on the surface temperature, ice thickness, and ablation rates. In SAMSIM the gas volume fraction could also be used to parametrize the optical properties because the number of air bubbles has a large impact on the optical properties of the ice (Light et al., 2008). However, because the focus of this paper is on the salinity evolution, we will use constant values of "alb", "pen", and κ for sea ice to remove a source of variability in the model results (values shown in Table 1).

2.2 Modified gravity drainage

We have implemented a slight change to the calculation of the Rayleigh number of the layer i which is used in the gravity drainage parametrizations introduced in Griewank and Notz (2013) as

$$R^i = \frac{g \triangle \rho^i \, \widetilde{\Pi}^i h^i}{\kappa \mu}. \tag{1}$$

The terms that enter the equation are the standard gravity g, the density difference between the brine in layer i and the lowest layer $\triangle \rho^i$, the distance from the layer i to the ocean h^i, the thermal diffusivity κ, the dynamic viscosity μ, and the permeability term $\widetilde{\Pi}^i$. In Griewank and Notz (2013), the minimal permeability $\widetilde{\Pi}^i = \min(\Pi^i, \Pi^{i+1}, \ldots, \Pi^n)$ was used as a simplification of the harmonic mean. However, Vancoppenolle et al. (2013) demonstrated that using the minimal permeability instead of the harmonic mean leads to substantially different Rayleigh numbers. Accordingly, we replace the minimal permeability with the harmonic mean in the definition of the Rayleigh number. So instead of $\widetilde{\Pi}^i$ we use the bulk permeability $\bar{\Pi}^i$ for a Darcy flow through a stack of layers, which is given by the harmonic mean overall layers from i to the lowest layer n:

$$\bar{\Pi}^i = \frac{\Sigma_{k=n}^i \triangle z^k}{\Sigma_{k=n}^i \frac{\triangle z^k}{\Pi^k}}, \tag{2}$$

where Π^i is the permeability and $\triangle z^i$ is the thickness of the layer i.

Changing the definition of the Rayleigh number requires the free parameters α and R_{crit} which link the amount of brine to be readjusted, leaving each layer $br_\downarrow^i$ to the Rayleigh number, time step dt, and layer thickness $\triangle z$ via

$$br_\downarrow^i = \alpha(R^i - R_{crit})\triangle z^i \cdot dt. \tag{3}$$

To readjust α and R_{crit}, the same procedure is used as that which initially determined the free parameters in Griewank and Notz (2013). The procedure numerically derives values which lead to the best agreement between modelled salinity and the laboratory measurements of Notz (2005). Two separate sets of measurements and the mean of the two sets are used, resulting in the following free parameter pairings: $\alpha = 0.000510$, $R_{crit} = 7.10$; $\alpha = 0.000681$, $R_{crit} = 3.23$; $\alpha = 0.000584$, $R_{crit} = 4.89$. As in Griewank and Notz (2013) we will use the values optimized to fit the mean of the two measurement sets as the default values: $\alpha = 0.000584$, $R_{crit} = 4.89$. In Sect. 4.3.3 the effect of the parameter uncertainty of α and R_{crit} on the multiyear salinity profile is addressed.

Updating the Rayleigh number definition has a noticeable effect on the modelled salinity evolution of both the complex and simple gravity drainage parametrizations. However, the qualitative conclusions of Griewank and Notz (2013) and this paper are unaffected by the changed definition of the Rayleigh number. That the qualitative results are unaffected by the change in Rayleigh number definition can be seen by comparing this paper to the results of Griewank (2014), which uses the same simulations but the original Rayleigh number definition of Griewank and Notz (2013).

2.3 Surface melt

There are two main difficulties which complicate simulating surface melt in a 1-D thermodynamic sea-ice model. The first is the strong spatial heterogeneity of melting sea ice. Although certain aspects such as melt ponds can be parametrized, there is no way to overcome the fact that a 1-D approximation is less valid for melting sea ice than for growing sea ice. The second major difficulty is that many physical processes which occur at the surface during sea-ice melt are poorly understood. This is especially true for processes which occur at the snow–ice boundary and processes which involve capillary forces in snow or ice.

We have decided against separating the 1-D column into a ponded and non-ponded fraction because this is impossible without violating the core assumption of SAMSIM that each layer is horizontally and vertically homogeneous. A possible compromise would be to couple a 1-D column with a melt pond cover to another 1-D column with no pond, which would come with its own issues of how these columns interact with each other. The classic approach is to implement a melt pond and albedo parametrization which is applied evenly to the column surface without taking any horizontal variability into account. However, we have decided to not introduce such an albedo parametrization for two reasons. Firstly, most albedo parametrizations are not suitable for SAMSIM. For example, some parametrizations change the albedo as an empirical function of surface temperature. If the parametrization assumes that the surface layer is salt free, the parametrization will assume that the surface temperature during melt will always be at $0\,°C$. However, in SAMSIM the surface temperature varies during melt depending on the salinity of the top ice layer. Other parametrizations rely on the surface melt speed, which is not a variable in SAMSIM. Instead, SAMSIM has meltwater formation and surface ablation, which are linked but not identical to the definition of surface ablation used by Bitz and Lipscomb (1999). Secondly, slight albedo changes would overshadow the effects of the sea-ice salinity. If the albedo parametrization were fully physically consistent with SAMSIM then this would be acceptable. However, albedo parametrizations mostly rely on empirical measurements, are intended to improve large-scale models, and are ill-suited to determine how the albedo would react to a 5 % increase of gas volume fraction or a $0.1\,°C$ increase of temperature in the top ice layer of SAMSIM. Including an albedo parametrization would result in a large non-physical source of variability which would greatly complicate interpreting the results. Extending SAMSIM by an albedo parametrization that is compatible with SAMSIM physics remains desirable, however, and will be the subject of future work. For now we simply use a constant value for the ice albedo.

From the measurements taken at the Surface Heat Budget of the Arctic Ocean Project (SHEBA) site, Eicken et al. (2002) identified three stages of melt for Arctic multiyear ice. During stage I melt ponds form, fed by the horizontal transport of melting snow. The snow cover still persists and, while most of the meltwater movement is horizontal, some meltwater drains to the bottom of the ice through cracks and flaws in the ice. Stage II begins when the snow cover has completely melted away. During stage II meltwater moves horizontally until it reaches flaws as well as vertically through the ice. In stage III the flaws have enlarged to the point of ice disintegration. Meltwater moves vertically through the ice and horizontally until it reaches cracks and the edge of the ice

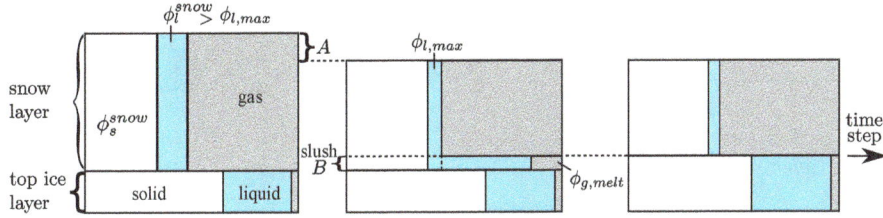

Figure 2. Sketch of snow melt by snow-to-slush conversion as described in Sect. 2.3.1. Snow-to-slush conversion occurs when the liquid fraction exceeds $\Psi_{l,\,max}$ as shown in the left sketch. A slush layer of thickness B is formed, which is instantaneously added to the top ice layer. A is the thickness lost by snow-to-slush conversion. The top ice layer thickness increases by B while the snow layer thickness is reduced by $A + B$. The white, blue, and grey areas represent the solid, liquid, and gas volume fractions of the model layers. The combined solid and liquid volumes of the snow and top ice layer are conserved during the conversion.

flows, and convective overturning occurs in the ice close to the ice–ocean interface.

In SAMSIM, surface melt is implemented by separating melt into two separate stages. The first stage is snow melt, in which snow is converted to slush. This process thins the snow layer by transforming a fraction of the snow into slush, which is then added to the top sea-ice layer as described in Sect. 2.3.1. The second stage is surface ablation, in which a fraction of the liquid volume of the top ice layer is designated as meltwater as described in Sect. 2.3.2. This meltwater is either transported directly into the ocean or flows through the ice and cracks according to the flushing parametrization introduced in Sect. 2.4.2.

2.3.1 Snow melt

The physics of snow is very complex. The snow layer in SAMSIM is intended to simulate only the most basic aspects of snow on sea ice. In contrast to the widely used 1-D thermodynamic sea-ice model of Bitz and Lipscomb (1999), which is implemented in both the Los Alamos (CICE) and the Louvain-la-Neuve sea-ice models, snow does not turn directly into meltwater in SAMSIM. Instead, melted snow from the snow surface percolates downward and accumulates on the sea-ice surface, forming a slush layer of depth B as illustrated in Fig. 2. This snow-to-slush conversion in SAM-SIM is based on two core assumptions. The first assumption is that the snow can only retain a maximum liquid mass fraction ($\psi_{l,\,max}$) which is a function of the snow solid mass fraction. The function we use is

$$\psi_{l,\,max} = 0.057 \frac{(1 - \psi_s^{snow})}{\psi_s^{snow}} + 0.017, \qquad (4)$$

which we take from the laboratory study of Coleou and Lesaffre (1998). In Fig. 2 the volume fractions are shown instead of the mass fractions because the volume fractions are proportional to the area depicted.

The second core assumption is that when the liquid water content surpasses the retainable amount, the excess water pools at the bottom of the snow layer, forming a layer of slush. At each time step the depth of the slush layer is deter-mined and then the slush layer is added to the top ice layer. Since the slush layer is merged with the top ice layer as soon as it forms, there is never a slush layer present at the beginning of the following time step. As such, the slush layer is not a physical representation of any physical material but instead a means to transform the model definition of snow into the model definition of sea ice. However, as the model definition of sea ice does not limit the liquid fraction, the sea ice can be in a condition which could be referred to as slush.

Two additional assumptions are required to determine the slush depth which is marked as B in Fig. 2: the gas fraction of the slush $\phi_{g,\,melt}$ and the solid fraction of the slush layer and remaining snow layer. We assume that the solid volume fraction equals the solid fraction of the previous time step and that $\phi_{g,\,melt}$ is a constant. In this paper we set $\phi_{g,\,melt}$ to 20 %, which we base on the measured surface sea-ice densities of Eicken et al. (1995).

Following these assumptions, when the liquid volume fraction of the snow layer exceeds $\phi_{l,\,max}$, the slush depth B is calculated from the snow solid fraction of the last time step (ϕ_s^{snow}) and the gas content as

$$B = \Delta z \frac{\phi_l^{snow} - \phi_{l,\,max}}{1 - \phi_{l,\,max} - \phi_s^{snow} - \phi_{g,\,melt}}. \qquad (5)$$

As a result, the top ice layer grows thicker by B, and mass and enthalpy are transferred according to the composition of the slush layer. To maintain the solid fraction of the last time step, the snow needs to be reduced in thickness by A as illustrated in Fig. 2. In total, the snow-to-slush conversion shrinks the snow layer by $A + B$, the total snow and ice column shrinks by A, the top ice layer grows by B, and the snow layer retains its density.

To our current knowledge, the approach of converting snow into slush before it can run off as meltwater is unique. Compared to the standard approach, in which snow melts at the top of the snow layer and immediately runs of as meltwater, our approach leads to a slight delay in the onset of flushing. This delay is because our approach requires the whole snow layer to convert to the model definition of sea-ice via slush formation before runoff occurs.

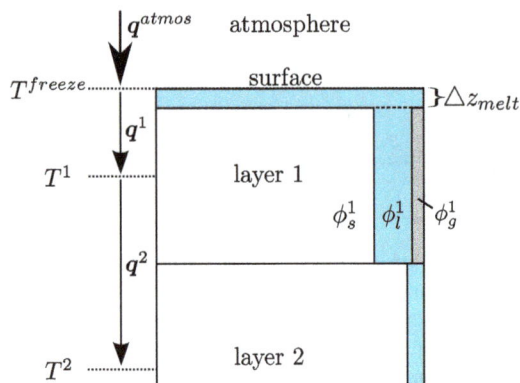

Figure 3. Sketch of meltwater formation caused by surface melting as described in Sect. 2.3.2. The white, blue, and grey areas represent the solid, liquid, and gas volume fractions of each model layer (ϕ_s, ϕ_l, and ϕ_g). The meltwater is located in a film which is $\triangle z_{melt}$ thick and located below the surface of the top layer. $\triangle z_{melt}$ is determined by the amount of latent heat release necessary to balance the energy difference between the atmospheric heat flux to the surface q^{atmos} and the flux from the surface into the top ice layer q^1.

In reality, sea ice has a varying surface height, which causes the meltwater in the slush to flow into melt ponds. In SAMSIM, by the time the snow layer has melted away, the top model layers that were formed by snow-to-slush conversion are predominantly liquid and salt free but also contain the solid fraction of the meltwater-soaked snow. These top ice layers can be interpreted as a spatial average over melt ponds and snow remnants. As a result the snow melt stage of SAMSIM is shorter than the first melt stage of Eicken et al. (2002) because not all of the latent heat which resided in the snow layer before the onset of melt needs to be released before the snow layer disappears in the model. Although the implemented snow-to-slush conversion neglects many of the finer aspects of snow physics, our approach, by having some interaction between the meltwater which forms at the snow surface with the underlying snow, captures snow melt somewhat more realistically than the standard approach of turning snow directly into meltwater.

Two additional processes also convert snow to slush: flooding as introduced in Sect. 2.4.5 and meltwater wicking. Wicking occurs when the top ice layer is so liquid that excess brine seeps into the snow. This process is incorporated into the model as introduced in the following subsection.

2.3.2 Surface ablation

Surface ablation in general refers to an ice-thickness decrease at the surface. Surface ablation is by necessity linked to a flux of melted ice away from the ice surface. In SAMSIM, surface ablation occurs when liquid from the top ice layer is removed via flushing. In this subsection we describe how SAMSIM determines how much liquid is available to be removed from

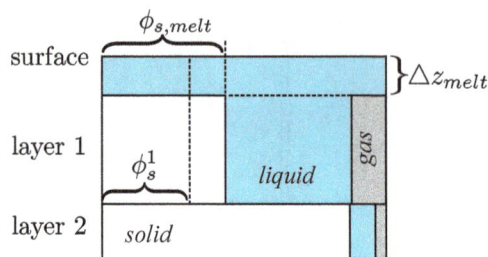

Figure 4. Formation of meltwater in the top ice layer when $\phi_s^1 < \phi_{s,\,melt}$ as described in Sect. 2.3.2. The thickness of the layer of meltwater ($\triangle z_{melt}$) is determined by how much the solid fraction has to be raised to equal $\phi_{s,\,melt}$. The white, blue, and grey areas represent the solid, liquid, and gas volume fractions of each model layer.

the top layer, what the properties of this liquid is, and how this liquid interacts with the snow layer and the top ice layer.

To describe this clearly we must first clarify how meltwater is defined in SAMSIM. The model definition of meltwater is the liquid in the top layer with the ability to leave the top layer. This ability distinguishes meltwater from the rest of the liquid in the top layer. Otherwise meltwater is identical to the remaining liquid in the top layer (i.e. temperature, salinity, density). Meltwater is assumed to be located on the ice surface in the top sea-ice layer as a thin film. The meltwater film is a part of the top layer, and its thickness is $\triangle z_{melt}$ as shown in Figs. 3 and 4.

The amount of meltwater which is present in the top layer is a diagnostic variable which is computed at each time step independently of the amount of meltwater in the previous time step. As the amount of meltwater determines the meltwater film thickness, the thickness is also calculated anew at each time step.

Meltwater can leave the top layer via two processes. The first process is via parametrized flushing, which is detailed in Sect. 2.4.2 and 2.4.4. Flushing leads to surface ablation because the thickness of the top layer is reduced by the thickness of the meltwater film when the water flushes away. The second process by which meltwater can leave the top layer is via wicking into the snow layer, as explained at the end of Sect. 2.3.1.

SAMSIM relies on three assumptions to diagnose meltwater amount and thickness. The first is that ice melted at the surface of the top sea-ice layer instantly turns into meltwater. The second is that if the solid fraction of the top ice layer sinks below a minimal low value, excess brine turns into meltwater. The third is that over time the gas fraction increases until it reaches the value of $\phi_{g,\,melt}$. The first two assumptions determine how much meltwater is available in the top layer, while the third assumption influences how thick the melt film is.

SAMSIM determines if melting occurs at the ice surface by analyzing the heat fluxes at the surface. As soon as the surface temperature surpasses the freezing temperature given by the bulk salinity of the top ice layer, meltwater can form. The amount of meltwater formed is determined by the amount of latent heat release necessary to balance the energy difference between the atmospheric heat flux to the surface and the flux from the surface into the top ice layer (depicted in Fig. 3). This approach is commonly used in sea-ice thermodynamic models (e.g. Bitz and Lipscomb, 1999) but needs to be adapted to incorporate the varying density and gas fraction of SAMSIM. The discretized diffusive heat flux from the ice surface into the top ice layer is

$$q^1 = -k^1 2 \frac{T^{\text{freeze}} - T^1}{\triangle z^1}. \tag{6}$$

The thermal conductivity of the top ice layer k^1 is a linear combination of the liquid and solid phases, while the gas phase is treated as an insulator. The depth of the meltwater film for a given atmospheric energy flux q^{atmos} is then

$$\triangle z_{\text{melt}} = \frac{q^{\text{atmos}} - q^1}{\phi_s^1 \rho_s L}. \tag{7}$$

The second way meltwater can form is when the solid fraction of the top ice layer ϕ_s^1 falls below a minimal low value $\phi_{s,\,\text{melt}}$. When this occurs the solid fraction ϕ_s^1 is rearranged by $\triangle z_{\text{melt}}$ until ϕ_s^1 reaches $\phi_{s,\,\text{melt}}$, as shown in Fig. 4. From volume conservation it follows that

$$\triangle z_{\text{melt}} = \triangle z^1 \left(1 - \frac{\phi_s^1}{\phi_{s,\,\text{melt}}} \right). \tag{8}$$

This second way of forming meltwater ensures that meltwater forms before the top ice layer is fully liquid. Not shown in the Fig. 4 is that a similar limit exists on the gas fraction which arises from our third assumption that the gas fraction increases to a specific value over time. If the gas fraction exceeds $\phi_{g,\,\text{melt}}$ then the top ice layer is compacted to reduce ϕ_g^1 to $\phi_{g,\,\text{melt}}$, which also slightly increases the density of the top layer. $\phi_{g,\,\text{melt}}$ is the same parameter which determines the amount of air captured in the slush during snow melt and is set to 0.2 based on density measurements at the surface of Eicken et al. (1995). To our knowledge there are no measurements from which to estimate $\phi_{s,\,\text{melt}}$. As first guess we assume a value of 0.4, which is slightly above the solid fraction assigned to fresh snow in SAMSIM. If meltwater forms primarily due to low solid fractions, the top ice layer will approach the given values of $\phi_{g,\,\text{melt}}$ and $\phi_{s,\,\text{melt}}$ over time.

If the meltwater forms due to a low solid fraction while snow is present, the meltwater is assumed to wick up into the snow and creates a slush layer which is then added to the top ice layer again. We refer to this as wicking, and it is similar to snow melt (Fig. 2). The difference between wicking and snow melt is that in wicking the amount of water available to form slush is given by the amount of meltwater present in the top ice layer, while in snow melt the amount is given by how far the liquid fraction of snow exceeds the threshold limit.

2.4 Salinity parametrizations

There are three known relevant desalination processes in sea ice: gravity drainage, flushing, and flooding (Notz and Worster, 2009). We addressed how gravity drainage is implemented in SAMSIM in our previous publication (Griewank and Notz, 2013). In this subsection we introduce parametrizations for flushing and flooding, making SAMSIM the second published 1-D model capable of capturing the full salinity evolution. The first model capable of capturing the full salinity cycle is the 1-D LIM sea-ice model of Vancoppenolle et al. (2006).

Parametrizing flushing faces the same challenges as modelling surface melting, namely high horizontal heterogeneity, insufficient data, and a lack of theoretical understanding. No quantitative laboratory studies of flushing have been published to this date and, due to sampling issues and challenging conditions, field studies have been limited to studies of dye dispersion and ice-core salinity (Eicken et al., 2002). The understanding of flooding is even poorer and is limited to the analysis of ice cores which contain flooded snow ice.

2.4.1 Flushing

The first and only published flushing parametrization incorporated in a full thermodynamic sea-ice model by Vancoppenolle et al. (2006) assumes that once the ice reaches a certain permeability, a fraction of the meltwater flows downward through the sea ice and into the ocean below. Although this approach neglects many aspects of flushing, it is able to reproduce field measurements of salinity (Vancoppenolle et al., 2007). In this subsection we will introduce two parametrizations. The complex parametrization attempts to model flushing as a physically consistent hydraulic system, and the simple parametrization is a numerically cheap alternative based on the assumption that the liquid fraction increases towards the surface during surface melt.

2.4.2 Complex flushing

It is known from the field observations of Eicken et al. (2002) that much of the brine movement during flushing occurs horizontally in the upper layers. Once the horizontally flowing meltwater reaches a flaw or crack it drains below the sea ice, which can lead to underwater ice formation (Eicken et al., 1995; Notz et al., 2003). These cracks can also be situated below melt ponds as discussed by Polashenski et al. (2012), who refer to them with the term macroscopic holes. The parametrization of Vancoppenolle et al. (2006) has no explicit treatment of horizontal fluxes. Our goal is to design a flushing parametrization which is as physically consistent as possible in a 1-D model and includes horizontal brine fluxes which are highest when close to the ice surface. Additionally the parametrization should have as few free parameters as possible. The resulting parametrization (sketched in Fig. 5)

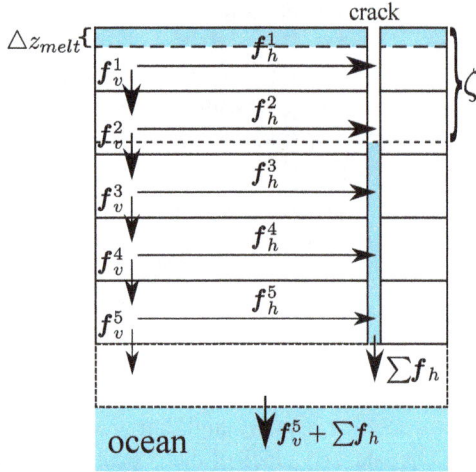

Figure 5. Brine fluxes of the complex flushing parametrization resulting from meltwater formation at the surface as described in Sect. 2.4.2. The horizontal fluxes f_h transport heat and salt to the lowest layer directly via cracks in the ice, while the vertical fluxes f_v advect heat and salt from layer to layer. ζ is the freeboard of the ice and $\triangle z_{melt}$ is the depth of the meltwater.

treats sea ice as a hydraulic network in which each model layer has a vertical and a horizontal hydraulic resistance (R_v and R_h). The assumptions on which the parametrization is based are as follows:

1. Cracks always exist in the ice.

2. As we have no data from which to deduce the frequency of these cracks, as a zero-order first guess we assume average horizontal distance between these cracks grows linearly with ice thickness.

3. Once brine reaches such a crack it drains away to the ice–ocean interface without interacting with the underlying ice layers.

4. The vertical resistance represents the resistance to brine flowing from the top to the bottom of a layer. The horizontal resistance represents the resistance that brine needs to overcome to reach a crack.

5. Flushing meltwater flows vertically from layer to layer and horizontally to the cracks. The specific amount for each layer is determined by the hydraulic resistances and the hydraulic head.

6. The hydraulic head is assumed to be equal to the freeboard ζ, resulting in a pressure difference of $\triangle p = \zeta \rho g$ for the brine density ρ and gravitational constant g.

The resulting parametrization has only a single free parameter β which determines the average distance x to the next crack for a given ice thickness h through $x = \beta \cdot h$.

The Darcy flow in a porous medium with a hydraulic resistance of R leads to a mass flux f of

$$f = \frac{\triangle p \cdot A}{R} \rho \qquad (9)$$

for the pressure difference $\triangle p$ and liquid density ρ. In SAMSIM, for each layer i the vertical hydraulic resistance

$$R_v^i = \frac{\mu}{\Pi(\phi_l^i)A} \triangle z^i \qquad (10)$$

is defined by the permeability Π, which is a function of the layer's liquid fraction ϕ_l^i, the brine viscosity μ, the column area A, and the layer thickness $\triangle z$. SAMSIM uses the permeability function of Freitag (1999), which was derived from measurements of vertical flows. We use it here for both horizontal and vertical permeability. This simplification should not adversely affect our results, since the major simplification lies in the underlying assumption that the permeability is only a function of solid fraction.

To define the horizontal hydraulic resistance we take the average distance to the next crack from our assumptions, resulting in

$$R_h^i = \frac{\mu}{\Pi(\phi_l^i)A_h^i} x. \qquad (11)$$

In contrast to the vertical flow area A, which is always $1\,\text{m}^2$ in the column model, the horizontal flow area A_h^i varies with layer thickness as well as with the geometry of the cracks and resulting flow field. We take A_h^i to be equal to the vertical layer surface with an area of $\triangle z^i \cdot 1\,\text{m}$.

The resulting horizontal and vertical brine fluxes (f_h and f_v as shown in Fig. 5) are then computed from hydraulic head and resistance. The total resistance over multiple layers is calculated as a sum of parallel and serial resistances, the same method used in resistor ladder circuits. To illustrate how the layers interact, refer to the sketch with six layers shown in Fig. 5 as an example. The lowest layer 6 has by definition no hydraulic resistance. The total resistance of the second lowest layer 5 (R_{total}^5) is

$$R_{total}^5 = \frac{R_v^5 R_h^5}{R_h^5 + R_h^5} \qquad (12)$$

because R_v^5 and R_h^5 are connected in parallel. The total resistance over layers 4 and 5 (R_{total}^4) is the parallel resistance of R_v^4 with the serially connected R_h^4 and R_{total}^5, resulting in

$$R_{total}^4 = \frac{(R_{total}^5 + R_v^4)R_h^4}{R_{total}^5 + R_v^4 + R_h^5}. \qquad (13)$$

Generalizing this for all layers results in

$$R_{total}^i = \frac{(R_{total}^{i+1} + R_v^i)R_h^i}{R_{total}^{i+1} + R_v^i + R_h^i}, \qquad (14)$$

which is true for any number of layers. The total amount of flushing brine through the whole ladder circuit shown in Fig. 5 is accordingly

$$f_v^5 + \sum_1^5 f_h = \frac{\Delta p \cdot A}{R_{total}^1} \rho. \tag{15}$$

The total amount of flushing brine can not exceed the amount of meltwater present in the top ice layer.

The calculated vertical fluxes advect salt and heat from layer to layer using the upstream method, while horizontal fluxes transport both salt and heat directly to the lowest model layer, i.e. the ice–ocean interface. As the thermal profile in melting ice is almost uniform and the brine salinity is linked to temperature, the vertical fluxes lead to a smaller desalination than the horizontal fluxes.

Although the top ice layer can accumulate meltwater faster than it can flush it away, a fully liquid top layer in the model is impossible with the complex flushing parametrization. As the top ice layer becomes more and more liquid, the permeability increases and the horizontal hydraulic resistance of the top ice layer decreases, resulting in a strong horizontal flushing in the top ice layer. This strong flushing removes water from the top layer and prevents the top layer from ever becoming fully liquid.

2.4.3 Complex flushing examples

To illustrate the fluxes which result from the complex flushing parametrization, we apply some numbers to a specific example with six layers as shown in Fig. 5. For this simple thought experiment, layers 1 through 5 are identical with the same permeability and 20 cm thick. Accordingly, the vertical and horizontal resistances of each layer are equal to each other: $R_h^1 = R_h^2 = \ldots = R_h$ and $R_v^1 = R_v^2 = \ldots = R_v$. The ratio of R_h and R_v is determined by the free parameter β, the layer thickness Δz, and the total ice thickness h. In our example we chose $\Delta z = 0.2$ m, which results in $h = 1$ m because we have five layers of ice. Combined, these result in $R_h = R_v \beta / 0.2^2 = 25 R_v \beta$. We can now calculate the ratios of the resulting fluxes for a given value of β, which are shown in Table 2.

For the default value of $\beta = 1$, 60 % of the flushing brine would penetrate vertically through all five layers of the ice while 40 % of the flushing brine would flow horizontally until falling through cracks and flaws (row (a) of Table 2). The horizontal fluxes are strongest in the top layer and decrease with depth. A lower value of β would favour horizontal fluxes. Reducing β to 0.2 results in only 18 % of the brine flushing vertically through all five layers, while over 50 % flushes horizontally in the three top layers (row (b) of Table 2).

In the previous example all layers have the same permeability. To illustrate how the complex flushing layer reacts if the lower ice is less permeable, we repeat the same scenario with a higher permeability close to the surface. Specifically, let us assume that the top two layers are 20 times more

Table 2. Horizontal and vertical fluxes of the thought experiment detailed in Sect. 2.4.3 and shown in Figure 5 to illustrate the complex flushing parametrization. All fluxes are given in percent of total flushing ($\sum f_h + f_v^5 = 100$). In (a) and (b), all five layers have the same permeability, while in (c) the top two layers are 20 times more permeable. The free parameter β is changed from the default value of 1.0 to 0.2 in (b).

Layer	1	2	3	4	5
	(a) $\beta = 1.0$				
f_h	14	11	8	5	2
f_v	86	75	68	63	60
	(b) $\beta = 0.2$				
f_h	35	22	14	8	4
f_v	65	43	29	21	18
	(c) $\beta = 1.0$				
f_h	42	39	2	1	1
f_v	58	19	17	16	15

permeable than the lower three. This reduces the percentage of brine that flushes vertically through the whole ice layer from 60 to 15 %, while over 80 % leaves the ice in the top two layers horizontally (compare row (c) to (a) in Table 2). Meanwhile, the horizontal flushing in layers three to five is very small. Less than 5 % of the total brine leaves the ice through cracks and flaws in the lower three layers. If the third layer were impermeable, all flushing would occur horizontally through the top two layers.

The scenario of higher permeable upper layers is slightly more realistic than the uniform permeability scenario; however, SAMSIM is run with many more layers and a correspondingly detailed vertical permeability profile. Idealized simulations illustrating how the complex flushing interacts with salinity and thermodynamics are discussed in Sect. 3.

2.4.4 Simple flushing

We propose a second, numerically cheaper, parametrization which we will refer to as the simple flushing parametrization. In contrast to the complex parametrization, which calculates brine fluxes that affect salinity via advection, the simple parametrization directly modifies the salinity to fulfill a stability criterion. This stability criterion is based on the simple assumption that the liquid fraction is highest in the top ice layer during melt and decreases into the ice. If this were not the case, the ice below the top layer could become fully liquid. Indeed, fully liquid pools inside the ice have, to our knowledge, never been observed, although rotten ice and slush layers seem to be common during the melt period. This stability criterion is only applied when surface melt occurs and has no affect on the rest of the year when solid fraction is high enough to prohibit liquid from running off as meltwater.

The implementation is as follows. At each time step the meltwater which forms in the top ice layer as explained in Sect. 2.3.2 is removed. The salinity of the meltwater is higher than the bulk salinity over the total layer because the solid fraction of the ice is salt free. Accordingly, meltwater removal leads to a reduction of the bulk salinity in the top ice layer. Over time this ensures that the top layer becomes less saline than the second layer. Given that the temperature difference between the top layers is small during surface melt, the second, saltier layer will gradually become more liquid than the fresher top layer.

To ensure that our assumption is fulfilled and the liquid fraction is highest in the top layer, SAMSIM checks each time step if $\phi_l^1 > \phi_l^2$. When this occurs, the salinity of the second layer is simply reduced by a fixed fraction ϵ. This increases the solid fraction while raising the temperature.

The same procedure is then applied to the third layer, to ensure that the second layer is not less liquid than the third layer, and after that to the fourth, fifth etc. until a layer is reached which is less liquid. As long as $\phi_l^i > \phi_l^{i+1}$, the salinity of layer $i+1$ will be reduced by the factor ϵ. For example if $\phi_l^1 < \phi_l^2 < \phi_l^3 > \phi_l^4 < \phi_l^5$, the salinity of the second and third layer are reduced while the fourth and fifth remain untouched.

2.4.5 Flooding

Flooding can occur when the weight of snow pushes the ice below the ocean surface, causing ocean water to well up and flood the snow. The resulting frozen mix of snow and ocean water, called snow ice, can be identified by various means in ice cores, from which we know that flooding occurs mainly in the Antarctic and contributes up to 25 % of ice production in certain areas (Jeffries et al., 2001; Maksym and Jeffries, 2001). We base our understanding and treatment of flooding on the work of Maksym and Jeffries (2000, 2001) and Jeffries et al. (2001). To readers interested in flooding we recommend the PhD thesis of Maksym (2001).

Although at first glance flooding seems to be the same process as flushing but with a reversed pressure gradient, there are a number of additional uncertainties. Field measurements have shown that a negative freeboard does not automatically lead to flooding although the lower the freeboard, the higher the chance of flooding is. Additionally, very little is known about what happens to the flooded brine once it reaches the ice surface. As flooding occurs at the bottom of the snow mantle, direct observations of flooding are extremely difficult to obtain. Snow metamorphism is in itself a complex process, but the interactions between flooding brine and snow are even more complex and little research has been devoted to this specific issue. Brine movement must occur at the ice surface after or during flooding, because otherwise snow–ice salinities would be higher than the measured values.

As for flushing and gravity drainage, we again developed two separate parametrizations for flooding. However, the two flooding parametrizations are rather similar. We will simply refer to the slightly more sophisticated parametrization as the *complex* parametrization and the simpler one as the *simple* flooding parametrization.

2.4.6 Complex flooding

The complex parametrization assumes that during flooding ocean water passes through cracks and channels in the ice to flood the snow layer. The flooding ocean water is assumed not to interact with the brine in the sea ice: Maksym and Jeffries (2001) showed that if flooding resulted in an upward brine displacement through the whole ice, the resulting desalination would quickly turn the ice impermeable. Experiments with SAMSIM reached the same conclusion as Maksym and Jeffries (2001) that upward brine displacement would quickly turn the ice impermeable (experiments not shown). Although the complex flushing parametrization consists partially of vertical flows that displace brine, these only seldom cause the ice to become impermeable for three reasons. Firstly, as a layer becomes less permeable the flushing brine is increasingly diverted horizontally. Secondly, the temperature gradients are much smaller in melting ice so that brine advection leads to less desalination. And thirdly, the ice is usually cooled by the atmosphere during flooding which can compensate the latent heat released during desalination.

The flux of ocean water to the surface is calculated as a Darcy flow driven by the negative freeboard and limited by the permeability of the least permeable model layer. Here we assume that the permeability function of Freitag (1999) provides a useful estimation regardless of the detailed pathways that the ocean water takes through the ice. Although this is a simplification, the major uncertainty stems from the uncertainty in permeability itself and the poor physical understanding of flooding.

Our approach of using the ice permeability to regulate the strength of flooding can lead to a large negative freeboard if the ice layer is impermeable. To avoid this a maximum negative freeboard ζ_{max} is defined. If the freeboard sinks below this threshold, the flux of ocean water necessary to raise the freeboard to the threshold is determined and applied.

The ocean water transported to the ice surface forms a slush layer which is immediately added to the top ice layer at each time step. This is the same approach SAMSIM uses to imitate snow melt and meltwater wicking into the snow layer (described in Sects. 2.3.1 and 2.3.2). However, given a snow solid volume fraction of approximately 30–40 %, this approach would result in the flooded slush layer having a very high salinity of roughly $20\,\mathrm{g\,kg^{-1}}$, which is inconsistent with measurements. To avoid this high salinity, we assume that the ocean water which floods the snow simultaneously wicks upward and dissolves additional snow into the slush which leads to a freshening of the slush. The ratio of dissolved to flooded snow is assumed to be constant and is defined by an additional free parameter δ.

In this paper we use a value of 5 cm for ζ_{max}, which is based on the freeboard measurements analyzed in Maksym and Jeffries (2000), and we use a value of 0.5 for δ as a preliminary best guess.

2.4.7 Simple flooding

The simple parametrization is the complex parametrization stripped of the permeability-dependent flooding speed and without snow dissolving into the slush layer. The simple parametrization is identical to the complex parametrization if the free parameters are set to $\zeta_{max} = 0$ m and $\delta = 0$. This means that as soon as a negative freeboard develops, flooding sets in right away and no snow is dissolved into the forming slush.

2.5 Salinity set-ups

In Sect. 2.4 we have presented four parametrizations, two for flushing and two for flooding. Together with the two gravity drainage parametrizations introduced in Griewank and Notz (2013), SAMSIM now has two complete sets of desalination processes. The first set consists of the complex flushing, the complex flooding, and the complex gravity drainage parametrization. The second set of parametrizations consists of the simple flushing, the simple flooding, and the simple gravity drainage parametrization. The parametrizations of the first set all compute brine fluxes which result in salt and heat advection. Accordingly, the rate of salinity change is determined by the strength of brine flow and the salinity gradients between layers. In contrast, the parametrizations of the second set directly adjust the salinity profile to fulfill defined stability criteria.

We will refer to the first set of parametrizations as the *complex* salinity approach because it consists of the more sophisticated parametrizations which were designed to be as close to reality as possible. The second set will be referred to as the *simple* approach because the parametrizations included were developed as simpler alternatives to the parametrizations of the complex approach.

The third and final salinity approach employed in this paper prescribes a depth-dependent salinity profile completely independent of the ice properties. The profile used is a crude approximation of measured multiyear ice salinity and is the same profile introduced and used in Griewank and Notz (2013). The profile consists of a linear decrease in bulk salinity from 34 g kg^{-1} at the ice–ocean interface to 4 g kg^{-1} at 15 cm above the bottom and a second linear decrease from the 4 g kg^{-1} at 15 cm above the ice–ocean interface to 0 g kg^{-1} at the surface. This approach is referred to as the *prescribed* approach. The prescribed profile is by choice highly idealized so that the *prescribed* approach provides a stark contrast to the *simple* and *complex* approaches. A more

realistic profile could have been derived from simulations using the complex approach but we prefer the idealized profile because it is independent of both SAMSIM and the chosen forcing.

An important aspect of the complex parametrization set is that the simulated brine fluxes result in heat fluxes both in the ice and into the ocean. This is most relevant during growth when gravity drainage continually moves colder brine to the ocean while taking up relatively warm ocean water, resulting in a small but steady increase of oceanic heat flux in our limited model domain. Because flushing mostly occurs in ice close to the freezing temperature, the energy lost due to flushing is small. However, these heat fluxes caused by brine flux lead to the *complex* approach having a different oceanic heat flux than the *prescribed* and *simple* approaches. To avoid this change in oceanic heat input when comparing the three salinity approaches against each other, the heat fluxes resulting from gravity drainage and flushing are subtracted from the lowest layer at each time step for the *complex* approach. This heat flux modification was already applied in Griewank and Notz (2013) to ensure that the various approaches can be compared to each other.

3   Idealized flushing experiments

In this section we take a closer look at the complex flushing parametrization to study how it interacts with temperature and salinity as well as how sensitively it reacts to various parameters. We prefer to use an idealized set-up, rather than a set-up based on field conditions, for two reasons. The first reason is that in the idealized experiment we can remove all feedbacks and processes not related to flushing. The second reason is that the idealized set-up allows us to chose conditions that highlight how the flushing parametrization interacts with the salinity and thermodynamics of the sea ice. As the full parameter space of all model parameters which interact with flushing in some way is too large to be fully explored in a useful way, we focus on the two parameters which have the strongest effect. The first of these two parameters is β, which determines the linear relationship of average horizontal flow distance to ice thickness in the complex flushing parametrization. The second parameter is the layer thickness $\triangle z_0$.

The idealized experiment begins with a 1 m thick homogeneous slab of ice with a bulk salinity of 5 g kg^{-1} and a temperature of roughly $-10\,°C$. The ocean below the ice is at 34 g kg^{-1} and 0 °C. A constant oceanic heat flux of 15 W m^{-2} is applied to the bottom while a constant heat influx of 380 W m^{-2} is applied to the surface. After subtracting the outgoing thermal radiation at 0 °C at the surface, the net heat input into the surface is slightly below 70 W m^{-2}. The heat fluxes were chosen such that the 1 m slab of ice melts over 1 month, which is the same order of magnitude found in reality. The cold initial temperature was chosen as it highlights the thermodynamic interactions of the flushing brine. All brine fluxes that occur in the experiment are caused by flushing as gravity drainage is deactivated and no flooding occurs.

We will first make some general observation of how flushing occurs in the idealized experiment in Sect. 3.1 before analysing how the flushing parametrization reacts to β and $\triangle z$ in Sect. 3.2 and 3.3.

3.1 General observations

In the idealized experiment the homogeneous sea-ice slab melts away over 1 month (Fig. 6). The constant surface heat input results in a constant rate of surface ablation. As the initial ice temperature of roughly $-10\,°C$ is well below the freezing temperature of the underlying $34\,g\,kg^{-1}$, water-bottom growth occurs over the first 3–4 days. This newly formed ice retains the $34\,g\,kg^{-1}$ salinity as no gravity drainage is activated in this simulation.

Flushing commences once the ice surface reaches melting temperature after a few days. The resulting desalination is clearly visible in the salinity profile as well as in the temperature profile (Fig. 6). The downward flushing meltwater quickly desalinates the upper ice, which causes a release of latent heat that warms the desalinated ice to $0\,°C$. However, after roughly 1 week the flushing stops penetrating downward into the ice and no further desalination occurs in the ice. By comparing the salinity and temperature profiles we can see that the kink in the $3\,g\,kg^{-1}$ salinity contour occurs when ice layers with zero salinity are below $0\,°C$. As freshwater ice below $0\,°C$ is a complete solid, it is impermeable and flushing can not penetrate below this level. This occurs because the temperature in the lower and saline ice cools the freshly desalinated ice layers, while the isothermal desalinated ice transports no heat via thermal diffusions.

Until the impermeable upper layers have melted away after half a month, flushing is restricted to the top layers. Once the impermeable layers have melted away, flushing begins to penetrate into the ice again. As the interior of the ice is by now quite close to the freezing temperature, the newly desalinated layers do not refreeze, and after a few days the ice is fully desalinated.

Two noteworthy secondary effects of flushing occur in the idealized experiment. The first is that while flushing reduces the bulk salinity close to the surface, it also leads to an increase of salinity in the lower ice (visible in the $7\,g\,kg^{-1}$ contour of Figs. 6 and 7b–d). This is caused by the positive temperature gradient near the ice–ocean interface, which leads to the vertically flushing brine moving from colder to warmer layers. As the brine is saltier in the colder layers due to the liquidus relationship, salt advection leads to a bulk salinity increase in the lowest ice layers. This effect disappears if gravity drainage is activated (Fig. 7a), which explains why this salinity increase due to flushing has not been observed to our knowledge. To determine if flushing could in principle

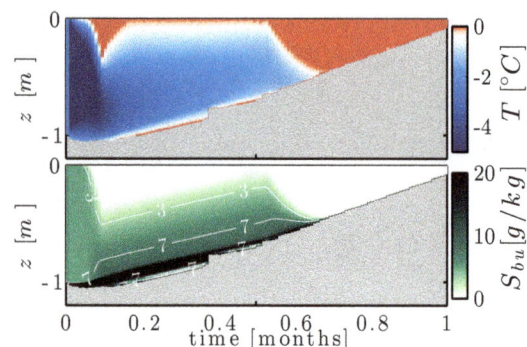

Figure 6. Temperature and bulk salinity evolution of the idealized flushing experiment using the default model set-up (experiment set-up in Sect. 3, model set-up in Table 1). Plot background is grey. The 3 and $7\,g\,kg^{-1}$ contour lines are included.

lead to such an increase in salinity if gravity drainage is absent, experiments with a multiphase material, in which both phases have a similar density to inhibit convection, would be required. An additional requirement needed to generate these high salinities close to the ice–ocean interface is that the oceanic heat flux is relatively small so that the flushing parametrization has sufficient time to transport salt into the lower layers before they melt away.

The other noteworthy secondary effect of flushing occurs at the ice–ocean interface. As described in Sect. 3, the fresh meltwater which drains through flaws and cracks flows into the lowest model layer. This results in a freshening of the lowest model layer (e.g. layer 6 in Fig. 5). If the lowest layer freezes after it has been freshened by flushing meltwater, it results in a thin layer of low-saline ice close to the ice–ocean interface. This effect is visible in the salinity plots of Figs. 6 and 7, where a thin line of low salinity at the ice–ocean boundary is outlined by the $7\,g\,kg^{-1}$ contour line from 0.2 to 0.4 months and once again briefly at 0.5 months. Because this thin layer of ice formed from meltwater, it is less saline than the ice above it. Since it is less saline, the thin layer has a higher solid fraction than the ice above at the same temperature. This leads to a thin sheet of solid freshwater ice below mostly liquid salty ice above. As a consequence, once the ice with low salinity (which is visible as an orange line in the temperature plot of Fig. 6) melts away, the ice above it melts away very quickly due to the low solid fraction. While this ice layer with its low salinity is similar to the false bottoms observed below summer ice, false bottoms occur in nature due to contact of fresh meltwater with sub-zero ocean water, which creates a negative oceanic heat flux. In the idealized experiment the oceanic heat flux is steady and positive, and the formation is dependent on the nonoccurrence of gravity drainage (compare to Fig. 7a).

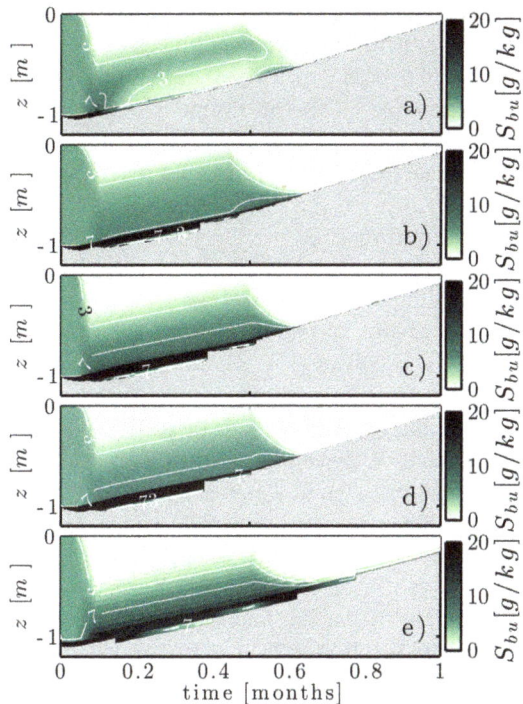

Figure 7. Salinity evolution of the idealized melting experiments in which one specific parameter or setting has been changed from the default values (default model results shown in Fig. 6, experiment description can be found in Sect. 3, default settings are listed in Table 1). The 3 and $7\,\mathrm{g\,kg^{-1}}$ contour lines are included. (**a**) Gravity drainage is included, which is otherwise disabled in the experiment. (**b**) The ratio of horizontal to vertical hydraulic resistance β is 0.2 instead of 1.0. In (**c**) the ratio of horizontal to vertical hydraulic resistance β is 5 instead of 1.0. In (**d**) the vertical spatial resolution $\triangle z_0$ is 2 mm instead of 1 cm, and in (**e**) the vertical spatial resolution $\triangle z_0$ is 5 cm instead of 1 cm.

3.2 Free parameter β

In this subsection we examine the importance of the single free parameter of the flushing parametrization β. We have no definitive physical or model limits on the possible value of β. Based on tracer studies of Eicken et al. (2002), we expect horizontal flows to be on the order of metres. Accordingly, we expect β to be in the single digits, and as a working assumption we set 1 as the default value. A value of 1 assumes the average horizontal travel distance to a crack equals the ice thickness, which implies that the cracks are on average roughly 4 times the ice thickness apart. However, the exact relationship of average travel distance to average crack spacing is a function of the geometric organization of the cracks and the 3-D flow path the meltwater follows. To test the parameter sensitivity around the default value of 1 we repeated the simulation with a value of 0.2 to 5.

Because a high β increases the horizontal hydraulic resistance the higher β is, higher values of β cause weaker horizontal fluxes and vice versa, as was shown in the thought ex-

periment of Sect. 2.4.3. In the idealized experiment the low value of $\beta = 0.2$ leads to a delayed onset and depth of flushing in contrast to $\beta = 5$ (Fig. 7b and c). The higher value of β increases the salinity at the ice–ocean interface, which results from more brine flushing completely through the ice. The results for $\beta = 0.2$, 1, and 5 differ only slightly, indicating that the complex flushing parametrization is much more dependent on the thermodynamics of the ice than the specific value of β. From the idealized experiment we conclude that changing β has the anticipated effect and that the parametrization has a low sensitivity to changes of β close to our default value of 1. This low sensitivity is an advantage for us because although we lack the data to derive the optimal value of β, having a non-optimal estimate of β should only impact our results slightly.

3.3 Vertical resolution

Changing the vertical resolution influences the complex flushing parametrization in many ways. The thickness of the top layer has an impact on how SAMSIM calculates meltwater formation, the grid spacing influences heat diffusion and tracer advection, and higher resolution allows more vertical variability of layer properties such as permeability.

The default value for $\triangle z_0$ in the model is 1 cm. As for β we repeated the idealized experiment with a value 5 times lower (i.e. 2 mm) and 5 times larger (5 cm). These values encompasses the practical range of values usable in SAMSIM.

In the idealized experiment, changing the resolution has only a minor effect (see Figs. 6b and 7d, e). The simulations with layer thicknesses of 5 cm and 2 mm are remarkably similar despite the higher resolution run using 25 times more layers. As a result, we do not expect the flushing parametrization to respond strongly to slight changes in vertical resolution.

3.4 Summary

The complex flushing parametrization responds weakly to changes of the parametrization parameter β and the model resolution $\triangle z$. Changing β has the expected effect, but no theoretical expectations or data are available to determine the optimal value. Accordingly, the chosen default value of 1.0 is uncertain and may be off by 1 order of magnitude. However, given the low sensitivity to β, even a change of magnitude would not qualitatively change our results. The vertical model resolution has little influence on the parametrized flushing beyond the change in underlying numerics. It is possible that the complex parametrization performs most realistically at a specific layer thickness or that the optimal value of β is resolution dependent, but this can not be determined until more precise data are available.

4   Arctic sea ice

In this section we study how SAMSIM simulates the salinity evolution in the Arctic using the complex salinity approach and compare the model output with ice-core data.

We have decided to limit the study to the Arctic because flooding and the corresponding snow-ice formation play a large role in the Antarctic. As explained in Sect. 2.4.5, we treat the flooding parametrizations currently implemented in SAMSIM as ad hoc solutions only suitable for dealing with isolated and sporadic flooding events. Accordingly, we will refrain from studying Antarctic ice until flooding is better understood.

Although a basic understanding of the salinity evolution has existed for many decades, the main processes driving this desalination still pose many unanswered questions. Using a model has the major advantage of being able to track the evolution consistently over long periods of time, while sea-ice cores can only provide snapshots. Simulating the salinity evolution with SAMSIM is an exercise in reproducing a vaguely known result of poorly understood origin. We aim to understand the impact and interactions of the various processes better while at the same time discovering the limitations of the developed parametrizations or the existence of neglected relevant processes.

4.1   Model set-up

To imitate Arctic conditions we use 3-hourly ERA-interim radiative fluxes and precipitation to provide the surface conditions for SAMSIM. Nine simulations, each forced with ERA-interim reanalysis data taken from one of nine locations spread over the Arctic, are run from July 2005 until December 2009. The coordinates of the chosen locations from south to north are: 70° N, 0° W; 72° N, 155° E; 75° N, 180° E; 75° N, 0° E; 75° N, 145° W; 80° N, 0° E; 80° N, 90° E; 85° N, 180° E; and 90° N. A simulation period of 4.5 years was chosen because it covers four yearly cycles of growth and melt, which covers the age of most Arctic sea ice (Lietaer et al., 2011).

SAMSIM also requires oceanic boundary conditions in the form of ocean salinity and oceanic heat flux. Due to the scarcity of oceanic heat flux measurements and for simplicity's sake, all runs share the same prescribed yearly heat-flux cycle, which is sinusoidal and based loosely on the heat fluxes (Huwald et al., 2005a) derived from the SHEBA measurements. The oceanic heat flux is highest in autumn ($14\,\mathrm{W\,m^{-2}}$) and lowest in spring ($0\,\mathrm{W\,m^{-2}}$). Similarly, a standard ocean salinity of $34\,\mathrm{g\,kg^{-1}}$ is used for all runs. The model settings and parameters used are listed in Table 1.

It is important to state that the boundary conditions we use are not necessarily a realistic approximation of the true conditions at the specific locations and time from which we chose the reanalysis data. Not only are the oceanic heat fluxes a strong approximation, the precision of the reanalysis data is limited by the lack of observations in the Arctic. Additionally, the influences of dynamic processes such as frazil formation, lead opening, melt ponds, and ice drift can not be accounted for in the 1-D SAMSIM model. Given the lack of melt pond formation and lead openings SAMSIM will tend to underestimate the amount of melt compared to reality.

4.2   Sample output

To give an example of the model output we have included the salinity evolution of one of the nine simulations for all three salinity approaches (Fig. 8). We chose the simulation forced with reanalysis data from 75 and 145° W as it has the same forcing during the first growth season as the growth season analyzed in Griewank and Notz (2013). Note that due to the modification to the Rayleigh number (see Sect. 2.2) the salinity evolution of the first growth season shown in Fig. 8 is not identical to the simulated salinity shown in Fig. 9 of Griewank and Notz (2013).

In the sample output the first-year ice survives the first melt season and is followed by 3 years of multiyear ice. The yearly cycle in sea-ice thickness is clearly visible, with strong interannual variations in minimum and maximum ice thickness due to interannual variations in the forcing data, such as snowfall. The complex and simple approaches (Fig. 8a and b) both create a detailed salinity profile which evolves during growth and melt with large differences from year to year. In contrast, the prescribed approach (Fig. 8c) has neither interannual variability nor a seasonal evolution. As noted in Griewank and Notz (2013), the simple parametrization desalinates slightly stronger during growth, but during the melt season the complex approach loses more salt. In contrast, the prescribed salinity profile results in an increase of bulk salinity over the ice column during melt.

4.3   Ice-core data

We begin analyzing the SAMSIM salinity evolution by comparing the output against salinity characteristics derived from ice-core measurements. Despite its drawbacks, taking ice cores is by far the most widespread method of measuring sea-ice salinity. Gough et al. (2012) provide a thorough overview of statistical and physical sampling issues associated with ice-core salinity measurements. Due to the high horizontal heterogeneity of sea ice we will only use means over multiple ice cores. It is to be expected that the core measurements underestimate the salinity near the ocean interface due to brine loss (Notz and Worster, 2008).

After over a century of sporadic measurement campaigns beginning with Nansen's Fram expedition, the observational record of Arctic sea-ice salinity is sparse in time and space and no comprehensive compilation of the conducted measurements has been published in the last decades (e.g. Weeks and Lee, 1958; Cox and Weeks, 1974; Nakawo and Sinha, 1981; Eicken et al., 1995). We do not attempt to provide a

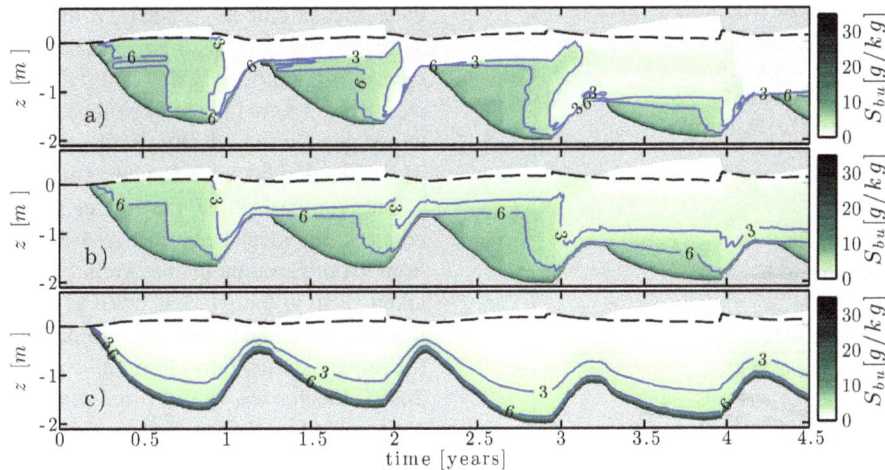

Figure 8. Salinity evolution of the (**a**) complex, (**b**) simple, and (**c**) prescribed salinity approach for one of the nine Arctic simulations forced with ERA-interim data from 75° N, 145° W. The salinity approaches are described in Sect. 2.5, and the model set-up is described in Sect. 4.1. The simulation time (x axis) begins on 1 July 2005. The dashed line marks the snow–ice boundary. The water surface is at $z = 0$.

rigorous comparison of model versus field data in this paper. Instead, we select three characteristic traits of sea-ice salinity to compare SAMSIM's results against. The three traits we compare against are the link between bulk salinity and ice thickness, the first-year salinity evolution from January to June, and the mean multiyear salinity profile from May to September.

4.3.1 Bulk salinity against thickness

The first trait we selected is the link between salinity and thickness which was studied by Cox and Weeks (1974) and Kovacs (1997). For the single growth season studied in Griewank and Notz (2013) the model results agreed well with the fit of Kovacs (1997) for first-year ice up to 2 m.

We separate first-year from multiyear ice before comparing the bulk salinity against thickness (Fig. 9). One simulation was singled out and highlighted, allowing the reader to track the progress over 4 years as the first-year ice turns into multiyear ice and becomes less saline over time. The simulation which was singled out is the same simulation as shown in Fig. 8.

Both first-year and multiyear ice show a distinctly different behaviour during growth and melt. The gradual transition from growth to melt is visible as a drop in bulk salinity at a constant thickness. A closer examination reveals that a slight thickness increase is visible in many simulations before ablation sets in. This bump in ice thickness arises from SAMSIM's definition of sea ice, which includes melting snow that has turned into slush (for details see Sect. 2.3.1). That this little bump appears at the end of the downward drop signals that until then no flushing has occurred. From that we can conclude that gravity drainage causes the drop in salinity.

Ice thinner than 20 cm has a wide spread in bulk salinity caused by melting and flooding at the onset of the growth

season. The simulated first-year ice thicker than 20 cm agrees well with the empirical results of Cox and Weeks (1974) and Kovacs (1997) during growth, with the model having only a slightly higher salinity. This bias is especially high for ice thinner than 0.5 m, which may be partially due to the fact that the underestimation of bulk salinity due to brine loss is higher for thin cores. After the onset of melt the bulk salinities are comparable to the estimates of Cox and Weeks (1974), which were based on a limited amount of cores that were at least 1 m thick.

As expected, multiyear sea ice shows a much smaller range of bulk salinities (Fig. 9b). During growth the bulk salinities show no coherent dependence on thickness, but during melt there appears to be a slight linear dependence on thickness. This is not far off from the estimation of Cox and Weeks (1974).

Both the modelled first-year and multiyear profiles are almost completely salt free at the end of the melt season. Neither Cox and Weeks (1974) or Kovacs (1997) included ice cores of such thin ice during melt, so we can not conclude from our comparison if this model behaviour agrees with reality. However, it is plausible that the 1-D nature of SAMSIM, which is built on the assumption that ice layers are totally homogeneous and all brine pockets are connected, would lead to an overestimation of desalination during flushing.

In conclusion, the modelled thickness–salinity relationship of growing first-year ice agrees well with the empirical fits to measurements of both Cox and Weeks (1974) and Kovacs (1997). For growing multiyear ice there is no one-to-one relationship between thickness and salinity, though growing multiyear ice tends to to be less salty the thicker it gets. The transition from growing to melting ice leads to a loss in bulk salinity at a constant thickness which is caused by

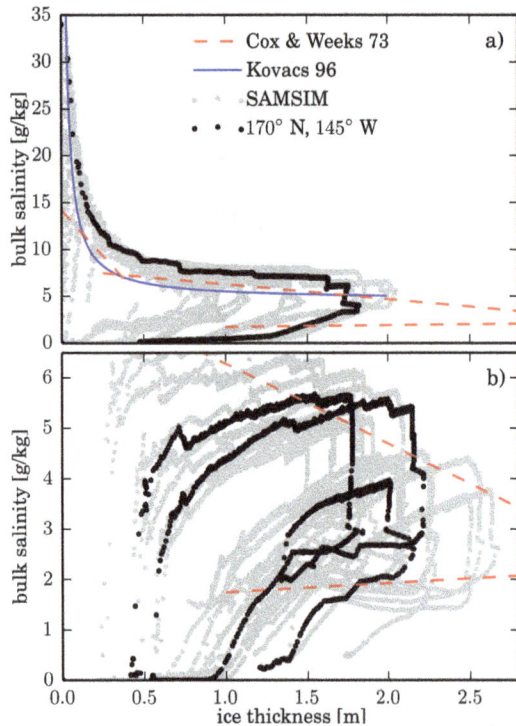

Figure 9. The vertically integrated vertical bulk salinity as a function of ice thickness for all reanalysis forced runs as described in Sect. 4.1. Each grey dot represents a 12-hourly snapshot. (**a**) Contains all 15 years of first-year ice and (**b**) contains all 21 years of multiyear ice in grey. Of all nine simulations, a single simulation is plotted in black (80° N, 90° E) to enable tracking the evolution over time. The blue curve in (**a**) is the empirical relationship for first-year ice published by Kovacs (1997) for ice up to 2 m. The red dashed lines mark the empirical linear relationships found by Cox and Weeks (1974) for growing (upper lines) and melting Arctic ice (lower line).

gravity drainage in the warming ice. Both melting first-year and multiyear ice show a weak linear dependence of salinity on thickness. In our simulations, the ice loses almost all its salt during melting; hence its mean salinity after the re-onset of growth is strongly affected by the salinity evolution of the newly forming ice.

4.3.2 First-year salinity evolution

The second trait of the modelled salinity we evaluate with core data is the evolution of first-year ice salinity from January until June. A longer time frame was not possible due to data availability; the period nonetheless allows us to study the salinity changes after gravity drainage is mostly restricted to the lower layers. We use the ice-core data taken as part of the Seasonal Ice Zone Observing Network and the Alaska Ocean Observing System by the sea-ice research group at the Geophysical Institute at the University of Fairbanks from 1999 to 2011 (Eicken et al., 2012). The great advantage of

these measurements, other than the sheer number of cores taken, is that by measuring repeatedly over a decade a large spread of conditions were captured. After rejecting all cores which did not include an ice thickness measurement or contained gaps in the salinity profile, a total of 86 first-year profiles remained between January and June.

The comparison of the model salinity with the Barrow cores is not ideal because SAMSIM is forced with conditions from throughout the Arctic, while the cores were all taken close to the Alaskan coast as part of an ongoing effort to understand and alleviate the impact of changing sea-ice on the human settlements along the coast (Druckenmiller et al., 2009). Ideally we would force our model with the forcing experienced by the ice measured at Barrow. This is not possible for a number of reasons. Firstly, we have no measurements of the oceanic heat flux. Secondly, although the ice was measured in Barrow we do not know where it was before the core was extracted. Most of the ice will likely have formed near the extraction points, but as illustrated by the multiyear ice cores taken in a region which is ice free in summer, there is a substantial amount of drift. Thirdly, we do not know when the ice was formed. The ice could have been formed during the initial freeze-up in fall, or later on in a lead or polynya. And lastly, due to the uncertainty in reanalysis data and the high variability in snow depth, we could not be certain that applying reanalysis forcing taken from the exact point where the cores grew would be correct. However, we do have a number of reasons to believe that the model-data comparison is useful. Firstly, the cores are taken over 12 years. This means that interannual variability will ensure that ice grown under a range of conditions was measured. Secondly, we show in the subsection on interannual salinity variability that the salinity variations resulting from atmospheric conditions are strongest in the uppermost 20 cm (Sect. 4.5). Because of this, we believe that the comparison should work well for the lower 80 % of the ice.

To compare the core profiles against the model profiles, both are first normalized to a depth of 0 to 1 before averaging over time. Often the salinity measurements did not extend all the way to the bottom of the ice, in which case the lowest measurement was extrapolated downwards. This extrapolation will contribute to the underestimation of salinity at the ice–ocean interface common to ice cores. We group the 86 core measurements into three bins of similar size based on the dates they were taken. The first bin spans from January to March (27 cores), the second from April to May (29 cores), and the final bin contains the remaining 29 cores taken in June.

As expected, even though the core profiles have a sharp increase of salinity at the ice–ocean interface they are still less saline at the ice–ocean boundary than SAMSIM (Fig. 10). Other than the top and bottom 10 % of the ice thickness, the simulated salinity profiles and the Barrow cores never differ by more than $2\,\mathrm{g\,kg^{-1}}$, which is in itself a mentionable model feat.

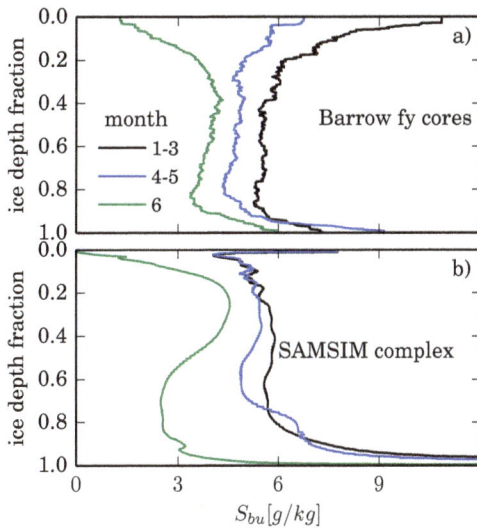

Figure 10. Time-averaged and vertically normalized salinity profiles from first-year ice cores (described in Sect. 4.3 and shown in **a**) and first-year ice from reanalysis forced simulations using the complex brine dynamic parametrizations (**b**). Both were averaged from January to March (1–3), April to May (4–5), and over June (6).

Other than the general agreement, this comparison highlights some limitations of SAMSIM's complex salinity approach. One of these limitations is that flushing and snow melt by design lead to a zero salinity at the surface once surface melt commences. Accordingly, the June SAMSIM profile is completely salt free at the surface while the core data show a salinity of roughly 1 g kg^{-1} at the surface (Fig. 10).

This total desalination at the surface is rooted in two of SAMSIM's design choices. The first design choice is that the snow layer in SAMSIM has zero salinity and that melting snow forms slush which is treated as sea ice. Accordingly, when snow melts the top ice layer will consist of melted snow slush and be absolutely salt free. The second design choice which leads to zero salinity at the surface is the implementation of flushing in SAMSIM. One of the core assumptions of the complex flushing parametrization is that the meltwater leaving the top ice layer has a brine salinity determined by the liquidus relationship. Accordingly, as the brine salinity of the top ice layer is by definition always higher than the bulk salinity of the top ice layer, flushing always results in a salinity decrease at the surface. This desalination quickly desalinates the surface once flushing commences as shown by the idealized flushing experiments (Sect. 3). While the freshly desalinated ice can freeze solid and thus inhibit any further flushing, this can only occur in the ice if the underlying ice is sufficiently cold as in the idealized example. At the surface this could also occur but only if there were a negative atmospheric heat flux to remove sufficient energy from the top layer to overcome the latent heat released during freezing.

The second distinct difference between model and core salinity is that SAMSIM has a high surface salinity with a very strong salinity gradient before the onset of melt (profiles from January to May, Fig. 10b). The sharp salinity gradient which occurs in the top few model layers could be a numerical artifact arising from SAMSIM's semi-adaptive grid. No matter which resolution is used, the initial ice growth occurs when only a few layers are active. This issue was investigated in our previous paper when comparing to freezing plate experiments conducted in the lab, but available data were insufficient to make any conclusions (Griewank and Notz, 2013). A different explanation is snow wicking, a process which transfers some of the surface salinity into the snow layer. In the model wicking only occurs when meltwater forms in the top ice layer beneath snow. The discrepancy between model and data at the surface could also arise from the neglect of frazil or pancake ice formation in SAMSIM. In frazil and pancake ice the wave motion and turbulence cause brine motion not captured by the gravity drainage parametrization which could desalinate the initial ice before it freezes into a static structure.

The third discrepancy between the cores and SAMSIM is that the bulk salinity in the upper 40 % does not change substantially from the period of January–March to that of April–May in the model. There are many possible explanations for this discrepancy, such as the non-ideal comparison itself (see second paragraph Sect. 4.3.2), insufficient simulations or core measurements, and errors of the core-salinity measurements. Another explanation is that the model is unable to simulate the salinity evolution correctly close to the surface during winter. A likely candidate to explain that the salinity remains constant near the surface is that the gravity drainage parametrization desalinates too quickly during growth. The modelled salinity is quickly reduced to 5 g kg^{-1} after which it stabilizes, instead of a weaker initial desalination followed by a gradual desalination over time (Fig. 10). The neglect of frazil and pancake ice formation in SAMSIM could again be an issue since turbulent conditions during the initial freeze-up would influence both the microstructure and permeability of the surface ice. It is also possible that the freeboard plays an important role, and that brine from above the waterline drains away by an unknown mixture of gravity drainage or flushing. The differences between the cores and SAMSIM as well as our poor understanding of what happens during flooding indicate that unknown, yet relevant, brine movements may occur at the ice–snow interface. It is also possible that the gravity drainage parametrization has some limitations. Despite the indirect model-to-data comparison and the three discrepancies in the salinity evolution discussed, SAMSIM successfully captures the general shape and magnitude of the three core-derived salinity profiles.

4.3.3   Multiyear salinity profile

The final and best-documented trait we select to compare is the mean multiyear salinity profile. The most widely used multiyear profile in the sea-ice modelling community is

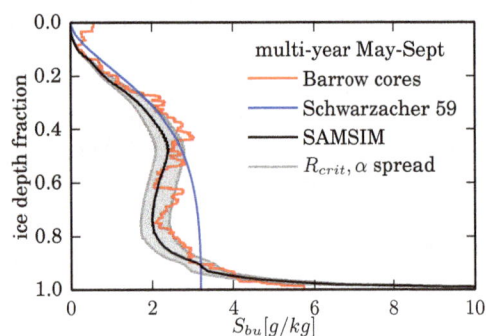

Figure 11. May to September mean of vertically normalized multiyear salinity profiles of reanalysis forced simulations using the complex brine dynamic parametrizations. Schwarzacher 59 refers to the fitted profile of Schwarzacher (1959), and Barrow cores refers to the multiyear ice cores taken by the Alaska Ocean Observing System from 1999 to 2011 (Eicken et al., 2012). The R_{crit}, α spread shows the SAMSIM profile using the two non-default values of the gravity drainage parameters obtained from the optimization process (see Sect. 2.2). Left line: $\alpha = 0.000681$, $R_{crit} = 3.23$. Right line: $\alpha = 0.000510$, $R_{crit} = 7.10$. The area between the two simulations is shaded in light grey.

based on 40 ice cores taken at the drifting ice station A in 1958 (Schwarzacher, 1959) from May to September. Although later studies have incorporated additional measurements (e.g. Cox and Weeks, 1974; Eicken et al., 1995), the basic shape has remained similar. The fitted bulk salinity profile of Schwarzacher (1959) on a normalized vertical coordinate z from zero to one,

$$S_{bu}(z) = 1.6(1 - \cos)(\pi z^{\frac{0.407}{0.573+z}}), \qquad (16)$$

is used in the 1-D models of Maykut and Untersteiner (1971) and Bitz and Lipscomb (1999). As the Schwarzacher cores were all taken from May to September, and the eight multi-year cores from Barrow were also taken in summer, we compare the normalized Barrow cores and Schwarzacher profile against the mean of SAMSIM from May to September. We did not compare directly to the Schwarzacher data as they were not easily available and the data displayed in Schwarzacher (1959) were not regularized before averaging. Although the fitted Schwarzacher profile has a $3.2\,\mathrm{g\,kg^{-1}}$ salinity at the ice–ocean interface (Fig. 11), an increase is clearly visible in the measurements, similar to the salinity increase of the eight multiyear salinity cores taken at Barrow. Due to this ignored increase and the repeatedly mentioned salinity loss in cores, we only compare to the upper 90 % the Schwarzacher profile. Although this comparison of SAMSIM to field data is far from perfect, it is the closest we can come to evaluating the flushing parametrization until controlled laboratory measurements are available.

We compare the May-to-September mean of all normalized multiyear SAMSIM profiles to the profile of Schwarzacher (1959) and the Barrow cores, which all share

a similar magnitude and shape in the upper 90 % (Fig. 11). Both SAMSIM and the Barrow cores have a slight maximum at a depth of 40 %, which indicates that the complex flushing parametrization predicts the desalination depth reasonably correctly. The good agreement between SAMSIM and ice-core data is a very positive result given that the complex flushing parametrization contains large parameter uncertainties and was developed from scratch without any data available to tune the free parameter β.

Between the depth fractions of 0.5 and 0.8, SAMSIM and the Barrow cores show a slight salinity decrease with depth while the Schwarzacher profile has a slight increase (Fig. 11). The differences between the model and the ice-core data are of similar magnitude to the differences caused by different values of α and R_{crit} obtained from the optimization process mentioned in Sect. 2.2. The Barrow cores and SAMSIM both have a sharp salinity increase in the lowest 10 %. That the model is saltier at the ice–ocean boundary is expected due to brine loss during coring and a lower spatial resolution of the measurements compared to the model.

4.3.4 Summary

According to SAMSIM there is a clear link between ice thickness and bulk salinity in growing first-year ice as described by Kovacs (1997). However, after the ice stops growing, gravity drainage in the warming ice causes a thickness independent desalination. Both melting first-year and multiyear ice show an approximately linear dependence of bulk salinity on ice thickness as suggested by Cox and Weeks (1974). The modelled ice loses almost all its salinity, a feature against which we do not have any core data to evaluate. The mean multiyear salinity profile of SAMSIM from May to September agrees well with the core data of Schwarzacher (1959) and from Barrow. The salinity evolution in first-year ice in SAMSIM is comparable to ice-core measurements at Barrow (Eicken et al., 2012). However, in contrast to the Barrow core data, the modelled salinity close to the ice surface remains constant from the period of January–March to that of April–May, indicating that in reality, brine fluxes occur close to the surface and are poorly captured by the complex set of parametrizations.

All comparisons between SAMSIM and ice-core data show that SAMSIM captures the general salinity evolution well, both qualitatively and quantitatively. Keep in mind that no tuning was used to reach these results and that all parametrizations were developed without any field data. Additionally, all parametrizations were developed separately, with no regard to possible interactions.

So far we have only evaluated characteristics of the Arctic simulations that we could compare against ice cores. From the comparison to ice cores we conclude that our parametrizations and understanding of desalination processes are sufficient to use SAMSIM as a tool to study Arctic sea ice beyond reproducing ice-core salinity.

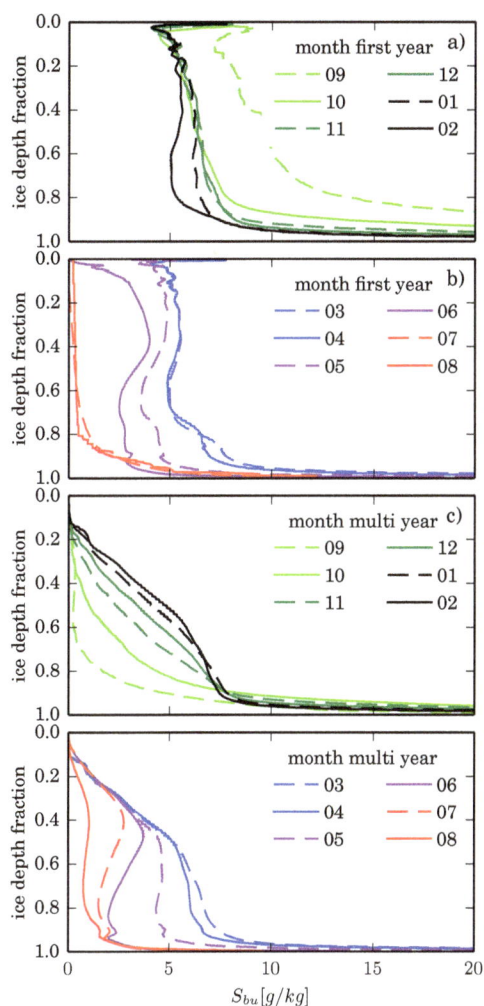

Figure 12. Monthly mean of vertically normalized salinity profiles of reanalysis forced simulations using the complex brine dynamic parametrizations as described in Sect. 4.1. The simulations were split into annual cycles beginning in September (month 9) and sorted into 15 years of first-year ice (**a** and **b**) and 21 years of multiyear ice (**c** and **d**). The corresponding ice thickness of the monthly means are shown in Fig. 13.

4.4 Mean salinity evolution

In this subsection we analyze the mean salinity evolution of the complex approach. In total, the model simulations yield 36 years of sea-ice growth and melt. Of those 36 years, 21 years are multiyear ice and 15 are first-year ice. Of the 15 years of first-year ice, 8 years end in open water while 7 form multiyear ice in the following year.

To process and visualize the salinity evolution we first normalize the depth of all salinity profiles of the model output between 0 and 1. This allows averaging over multiple normalized profiles and it simplifies comparing profiles of varying thicknesses. To resolve the mean annual cycle we sort all first-year and multiyear profiles into monthly bins beginning

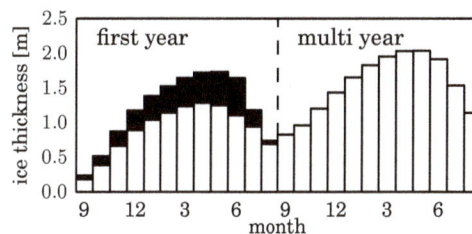

Figure 13. The white columns show the thickness of all monthly mean salinity profiles shown in Fig. 12. The black columns represent only first-year ice which evolves into multiyear ice the following year. To be included in the monthly average ice must be present, meaning that model output of ice-free water with an ice thickness of zero is excluded from the mean.

in September, which we then average (Fig. 12). A side effect of this averaging approach is that when there is no ice in the model output, this output does not affect the mean salinity profile. As a consequence, the mean August profile consists mostly of first-year ice which will turn into multiyear ice the following year, and there is a smooth transition from the August first-year profile to the September multiyear profile. This selection effect is clearly visible when comparing the mean ice thickness of all first-year simulations excluding ice-free output against the mean thickness of first-year ice which turns into multiyear ice next September (Fig. 13).

During the growth season the salinity of the first-year ice decreases to $5\,\mathrm{g\,kg^{-1}}$ after about 2 months with a sharp increase to $10\,\mathrm{g\,kg^{-1}}$ in the upper 5 % of the ice thickness (Fig. 12a). The salinity profile remains pretty stable between November and April, followed by a slight desalination in May at the onset of melt. The desalination accelerates during June and July until the upper 80 % of the ice has a very low salinity below $2\,\mathrm{g\,kg^{-1}}$ (Fig. 12b). The influence of flushing is clearly visible in the almost total loss of salt at the surface from June onwards. Although there is only little and indirect experimental evidence of gravity drainage occurring as the ice warms (e.g. Widell et al., 2006; Jardon et al., 2013) the salinity reduction in the lower half of the ice from April to June shows that gravity drainage is active in SAMSIM during the onset of melt. This desalination is consistent with results from idealized experiments we conducted that show a reduction of bulk salinity from above 5 to below $3\,\mathrm{g\,kg^{-1}}$ from gravity drainage when sea-ice begins to warm (Griewank and Notz, 2013).

At the end of the melt season the multiyear ice salinity is lowest (Fig. 12c). While the surface salinity remains low the newly formed ice at the bottom retains over $5\,\mathrm{g\,kg^{-1}}$. During the melt season the lower half of the ice is desalinated by gravity drainage while flushing maintains the low surface salinity. That this desalination is not only due to the loss of the saltier lower layers through melt is visible in the curve that develops in the lower half of the normalized profile as flushing by itself would lead to an increase in salin-

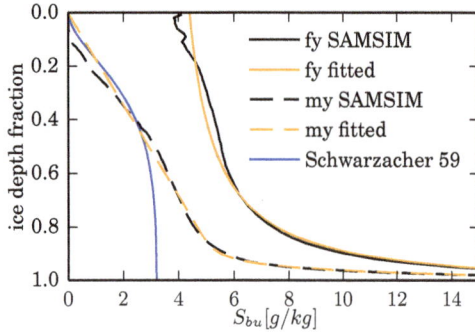

Figure 14. Yearly mean first-year (fy) and multiyear (my) sea-ice salinity profiles of SAMSIM using the complex parametrization. The fitted analytical functions of the profiles listed in Sect. 4.4 are added in orange. Although the profile of Schwarzacher (1959) is summer biased (see Sect. 4.3.3), we have included it as a reference.

ity (Fig. 6). That gravity drainage can act in such a manner is visible in the idealized experiment in which gravity drainage was enabled (Fig. 7a). This curve is also visible in the Barrow core data shown in Fig. 11. With the exception of the gravity drainage during melt, the overall multiyear salinity agrees well with expectations already voiced by Cox and Weeks (1974).

For readers interested in analytical approximations of the mean first-year and multiyear profiles we offer two functions, $S_{bu, fy}(z)$ and $S_{bu, my}(z)$. Both are a function of the normalized ice depth $0 \leq z \leq 1$ and are shown in Fig. 14 along with the mean SAMSIM profiles. The fitted first-year ice profile is

$$S_{bu, fy}(z) = \frac{z}{a + bz} + c \tag{17}$$

for $a = 1.0964$, $b = -1.0552$, and $c = 4.41272$, and the fitted multiyear ice profile is

$$S_{bu, my}(z) = \frac{z}{a} + \left(\frac{z}{b}\right)^{\frac{1}{c}} \tag{18}$$

with $a = 0.17083$, $b = 0.92762$, and $c = 0.024516$.

The transition from first-year to multiyear ice over the melt season can be approximated by a time-dependent combination of the two profiles in the form

$$S_{bu}(z, t) = (1 - t) \cdot S_{bu, my}(z) + t \cdot S_{bu, my}(z), \tag{19}$$

where $t = 0$ at the beginning of the melt season in June and $t = 1$ at the onset of growth in September.

4.5 Variability

While the previous subsection studied the mean salinity properties, in this subsection we will take a brief look at the salinity variability in SAMSIM using the complex approach. The model variability arises from two sources, the

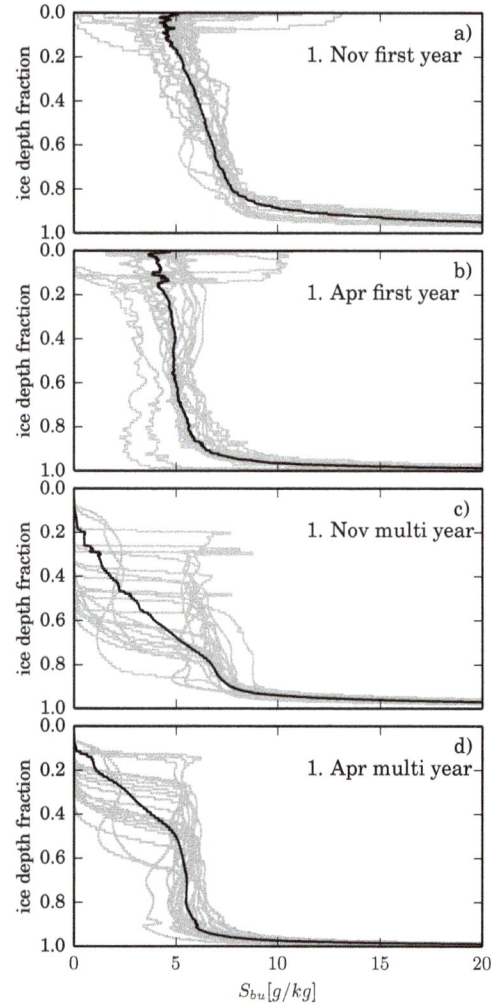

Figure 15. Vertically normalized salinity profiles of the reanalysis forced simulations (described in Sect. 4.1) using the complex salinity parametrizations on 1 November (**a** and **c**) and 1 April (**b** and **d**). First-year ice (**a** and **b**) and multiyear ice (**c** and **d**) are shown separately. The grey lines are the individual model realizations and the black line is the average overall profiles.

main one being the atmospheric forcing. Although the location at which the reanalysis data was selected has the largest impact, interannual variability ensures that all 36 years of simulated sea ice have a unique forcing. The second source for variability is the initial ice conditions at the beginning of the growth season. This second source only applies to the 21 years of multiyear ice since all first-year ice grows from ice-free water. The variance of the model can not be directly compared to ice-core variability, because the variability in ice cores additionally contains a large amount of variability due to small-scale horizontal heterogeneity (Gough et al., 2012).

To visualize the variability we have plotted all normalized salinity profiles at two dates in time as well as the mean overall profiles at that time point in Fig. 15. With few exceptions the first-year ice only deviates a few g kg^{-1} from the mean in the lowest 80 % of the ice. However, at the surface the spread is much higher, with values reaching from 0 to above $10 \, g \, kg^{-1}$ (see Fig. 15a and b). There are two main reasons for the higher variability at the surface. The first is that after 10–20 cm of ice has formed, the variability of the atmospheric forcing is severely dampened before it reaches the ice–ocean interface. As a result, the ice formed after the initial 10–20 cm grows under roughly similar conditions in all simulations. The second reason is that flooding and flushing both occur mainly at the surface of the ice. That such a similar high variability near the surface is not visible in the multiyear ice is because both processes are far less likely to occur in multiyear ice during the winter than in first-year ice. Farther south, where first-year ice seldom survives the melt season, rainfall and above-freezing surface temperatures occur during the growth season, both of which can cause flushing. As the first-year ice is less thick, strong snowfall that slows ice growth can lead to flooding more easily than in multiyear ice.

As all multiyear ice has experienced at least one melt season, it is not surprising that multiyear simulations have a salinity of zero at the surface (Fig. 15c and d). That all 21 years have zero surface salinity shows that flooding of multiyear ice does not occur in any of the simulations. Most of the variability in multiyear ice arises from the different ice thickness and salinity of the ice at the end of the melt season. The sudden salinity increases with depth arise from sudden quick growth in the beginning of the growth season beneath almost completely desalinated ice for many simulations between 0.2 and 0.6 in the November profiles (Fig. 15c). This growth can be quicker than in first-year ice of similar thickness due to the following reasons. The first reason is that by the time first-year ice reaches the same thickness, it has likely accumulated an insulating snow layer which slows ice growth. Secondly, the fresher multiyear ice has a higher thermal conductivity and lower thermal capacity which enhances heat transport from the ice–ocean interface to the ice–atmosphere boundary.

Over the next half-year the profiles are smoothed out and the salinity sinks to $7 \, g \, kg^{-1}$ or lower except in the lowest 10 % (Fig. 15d). Visible in both first-year ice and multiyear ice is that the salinity in the lowest layers is higher in November during ice growth than in April.

In conclusion, the variability in first-year ice is strongest at the surface and arises from the atmospheric forcing, while the variability in multiyear ice is mostly due to the thickness of the ice at the beginning of the growth season. A third possible source of variance is the variation in the oceanic heat flux. This is not included in this study as all simulations share the same prescribed annual cycle of oceanic heat flux.

5 Impact of parametrizing salinity

While the previous section focused on the salinity evolution and the processes which drive it, this section aims to quantify how parametrizing salinity affects sea-ice properties relevant to the climate system. We address this question, which is highly relevant to modellers seeking to improve climate models, by using the same runs used in the previous section (see Sect. 4.1). In this paper we only study the physical properties of sea-ice. Biogeochemical properties and feedbacks can not be assessed with SAMSIM currently.

To asses the total impact of parametrizing salinity in a climate model it is not sufficient to quantify the impact on the sea ice itself. It is also necessary to determine resulting feedbacks with the ocean and atmosphere. So far the only coupled model featuring a partially parametrized salinity is the NEMO-LIM model, which uses a prescribed atmospheric forcing. Using the NEMO-LIM model, Vancoppenolle et al. (2009) found that the large-scale sea-ice mass balance and the upper-ocean characteristics are quite sensitive to sea-ice salinity. Salinity variations introduced to NEMO-LIM increased sea ice volume by up to 28 % in the Southern Hemisphere because changes to ice–ocean interactions stabilized the ocean, leading to a reduced oceanic heat flux. In the Arctic the ocean stratification was not influenced by the implemented sea-ice variations; however, Vancoppenolle et al. (2009) discovered increases in ice thickness of up to 1 m due to changes of the sea-ice thermal properties.

From Vancoppenolle et al. (2009) we conclude that in the Arctic the oceanic feedbacks will be small due to the stable stratification of the Arctic Ocean. Although the atmospheric feedbacks remain unknown, we can use SAMSIM's more advanced salinity parametrizations with a much higher spatial and temporal resolution to take a more detailed look at how the salinity evolution affects the sea ice.

To quantify the impact of parametrizing salinity we compare quantities of the nine reanalysis forced simulations using the three salinity approaches introduced in Sect. 2.5. The specific quantities we use based on their importance for the climate system are the same four used in Griewank and Notz (2013). These are the ice thickness, the freshwater column stored in the ice and snow, the thermal resistance R_{th}, and the total enthalpy H integrated over the whole ice and snow column. Each of the nine runs is evaluated separately over the full 4.5 simulation years to ensure that opposing biases at different locations do not average out.

The metrics we use to compare the time-dependent quantities against each other are a time-integrated ratio and a time-integrated, weighted absolute difference. The ratio r of the quantity $x_i(t)$ using the salinity approach i to the same quantity using the different salinity approach $x_j(t)$ over the simulated 4.5 years is calculated as

$$r = \frac{\int_{t=0}^{t=4.5} {}^a x_i(t) \, \mathrm{d}t}{\int_{t=0}^{t=4.5} {}^a x_j(t) \, \mathrm{d}t}. \tag{20}$$

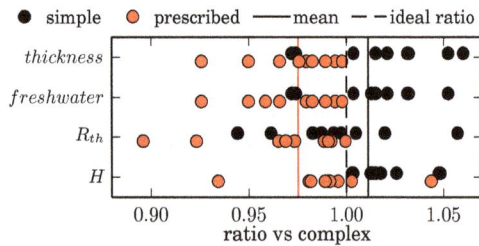

Figure 16. Ratios of the time-integrated ice thickness, freshwater column, thermal resistance R_{th}, and enthalpy H of the simple and the prescribed SAMSIM salinity approach compared to the complex approach (details in Sect. 5). The ratios were calculated separately for each of the nine reanalysis forced simulations over 4.5 years. Each dot shows the ratio of a specific simulation, while the lines show the mean overall runs and quantities.

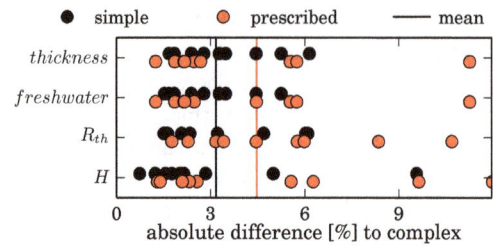

Figure 17. Time-integrated absolute differences of the ice thickness, freshwater column, thermal resistance R_{th}, and enthalpy H of simulations using the simple and prescribed SAMSIM salinity approach compared to simulations using the complex approach (details in Sect. 5). The absolute differences were calculated separately for each of the nine reanalysis forced simulations over 4.5 years. Each dot shows the ratio of a specific simulation, while the lines show the mean overall runs and quantities.

The second metric used, the weighted absolute difference "d", is determined by

$$d = \frac{\int_{t=0}^{t=4.5a} x_i(t) - x_j(t) \, dt}{\int_{t=0}^{t=4.5a} x_j(t) \, dt} \tag{21}$$

and is a measure of how large the differences are between the two quantities at each time step compared to the total value of the second quantity. The ratio is chosen to indicate if and by how much x_i is greater or smaller than x_j over time, while the absolute difference is chosen to detect compensating errors not apparent in the ratio. We quantify the impact by comparing the simple and prescribed approach against the complex approach.

The computed ratios for each simulation reveal that the prescribed approach with few exceptions leads to a lower ice thickness, freshwater column, thermal resistance, and total enthalpy than the complex approach (Fig. 16). Ratios range from 0.90 to 1.05. The mean of all ratios and quantities of the prescribed approach is 0.975; accordingly, the quantities of the complex approach are 2.5 % higher on average. The ratios of the simple approach have a slightly lower spread and are on average higher with a mean of 1.012. So on average the simple approach overestimates half as much as the prescribed approach underestimates.

The absolute differences paint a similar picture, with the prescribed approach having a slightly larger spread with differences up to 12 % (Fig. 17). On average the simple approach has lower differences with a mean of 3.3 % in comparison to the prescribed mean of 4.5 %. Because the absolute differences are roughly 2 times larger than the ratios, we can deduce that roughly half of the discrepancy between two simulations stems from a bias in one direction.

Given that the prescribed approach does not distinguish growing from melting ice and that the prescribed profile was not optimized or tuned in any way, the simulated ice properties using the prescribed approach are unexpectedly close to the complex approach. We also expected the prescribed ap-

proach to have a wider spread when compared to the complex approach, because the prescribed approach treats all ice the same regardless of its history while the complex approach is dependent on previous conditions (as visible in Fig. 8).

6   Summary and conclusions

We have incorporated surface melt, flooding, and flushing into SAMSIM. In contrast to the thermodynamic models derived from Maykut and Untersteiner (1971), such as Bitz and Lipscomb (1999) and Huwald et al. (2005b), surface melt in SAMSIM is implemented as a two-stage process. The first stage is the conversion of snow to slush followed by the second stage of surface ablation by meltwater runoff. All desalination processes are parametrized in two different ways in SAMSIM. The *complex* parametrizations calculate brine fluxes and are physically consistent, while the *simple* parametrizations attempt to imitate the effects of the complex parametrizations with less numerical overhead.

SAMSIM is the only 1-D thermodynamic sea-ice model other than the 1-D LIM model of Vancoppenolle et al. (2007) which has a fully prognostic salinity. In contrast to the flushing parametrization of Vancoppenolle et al. (2007), the complex flushing parametrization of SAMSIM explicitly includes both horizontal and vertical brine movements. A detailed discussion of why the complex gravity drainage parametrization of SAMSIM agrees better than the gravity drainage of LIM 1-D with both theoretical and numerical expectations is included in Griewank and Notz (2013). The complex flooding parametrization based on the results of Maksym and Jeffries (2000) is an ad hoc solution as the current understanding of flooding is insufficient to develop a more realistic parametrization. Nevertheless, SAMSIM is the first 1-D model to include flooding as well as flushing and gravity drainage, and the flooding parametrization does capture the basics of flooding and produces snow ice with reasonable salinities in a physically consistent manner.

Under idealized conditions, the complex flushing parametrization leads to an increase of salinity close to the ice–ocean interface if gravity drainage is deactivated. Although we do not have data available to determine optimal values of the ratio of vertical to horizontal hydraulic resistance β, our idealized experiments show that the flushing parametrization is only weakly sensitive to changes close to the default values. The vertical resolution of SAMSIM also only has a small impact on the flushing parametrization.

We study the salinity evolution of Arctic sea ice using 36 years of SAMSIM output. To imitate Arctic conditions we force SAMSIM with ERA-interim reanalysis precipitation and radiation fluxes from throughout the Arctic. The 36 years are separated into 15 years of first-year and 21 years of multiyear sea-ice and then compared against ice-core data. The mean multiyear salinity profile of Schwarzacher (1959) and the salinity evolution of first-year ice cores from Barrow, Alaska, agree well with SAMSIM simulations. However, while the first-year ice-core salinity at the surface decreases from January to May, the modelled salinity at the surface remains constant until the onset of melt. This discrepancy indicates that brine fluxes close to the ice–snow boundary are captured poorly by SAMSIM. Possible reasons for this discrepancy are discussed in detail in Sect. 4.3.2.

We deduce from the 36 years of simulated sea-ice that ice thickness is a good indicator of bulk salinity for growing first-year ice. The model results agree well with the empirical results of Cox and Weeks (1974) and Kovacs (1997). That the modelled bulk salinities of thin ice are higher than the ice-core data is at least partially due to the fact that brine loss during coring is especially high from thin and more saline ice. The transition from growth to melt is accompanied by a 1.5–4 $g\,kg^{-1}$ reduction of bulk salinity caused by gravity drainage before the onset of flushing. This onset of gravity drainage as the ice warms is consistent with earlier findings by Griewank and Notz (2013) and Jardon et al. (2013). The onset contradicts the general melt evolution depicted by Eicken et al. (2002) in which gravity drainage sets in at the end of the melt season. In general, thicker multiyear ice tends to be fresher, but during growth the bulk salinity increases with thickness. During melt both multiyear and first-year ice have a linear relationship of bulk salinity and thickness as Cox and Weeks (1974) hypothesized on a limited set of cores, but the slope of the linear relationship in the model is steeper than that proposed by Cox and Weeks (1974).

Our results show that the largest interannual variations of salinity occur at the surface of first-year ice and are caused by rain, surface melt, and flooding. In contrast, the lower 80 % of the salinity profile of first-year ice are similar to each other despite being forced with reanalysis data taken from different locations and years. The multiyear ice profiles vary depending on the ice thickness at the onset of growth and become more similar over the growth season.

We compare the ice thickness, freshwater column, thermal resistance, and total stored energy of the nine 4.5-year simulations of Arctic sea-ice using the three different salinity approaches against each other. Although certain quantities differ by up to 12 % for a specific simulation, on average the differences between the complex salinity approach and the other approaches are below 5 %. The simple approach has a roughly 30 % smaller difference compared to the complex approach than the prescribed approach (Fig. 17) and a roughly 50 % better ratio than the prescribed approach (Fig. 16).

Given that the strong arctic halocline should prohibit strong ice–ocean feedbacks, we expect that fully parametrizing the temporal sea-ice salinity evolution in the Arctic will not have a large effect on sea-ice thermodynamics in climate models. We expect that parametrized–prescribed hybrids, such as that proposed by Vancoppenolle et al. (2009), which parametrizes the evolution of the bulk salinity of the whole ice column and prescribes an empirical salinity profile based on the bulk salinity, will reproduce the dominant thermodynamic effects of the sea-ice salinity evolution. Prescribing age- and thickness-dependent salinity profiles such as those shown in Fig. 14 is also a viable alternative. The multiyear profile of Schwarzacher (1959) underestimates the mean salinity profile because it is based on cores taken from May to September, during which the salinity is lower (Fig. 12). A smooth transition from first-year to multiyear ice can be achieved by linearly transitioning from the first-year to the multiyear profile as discussed in Sect. 4.4. Further refinement can be achieved by taking into account the annual cycle (Fig. 12), the ice thickness (Fig. 9), and the sea-ice location.

Comparisons to laboratory and field salinity measurements have shown that the parametrized brine fluxes in SAMSIM are a reasonable approximation of reality. SAMSIM's semi-adaptive grid is convenient when studying processes which occur close to the ice–atmosphere or ice–ocean boundary, as it avoids numerical diffusion through layer advection in the surface and bottom layers. All dissolved tracers in brine can be easily advected similar to salt, and the gas volume fraction in each layer can be used to compute outgassing and uptake. Thanks to these properties SAMSIM is a valuable tool to study small-scale thermodynamic and other aspects of sea ice which are affected by brine dynamics such as sea-ice biology.

Acknowledgements. We would like to thank Thorsten Mauritsen, Jochem Marotzke, and our anonymous reviewers for commenting on the manuscript, the ECMWF for providing the ERA-interim reanalysis data freely for research, and the Barrow Sea Ice Observatory for hosting their ice-core data openly online.

Edited by: E. Larour

References

Bitz, C. M. and Lipscomb, W. H.: An energy-conserving thermo-dynamic model of sea ice, J. Geophys. Res., 104, 15669–15677, doi:10.1029/1999JC900100, 1999.

Coleou, C. and Lesaffre, B.: Irreducible water saturation in snow: experimental results in a cold laboratory, Ann. Glaciol., 26, 64–68, 1998.

Cox, G. F. and Weeks, W. F.: Salinity variations in sea ice, J. Glaciol., 13, 109–120, 1974.

Druckenmiller, M. L., Eicken, H., Johnson, M. A., Pringle, D. J., and Williams, C. C.: Toward an integrated coastal sea-ice obser-vatory: System components and a case study at Barrow, Alaska, Cold Reg. Sci. Technol., 56, 61–72, 2009.

Eicken, H., Lensu, M., Lepparanta, M., Tucker, W. B., Gow, A. J., and Salmela, O.: Thickness, Structure, and Properties of Level Multiyear Ice In the Eurasian Sector of the Arctic-ocean, J. Geo-phys. Res., 100, 22697–22710, doi:10.1029/95JC02188, 1995.

Eicken, H., Krouse, H., Kadko, D., and Perovich, D.: Tracer studies of pathways and rates of meltwater transport through Arctic summer sea ice, J. Geophys. Res., 107, 8046, doi:10.1029/2000JC000583, 2002.

Eicken, H., Gradinger, R., Kaufman, M., and Petrich, C.: Sea-ice core measurements (SIZONET), UCAR/NCAR – CISL – ACADIS, doi:10.5065/D63X84KG, 2012.

Flocco, D. and Feltham, D. L.: A continuum model of melt pond evolution on Arctic sea ice, J. Geophy. Res., 112, C08016, doi:10.1029/2006JC003836, 2007.

Freitag, J.: The hydraulic properties of Arctic sea ice – impli-cations for the small scale particle transport, Berichte zur Po-larforschung, Alfred-Wegener Institut füt Polar- und Meeres-forschung, Bremerhaven, 325, 1999 (in German).

Gough, A. J., Mahoney, A. R., Langhorne, P. J., Williams, M. J. M., and Haskell, T. G.: Sea ice salinity and structure: A winter time series of salinity and its distribution, J. Geophys. Res., 117, C03008, doi:10.1029/2011JC007527, 2012.

Griewank, P.: A 1D model study of brine dynamics in sea ice, PhD thesis, University of Hamburg, Germany, 2014.

Griewank, P. J. and Notz, D.: Insights into brine dynamics and sea ice desalination from a 1-D model study of gravity drainage, J. Geophys. Res., 118, 3370–3386, doi:10.1002/jgrc.20247, 2013.

Hunke, E. C., Notz, D., Turner, A. K., and Vancoppenolle, M.: The multiphase physics of sea ice: a review for model developers, The Cryosphere, 5, 989–1009, doi:10.5194/tc-5-989-2011, 2011.

Huwald, H., Tremblay, L. B., and Blatter, H.: Reconciling different observational data sets from Surface Heat Budget of the Arc-tic Ocean (SHEBA) for model validation purposes, J. Geophys. Res., 110, C05009, doi:10.1029/2003JC002221, 2005a.

Huwald, H., Tremblay, L.-B., and Blatter, H.: A multilayer sigma-coordinate thermodynamic sea ice model: Validation against Sur-face Heat Budget of the Arctic Ocean (SHEBA)/Sea Ice Model Intercomparison Project – Part 2 (SIMIP2) data, J. Geophys. Res., 110, C05010, doi:10.1029/2004JC002328, 2005b.

Jardon, F., Vivier, F., Vancoppenolle, M., Lourenco, A., Bouruet-Aubertot, P., and Cuypers, Y.: Full-depth desalination of warm sea ice, J. Geophys. Res., 118, 1–13, 2013.

Jeffery, N., Hunke, E. C., and Elliott, S. M.: Modeling the trans-port of passive tracers in sea ice, J. Geophys. Res., 116, C07020, doi:10.1029/2010JC006527, 2011.

Jeffries, M. O., Roy Krouse, H., Hurst-Cushing, B., and Maksym, T.: Snow-ice accretion and snow-cover depletion on Antarctic first-year sea-ice floes, Ann. Glaciol., 33, 51–60, 2001.

Kovacs, A.: Sea Ice – Part 1: Bulk Salinity Versus Ice Floe Thick-ness, CRREL Report, 96, 1–17, 1997.

Lietaer, O., Deleersnijder, E., Fichefet, T., Vancoppenolle, M., Comblen, R., Bouillon, S., and Legat, V.: The vertical age pro-file in sea ice: Theory and numerical results, Ocean Modell., 40, 211–226, 2011.

Light, B., Grenfell, T. C., and Perovich, D. K.: Transmis-sion and absorption of solar radiation by Arctic sea ice during the melt season, J. Geophys. Res., 113, C03023, doi:10.1029/2006JC003977, 2008.

Maksym, T.: Brine percolation, flooding and snow ice formation on Antarctic sea ice, PhD thesis, University of Alaska Fairbanks, USA, 2001.

Maksym, T. and Jeffries, M. O.: A one-dimensional percolation model of flooding and snow ice formation on Antarctic sea ice, J. Geophys. Res., 105, 26313–26331, 2000.

Maksym, T. and Jeffries, M. O.: Phase and compositional evolution of the flooded layer during snow-ice formation on Antarctic sea ice, Ann. Glaciol., 33, 37–44, 2001.

Maykut, G. A. and Untersteiner, N.: Some Results From A Time-dependent Thermodynamic Model of Sea Ice, J. Geophys. Res., 76, 1550, doi:10.1029/JC076i006p01550, 1971.

Nakawo, M. and Sinha, N. K.: Growth-rate and Salinity Profile of 1st-year Sea Ice In the High Arctic, J. Glaciol., 27, 315–330, 1981.

Notz, D.: Thermodynamic and fluid-dynamical processes in sea ice, PhD thesis, University of Cambridge, Cambridge, UK, 2005.

Notz, D. and Worster, M. G.: In situ measurements of the evo-lution of young sea ice, J. Geophys. Res., 113, C03001, doi:10.1029/2007JC004333, 2008.

Notz, D. and Worster, M. G.: Desalination processes of sea ice revisited, J. Geophys. Res., 114, C05006, doi:10.1029/2008JC004885, 2009.

Notz, D., McPhee, M., Worster, M., Maykut, G., Schlünzen, K., and Eicken, H.: Impact of underwater-ice evolution on Arctic summer sea ice, J. Geophys. Res., 108, 3223, doi:10.1029/2001JC001173, 2003.

Pedersen, C. A., Roeckner, E., Luthje, M., and Winther, J. G.: A new sea ice albedo scheme including melt ponds for ECHAM5 general circulation model, J. Geophys. Res., 114, D08101, doi:10.1029/2008JD010440, 2009.

Polashenski, C., Perovich, D., and Courville, Z.: The mechanisms of sea ice melt pond formation and evolution, J. Geophys. Res., 117, C01001, doi:10.1029/2011JC007231, 2012.

Rees Jones, D. W. and Worster, M. G.: Fluxes through steady chim-neys in a mushy layer during binary alloy solidification, J. Fluid Mech., 714, 127–151, 2013.

Saenz, B. T. and Arrigo, K. R.: Simulation of a sea ice ecosystem using a hybrid model for slush layer desalination, J. Geophys. Res., 117, C05007, doi:10.1029/2011JC007544, 2012.

Schwarzacher, W.: Pack-ice studies in the Arctic Ocean, J. Geophys. Res., 64, 2357–2367, 1959.

Semtner, A. J.: Model For Thermodynamic Growth of Sea Ice In Numerical Investigations of Climate, J. Phys. Oceanogr., 6, 379–389, doi:10.1175/1520-0485(1976)006<0379:AMFTTG>2.0.CO;2, 1976.

Tedesco, L., Vichi, M., Haapala, J., and Stipa, T.: A dynamic Biologically Active Layer for numerical studies of the sea ice ecosystem, Ocean Modell., 35, 89–104, 2010.

Tedesco, L., Vichi, M., and Thomas, D. N.: Process studies on the ecological coupling between sea ice algae and phytoplankton, Ecol. Model., 226, 120–138, 2012.

Turner, A. K., Hunke, E. C., and Bitz, C. M.: Two modes of sea-ice gravity drainage: A parameterization for large-scale modeling, J. Geophys. Res., 118, 2279–2294, 2013.

Vancoppenolle, M., Fichefet, T., and Bitz, C. M.: Modeling the salinity profile of undeformed Arctic sea ice, Geophys. Res. Lett., L21501, doi:10.1029/2006GL028342, 2006.

Vancoppenolle, M., Bitz, C. M., and Fichefet, T.: Summer landfast sea ice desalination at Point Barrow, Alaska: Modeling and observations, J. Geophys. Res., 112, C04022, doi:10.1029/2006JC003493, 2007.

Vancoppenolle, M., Fichefet, T., Goosse, H., Bouillon, S., Madec, G., and Maqueda, M. A. M.: Simulating the mass balance and salinity of Arctic and Antarctic sea ice. 1. Model description and validation, Ocean Modell., 27, 33–53, doi:10.1016/j.ocemod.2008.10.005, 2009.

Vancoppenolle, M., Goosse, H., de Montety, A., Fichefet, T., Tremblay, B., and Tison, J.-L.: Modeling brine and nutrient dynamics in Antarctic sea ice: The case of dissolved silica, J. Geophys. Res., 115, C02005, doi:10.1029/2009JC005369, 2010.

Vancoppenolle, M., Notz, D., Vivier, F., Tison, J., Delille, B., Carnat, G., Zhou, J., Jardon, F., Griewank, P., Lourenço, A., and Haskell, T.: Technical Note: On the use of the mushy-layer Rayleigh number for the interpretation of sea-ice-core data, The Cryosphere Discuss., 7, 3209–3230, doi:10.5194/tcd-7-3209-2013, 2013.

Weeks, W. F. and Lee, O. S.: Observations on the physical properties of sea-ice at Hopedale, Labrador, Arctic, 11, 134–155, 1958.

Wells, A. J., Wettlaufer, J. S., and Orszag, S. A.: Brine fluxes from growing sea ice, Geophys. Res. Lett., 38, L04501, doi:10.1029/2010GL046288, 2011.

Widell, K., Fer, I., and Haugan, P. M.: Salt release from warming sea ice, Geophys. Res. Lett., 33, L12501, doi:10.1029/2006GL026262, 2006.

Winton, M.: A reformulated three-layer sea ice model, J. Atmos. Oceanic Technol., 17, 525–531, 2000.

Influence of freshwater input on the skill of decadal forecast of sea ice in the Southern Ocean

V. Zunz[1,*] and H. Goosse[1]

[1]Université catholique de Louvain, Earth and Life Institute, Georges Lemaître Centre for Earth and Climate Research, Louvain-la-Neuve, Belgium
[*] *Invited contribution by V. Zunz, recipient of the EGU Young Scientists Outstanding Poster Paper Award 2013.*

Correspondence to: V. Zunz (violette.zunz@uclouvain.be)

Abstract. Recent studies have investigated the potential link between the freshwater input derived from the melting of the Antarctic ice sheet and the observed recent increase in sea ice extent in the Southern Ocean. In this study, we assess the impact of an additional freshwater flux on the trend in sea ice extent and concentration in simulations with data assimilation, spanning the period 1850–2009, as well as in retrospective forecasts (hindcasts) initialised in 1980. In the simulations with data assimilation, the inclusion of an additional freshwater flux that follows an autoregressive process improves the reconstruction of the trend in ice extent and concentration between 1980 and 2009. This is linked to a better efficiency of the data assimilation procedure but can also be due to a better representation of the freshwater cycle in the Southern Ocean. The results of the hindcast simulations show that an adequate initial state, reconstructed thanks to the data assimilation procedure including an additional freshwater flux, can lead to an increase in the sea ice extent spanning several decades that is in agreement with satellite observations. In our hindcast simulations, an increase in sea ice extent is obtained even in the absence of any major change in the freshwater input over the last decades. Therefore, while the additional freshwater flux appears to play a key role in the reconstruction of the evolution of the sea ice in the simulation with data assimilation, it does not seem to be required in the hindcast simulations. The present work thus provides encouraging results for sea ice predictions in the Southern Ocean, as in our simulation the positive trend in ice extent over the last 30 years is largely determined by the state of the system in the late 1970s.

1 Introduction

The sea ice extent in the Southern Ocean has been increasing at a rate estimated to be between 0.13 and 0.2 million km^2 per decade between November 1978 and December 2012 (Vaughan et al., 2013). The recent work of Eisenman et al. (2014) suggests that the positive trend in Antarctic sea ice extent may be in reality smaller than the value given in Vaughan et al. (2013). Indeed, an approximate continuation of the trends in sea ice extent corresponding to the version 1 of the Bootstrap algorithm provides a value around 0.1 million km^2 per decade between November 1978 and December 2012 (Fig. 1b of Eisenman et al., 2014). Nevertheless, even a slight expansion of the Antarctic sea ice is in clear contrast with the behaviour of its Arctic counterpart which is currently shrinking (e.g. Turner and Overland, 2009).

The processes that drive the evolution of the Antarctic sea ice and the causes of its recent expansion are still debated. The hypothesis that the stratospheric ozone depletion (Solomon, 1999) could have been responsible for the increase in sea ice extent is not compatible with the results of some recent model analyses (e.g. Sigmond and Fyfe, 2010, 2013; Bitz and Polvani, 2012; Smith et al., 2012) but the impact of ozone changes involves complex mechanisms that need to be further investigated (Ferreira et al., 2015). Besides, other studies have underlined the fact that the positive trend in sea ice extent could be attributed to the internal variability of the system (e.g. Mahlstein et al., 2013; Zunz et al., 2013; Polvani and Smith, 2013; Swart and Fyfe, 2013). Nevertheless, this explanation cannot be confirmed by present-day general circulation models (GCMs) involved in the Fifth Coupled Model Intercomparison Project (CMIP5,

Taylor et al., 2011). Indeed, because of the biases present in those models, they often simulate a seasonal cycle or an internal variability (or both) of the Southern Ocean sea ice that disagrees with what is observed (e.g. Turner et al., 2013; Zunz et al., 2013).

Hypotheses related to changes in the atmospheric circulation or in the ocean stratification (e.g. Bitz et al., 2006; Zhang, 2007; Lefebvre and Goosse, 2008; Stammerjohn et al., 2008; Goosse et al., 2009; Kirkman and Bitz, 2010; Landrum et al., 2012; Holland and Kwok, 2012; Goosse and Zunz, 2014; de Lavergne et al., 2014) have also been proposed. In particular, a link between the melting of the Antarctic ice sheet, especially the ice shelves, and the formation of sea ice has been recently proposed (e.g. Hellmer, 2004; Swingedouw et al., 2008; Bintanja et al., 2013). The meltwater input from the ice sheet leads to a fresher and colder surface layer in the ocean surrounding Antarctica. As a consequence, the ocean gets more stratified and there is less interaction between the surface and the warmer and saltier interior ocean, leading to an enhanced cooling of the surface. This negative feedback could counteract the greenhouse warming and could thus contribute to the expansion of the sea ice. Estimates of the Antarctic ice sheet mass imbalance are available thanks to satellite observations and climate modelling. These estimates report an increase in the melting of the Antarctic ice sheet over the past decade, mainly coming from West Antarctica (e.g. Rignot et al., 2008; Velicogna, 2009; Pritchard et al., 2012; Shepherd et al., 2012). According to Bintanja et al. (2013), incorporating realistic changes in the Antarctic ice sheet mass in a coupled climate model could lead to a better simulation of the evolution of the sea ice in the Southern Ocean. For past periods, this may be achieved using estimates of changes in mass balance but for future projections this requires a comprehensive representation of the polar ice sheets in models. Besides, Swart and Fyfe (2013) have shown that the freshwater derived from the ice sheet is unlikely to affect significantly the recent trend in sea ice extent simulated by CMIP5 models, when imposing a flux whose magnitude is constrained by the observations.

In addition to the studies devoted to a better understanding of the causes of the recent variations, models are also employed to perform projections for the changes at the end of the 21st century and predictions for the next months to decades. Such predictions are generally performed using GCMs. Unfortunately, as mentioned above, current GCMs have biases that reduce the accuracy of the simulated sea ice in the Southern Ocean. In addition, taking into account observations to initialise these models, generally through simple data assimilation (DA) methods, did not improve the quality of the predictions in the Southern Ocean (Zunz et al., 2013). However, two recent studies performed in a perfect model framework, i.e. using pseudo-observations provided by a reference simulation of the model instead of actual observations, underlined some predictability of the Antarctic sea ice (e.g. Holland et al., 2013; Zunz et al., 2014). According to

these studies, at interannual timescales, the predictability is limited to a few years ahead. Besides, significant predictability is found for the trends spanning several decades. Both studies have pointed out that the heat anomalies stored in the interior ocean could play a key role in the predictability of the sea ice. In particular, in their idealised study, Zunz et al. (2014) have described a link between the skill of the prediction of the sea ice cover and the quality of the initialisation of the ocean below it.

On the basis of those results, the present study aims to identify a procedure that could improve the quality of the predictions of the sea ice in the Southern Ocean at multi-decadal timescales. Unlike Holland et al. (2013) and Zunz et al. (2014), the results discussed here have been obtained in a realistic framework. It means that actual observations are used to initialise the model simulations as well as to assess the skill of the model. The results of Holland et al. (2013) and Zunz et al. (2013, 2014) encouraged us to focus on the prediction of the multi-decadal trends in sea ice concentration or extent rather than on its evolution at interannual timescales. Our study deals with two aspects that could influence the quality of the predicted trend in sea ice in the Southern Ocean: the initial state of the simulation and the magnitude of the freshwater input associated, for instance, with the Antarctic ice sheet mass imbalance. The initialisation procedure is based on the nudging proposal particle filter (NPPF, Dubinkina and Goosse, 2013), a data assimilation method that requires a large ensemble of simulations. Such a large amount of simulations cannot be afforded with GCMs because of their requirements in CPU time. We have thus chosen to work with an Earth-system model of intermediate complexity, LOVECLIM1.3. It has a coarser resolution and a lower level of complexity than a GCM, resulting in a lower computational cost. However, it behaves similarly to the GCMs in the Southern Ocean (Goosse and Zunz, 2014). It thus seems relevant to use this model to study the evolution of the Antarctic sea ice.

The climate model LOVECLIM1.3 is briefly described in Sect. 2.1, along with a summary of the simulations performed in this study. The data assimilation method used to compute the initial conditions of the hindcast simulations is presented in Sect. 2.2. Section 2.3 explains how the additional freshwater flux is taken into account in the simulations. Details about the estimation of the model skill are given in Sect. 2.4. The discussion of the results is divided into two parts: the simulations with data assimilation that provide the initial states (Sect. 3.1) and the hindcast simulations (Sect. 3.2). Finally, Sect. 4 summarises the main results and proposes conclusions.

2   Methodology

2.1   Model and simulations

The 3-D Earth-system model of intermediate complexity LOVECLIM1.3 (Goosse et al., 2010) used here includes representations of the atmosphere (ECBilt2, Opsteegh et al., 1998), the ocean and the sea ice (CLIO3, Goosse and Fichefet, 1999) and the vegetation (VECODE, Brovkin et al., 2002). The atmospheric component is a T21 (corresponding to a horizontal resolution of about $5.6° \times 5.6°$), three-level quasi-geostrophic model. The oceanic component consists of an ocean general circulation model coupled to a sea ice model with horizontal resolution of $3° \times 3°$ and 20 unevenly spaced vertical levels in the ocean. The vegetation component simulates the evolution of trees, grasses and desert, with the same horizontal resolution as ECBilt2. The simulations performed in this study span the period 1850–2009 and are driven by the same natural and anthropogenic forcings (greenhouse gases increase, variations in volcanic activity, solar irradiance, orbital parameters and land use) as the ones adopted in the historical simulations performed in the framework of CMIP5 (Taylor et al., 2011).

Three kinds of simulation are performed in this study and all of them consist of 96-member ensembles. First, a simulation driven by external forcing only provides a reference to measure the predictive skill of the model that can be accounted for by the external forcing alone (NODA in Table 1). This numerical experiment does not take into account any observation, neither in its initialisation nor during the integration. At the initialisation and every 3 months of simulation, the surface air temperature of each members of NODA is slightly perturbed, to have an experimental design as close as possible to the simulations with data assimilation (see below). Second, simulations that assimilate observations of surface air temperature anomalies (see Sect. 2.2 for details) are used to reconstruct the past evolution of the system, from January 1850 to December 2009, and to provide initial conditions for hindcast simulations. Third, the hindcast simulations are initialised on 1 January 1980 from a state extracted from a simulation with data assimilation and are not constrained by the observations during the model integration.

Two simulations with data assimilation, from 1850 to 2009, are analysed here: one without additional freshwater flux (DA_NOFWF in Table 1) and one that is forced by an autoregressive freshwater flux described in Sect. 2.3 (DA_FWF in Table 1), representing crudely the meltwater input to the Southern Ocean. The simulation DA_NOFWF provides the initial state of the first hindcast (HINDCAST_1 in Table 1). The three hindcasts HINDCAST_2.1, HINDCAST_2.2 and HINDCAST_2.3 (see Table 1) are initialised from a state extracted from DA_FWF. These three hindcasts differ amongst each other in the additional freshwater flux they receive during the model integration. No additional freshwater flux is applied for HINDCAST_2.1. HIND-

CAST_2.2 is forced by a time series resulting from the ensemble mean of the additional freshwater flux diagnosed in DA_FWF. The average over the period 1980–2009 of the ensemble mean diagnosed from DA_FWF is applied in HINDCAST_2.3 as a constant additional flux.

2.2   Data assimilation: the nudging proposal particle filter

Data assimilation consists of a combination of the model equations and the available observations, in order to provide an estimate of the state of the system as accurate as possible (Talagrand, 1997). The data assimilation simulations performed here provide a reconstruction of the past evolution of the climate system over the period 1850–2009. Such a long period appears necessary because of the long memory of the Southern Ocean. It allows the ocean to be dynamically consistent with the surface variables, constrained by the observations, over a wide depth range. The state of the system on 1 January 1980 is then extracted and used to initialise the hindcast. After the initialisation, the hindcast is driven by external forcing only and no further observations are taken into account.

In this study, observed anomalies of surface air temperature are assimilated in LOVECLIM1.3 thanks to a nudging proposal particle filter (Dubinkina and Goosse, 2013). The assimilated observations are from the HadCRUT3 data set (Brohan et al., 2006). This data set has been derived from in situ land and ocean observations and provides monthly values of surface air temperature anomalies (with regard to 1961–1990) since January 1850. Model anomalies of surface air temperature are computed with regard to a reference computed over 1961–1990 as well, from a simulation driven by the external forcing only, without data assimilation and additional freshwater flux.

The NPPF is based on the particle filter with sequential resampling (e.g. van Leeuwen, 2009; Dubinkina et al., 2011) that consists of three steps. First, an ensemble of simulations, the *particles*, is integrated forward in time with the model. These particles are initialised from a set of different initial conditions. Therefore, each particle represents a different solution of the model. Second, after 3 months of simulation, a weight is attributed to each particle of the ensemble based on its agreement with the observations. To compute this weight, only anomalies of surface air temperature southward of 30° S are taken into account. Third, the particles are resampled: the ones with small weight are eliminated while the ones with large weight are retained and duplicated, in proportion to their weight. This way, a constant number of particles is maintained throughout the procedure. A small perturbation is applied on the duplicated particles to generate different solutions of the model and the three steps are repeated until the end of the period of interest.

In the NPPF, a nudging is applied on each particle during the model integration. It consists of adding to the model

Table 1. Summary of the simulations analysed in this study.

Simulation	Number of members	Time period	Initialisation	Data assimilation	Additional freshwater flux during the simulation
NODA	96	Jan 1850–Dec 2009	on 1 Jan 1850	NO	NO
DA_NOFWF	96	Jan 1850–Dec 2009	on 1 Jan 1850	YES	NO
DA_FWF	96	Jan 1850–Dec 2009	on 1 Jan 1850	YES	Autoregressive FWF following Eq. (1).
HINDCAST_1	96	Jan 1980–Dec 2009	on 1 Jan 1980 from DA_NOFWF	NO	NO
HINDCAST_2.1	96	Jan 1980–Dec 2009	on 1 Jan 1980 from DA_FWF	NO	NO
HINDCAST_2.2	96	Jan 1980–Dec 2009	on 1 Jan 1980 from DA_FWF	NO	Ensemble mean of the FWF computed in DA_FWF between 1980 and 2009 (see Fig. 6).
HINDCAST_2.3	96	Jan 1980–Dec 2009	on 1 Jan 1980 from DA_FWF	NO	Ensemble mean of the FWF computed in DA_FWF, averaged over the period 1980–2009 ($= 0.01\,\text{Sv}$).

equations a term that pulls the solution towards the observations (e.g. Kalnay, 2007). The nudging alone, i.e. not in combination with another DA method, has been used in many recent studies on decadal predictions (e.g. Keenlyside et al., 2008; Pohlmann et al., 2009; Dunstone and Smith, 2010; Smith et al., 2010; Kröger et al., 2012; Swingedouw et al., 2012; Matei et al., 2012; Servonnat et al., 2014). In LOVECLIM1.3, the nudging has been implemented as an additional heat flux between the atmosphere and the ocean $Q = \gamma(T_{\text{mod}} - T_{\text{obs}})$. T_{mod} and T_{obs} are the monthly mean surface air temperature simulated by the model and from the observations respectively. γ determines the relaxation time and equals $120\,\text{W}\,\text{m}^{-2}\,\text{K}^{-1}$, a value similar to the ones used in other studies (e.g. Keenlyside et al., 2008; Pohlmann et al., 2009; Smith et al., 2010; Matei et al., 2012; Swingedouw et al., 2012; Servonnat et al., 2014). The nudging is applied on every ocean grid cell, except the ones covered by sea ice and the amplitude of the nudging applied on a particle is taken into account in the computation of its weight (Dubinkina and Goosse, 2013).

2.3 Autoregressive additional freshwater flux

As the freshwater related to the melting of the Antarctic ice sheet may contribute to the variability of the sea ice extent (e.g. Hellmer, 2004; Swingedouw et al., 2008; Bintanja et al., 2013), it appears relevant to check its impact on the data assimilation simulations as well as on the hindcasts. However, deriving the distribution of the freshwater flux from the estimate of the observed Antarctic ice sheet mass imbalance is not possible for the whole period covered by our simulations, because of the lack of data. Furthermore, the configuration of the model used in our study does not allow simulating this freshwater flux in an interactive way. We have thus chosen to apply a random freshwater flux, described in term of an autoregressive process as in Mathiot et al. (2013), on each particle during the data assimilation simulations DA_FWF (see Table 1 for details). This allows determining the most adequate value of the additional freshwater flux for the model

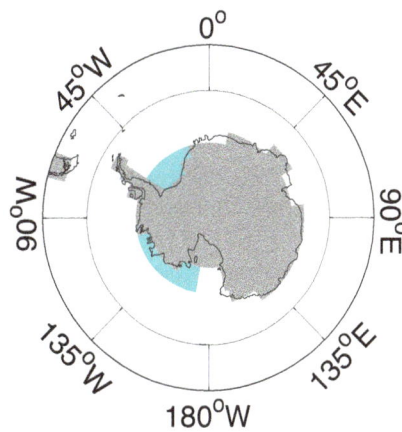

Figure 1. Spatial distribution of the additional freshwater flux included in model simulations (shaded blue). The shaded grey areas correspond to the land mask of the ocean model.

using the NPPF. Because of this additional freshwater flux, the parameters selected to define the error covariance matrix, required to compute the weight of each particle (see Dubinkina et al., 2011), are slightly modified in comparison to the values applied for these parameters in the data assimilation without additional freshwater flux (DA_NOFWF).

The freshwater flux is computed every 3 months, i.e. with the same frequency as the particle filtering. In DA_FWF, the additional freshwater flux is defined as

$$\text{FWF}(t) = 0.8\text{FWF}(t-1) + \epsilon_{\text{FWF}}(t), \tag{1}$$

where $\epsilon_{\text{FWF\_1}}$ is a random noise following a Gaussian distribution $N(0, \sigma_{\text{FWF\_1}})$, with $\sigma_{\text{FWF\_1}}$ equal to $40\,\text{mSv}$.

The parameters of the autoregressive processes described in Eq. (1) have been chosen with the goal to obtain a freshwater flux roughly compatible with the estimates of the current Antarctic ice sheet mass loss. The standard deviation of the resulting additional freshwater flux obtained from the simulation DA_FWF (see Fig. 6), computed from the averages over independent 6-year time periods between 1850 and

2009, equals 7 mSv ($\approx 218 \, \mathrm{Gt\,yr^{-1}}$). This value of the standard deviation is about 3 times larger than the changes in the freshwater input derived from the West Antarctic ice sheet melting between the periods 1992–2000 and 2005–2010 reported in the reconciled estimates of Shepherd et al. (2012) ($\approx 64 \, \mathrm{Gt\,yr^{-1}}$). Alternatively, we can also consider that the ice sheet mass imbalance is not the only contributor to the additional freshwater flux required by the model. For instance, variations in precipitation are also expected to impact the freshwater balance in the Southern Ocean and might not be simulated adequately by the model. A formulation of the additional freshwater flux that allows stronger variations of this freshwater flux and implies a larger impact has also been tested. The results of this additional simulation are discussed in Sect. S1 of the Supplement, along with three additional hindcast simulations.

The melting of the Antarctic ice sheet being particularly strong over West Antarctica (e.g. Rignot et al., 2008; Velicogna, 2009; Pritchard et al., 2012; Shepherd et al., 2012), we have chosen to distribute uniformly the freshwater flux in the ocean between 0 and 170° W, south of 70° S (area in blue on Fig. 1). Here, the distribution of the freshwater flux is thus not limited to the cells adjacent to Antarctica, unlike Bintanja et al. (2013); Swart and Fyfe (2013). This is based on the assumption that a part of the freshwater might be redistributed offshore by icebergs (e.g. Silva et al., 2006) or coastal currents not well represented in a coarse-resolution model. The spatial distribution of the additional freshwater flux likely impacts the model results. Here, we have chosen a spatial structure as simple as possible, consistent with the available observations, in order to limit the parameters associated with the additional freshwater flux. A detailed investigation of the impact of different spatial distributions of the additional freshwater input on the model solutions would probably provide insightful results but is out of the scope of the present study.

The additional freshwater flux increases the range of solutions reached by the particles and can randomly bring some of them closer to the observations. When a particle is picked up because of its large weight, it is duplicated and the copied particles inherit the value of the freshwater flux that possibly brought the particle close to the observations. This value keeps influencing the copied particles because the freshwater flux is autoregressive. It could thus improve the efficiency of the particle filter. Furthermore, by selecting the solutions that best fit the observations, the particle filter allows estimating the freshwater flux that is more likely to provide a state compatible with the observations.

2.4 Skill assessment

In order to measure the skill of the model combined with the assimilation of observations, the results of the data assimilation simulations and of the hindcasts are compared to observations of the annual mean sea ice concentration (the fraction of a grid cell covered by sea ice) and sea ice extent (the sum of the areas of all grid cells having a sea ice concentration above 15 %), between 1980 and 2009. This corresponds to the period for which reliable observations of the whole ice covered area are available. The sea ice concentration and extent data used here, unless specified otherwise, have been derived from the Nimbus-7 SMMR and DMSP SSM/I-SSMIS satellite observations through version 2 of the Bootstrap algorithm (Comiso, 1999). The impact of the uncertainty of those estimates on our conclusion is discussed in Sects. 3 and 4.

Particular attention is paid to the trend in sea ice concentration and extent. Significance levels for the trends are computed on the basis of a two-tailed t-test. The autocorrelation of the residuals is taken into account in both the standard deviation of the trend and in the number of degrees of freedom used to determine the significance threshold (e.g. Santer et al., 2000; Stroeve et al., 2012). This statistical test provides an estimate of the relative significance of the trend, but we have to keep in mind that the assumptions inherent to this kind of test are rarely totally satisfied in the real world (e.g. Santer et al., 2000).

The ensemble means computed for the results of the data assimilation simulations consist of weighted averages. The ensemble mean $X(y, m)$ of the variable x, for the month m in the year y is thus defined as

$$X(y,m) = \frac{1}{K} \sum_{k=1}^{K} x_k(y,m).w_k(y,m), \qquad (2)$$

where k is the member index, K is the number of members within the ensemble and $w_k(y,m)$ is the weight attributed to the member k during the data assimilation procedure. The ensemble means of each month of the year are then averaged over a year to obtain the annual mean.

The standard deviation of the annual mean of the ensemble cannot be computed explicitly because of the possible time discontinuity in the results of individual members, arising from the resampling occurring every 3 months. An estimate of this standard deviation is however assessed by multiplying the weighted standard deviation of each month of a year by a coefficient and averaging it over the year. These coefficients are introduced to take into account the fact that the standard deviation of the annual mean is not the mean of the standard deviation from every month. They are obtained here by computing the mean ratio between the ensemble standard deviation of the annual mean and the ensemble standard deviation of each month in the simulation NODA.

The ensemble means and standard deviations calculated for NODA and for the hindcast simulations correspond to classical values that do not include any weight as this procedure is only required when data assimilation is applied.

3 Results

In this section, the results of the various simulations (see Table 1 for details) are discussed. First, the reconstructions of the evolution of the sea ice between 1850 and 2009, provided by the simulations NODA, DA_NOFWF and DA_FWF, are presented in Sect. 3.1 and compared to observations. Second, the hindcasts initialised with a state extracted from a data assimilation simulation are analysed to measure the skill of the prediction system tested in this study (Sect. 3.2).

3.1 Data assimilation simulations

The observations of yearly mean sea ice extent, based on version 2 of the Bootstrap algorithm, display a positive trend between 1980 and 2009 equal to $19.0 \times 10^3 \, \text{km}^2 \, \text{yr}^{-1}$, significant at the 99 % level (Fig. 2). This trend in sea ice extent is the result of an increase in sea ice concentration in most part of the Southern Ocean, particularly in the Ross Sea (Fig. 3a).

When no data assimilation is included in the model simulation (NODA), the ensemble mean displays a decreasing trend in sea ice extent in response to the external forcing (Fig. 2a and b), similar to the one found in other climate models (e.g. Zunz et al., 2013). Consequently, for the ensemble mean, 30-year trends are negative during the whole period of the simulation without data assimilation (Fig. 2b). Over the period 1980–2009, the ensemble mean of the trend in sea ice extent equals $-15.5 \times 10^3 \, \text{km}^2 \, \text{yr}^{-1}$, with an ensemble standard deviation of $14.5 \times 10^3 \, \text{km}^2 \, \text{yr}^{-1}$, and the reduction of sea ice concentration occurs everywhere in the Southern Ocean (Fig. 3b), except in the Ross Sea and in the Western Pacific sector. This negative trend obtained for the ensemble mean is the result of a wide range of behaviours simulated by the different members belonging to the ensemble (light green shading in Fig. 2a and b) and, considered individually, the members can thus provide positive or negative values for the trend. This indicates thus that, for some members, the natural variability could compensate for the negative trend in sea ice extent simulated in response to the external forcing. Positive trends similar to the one observed over the last 30 years are however rare in NODA. For instance, only 14 of the 96 members have a positive trend over the period 1980–2009 and none of them have a trend larger than the observed one.

In NODA, the ensemble mean displays an increase in the heat contained in both the upper ocean, defined here as the first 100 m below the surface, and the interior ocean, considered to lie between -100 and -500 m (green solid lines in Fig. 4a and b). The correlation between these two variables equals 0.89 over the period 1980–2009 (Table 2). This warming of the ocean results directly from the increase in the external forcing and is consistent with the decrease in sea ice extent (Fig. 2a). Besides, the ocean salt content in the first 100 m decreases (Fig. 4c). This is likely due to the enhanced hydrological cycle in a global warming context and the inher-

ent increase in precipitation at high southern latitudes that freshens the ocean surface (e.g. Liu and Curry, 2010; Fyfe et al., 2012). Indeed, in NODA, the freshwater input resulting from precipitation integrated south of 60° S is about 365 mSv in the early 1850s and increases up to about 375 mSv in 2009. In the simulation NODA, the negative correlation of -0.94 between the ocean heat and salt content in the first 100 m below the surface over the period 1980–2009 (see Table 2) is linked to the response of these two variables to the external forcing. Nevertheless, this contribution of the external forcing can be masked in individual members by internal variability, leading to low correlations between the heat content at surface and in the interior or between heat and salt contents at surface on average over the ensemble (Table 2).

As the ocean heat content in ice-covered regions is related to the temperature of the freezing point, which is in turn determined by the salinity of the seawater, the co-variations of the ocean heat and salt contents may be constrained by the salinity dependance of the freezing point temperature. Nevertheless, in all our simulations, the variations in the sea surface salinity associated with the freshwater input imply very weak changes in the freezing point temperature (standard deviation = 0.001 °C over the period 1850–2009). Besides, the variations in the upper ocean heat content in NODA correspond to a standard deviation of the ocean temperature averaged over the first 100 m, south of 60° S, equal to 0.03 °C. Therefore, it can be reasonably assumed that the salinity dependance of the freezing point temperature has a negligible impact on the ocean temperature and heat content.

If observations of the anomalies of the surface air temperature are assimilated during the simulation, without additional freshwater flux (DA_NOFWF), the model is able to capture the observed interannual and multi-decadal variability of this variable, as expected (Fig. 5b). Consequently, the trend in the ensemble mean sea ice extent is more variable than in NODA. Over the period 1850–2009, the values of the 30-year trend in sea ice extent, computed from the ensemble mean, stand between -29.1×10^3 and $13.6 \times 10^3 \, \text{km}^2 \, \text{yr}^{-1}$ (Fig. 2d). Between 1980 and 2009, the trend in sea ice extent equals $-3.0 \times 10^3 \, \text{km}^2 \, \text{yr}^{-1}$. On average over the ensemble, the trend is thus less negative than in the case where no observations are taken into account during the simulation but it still has a sign opposite to the observed one. The difference with the estimates derived from version 2 of the Bootstrap algorithm between November 1978 and December 2009 is of the order of $20 \times 10^3 \, \text{km}^2 \, \text{yr}^{-1}$. The difference with the estimates from version 1 of the Bootstrap algorithm is slightly smaller, being around $15 \times 10^3 \, \text{km}^2 \, \text{yr}^{-1}$ (Eisenman et al., 2014). The trends in sea ice concentration display a pattern roughly similar to the observed one (Fig. 3a and c), with an increase in the eastern Weddell Sea, in the eastern Indian sector, in the Western Pacific sector and in the Ross Sea, the sea ice concentration decreasing elsewhere. The decrease in sea ice concentration occurring in the Bellingshausen and Amund-

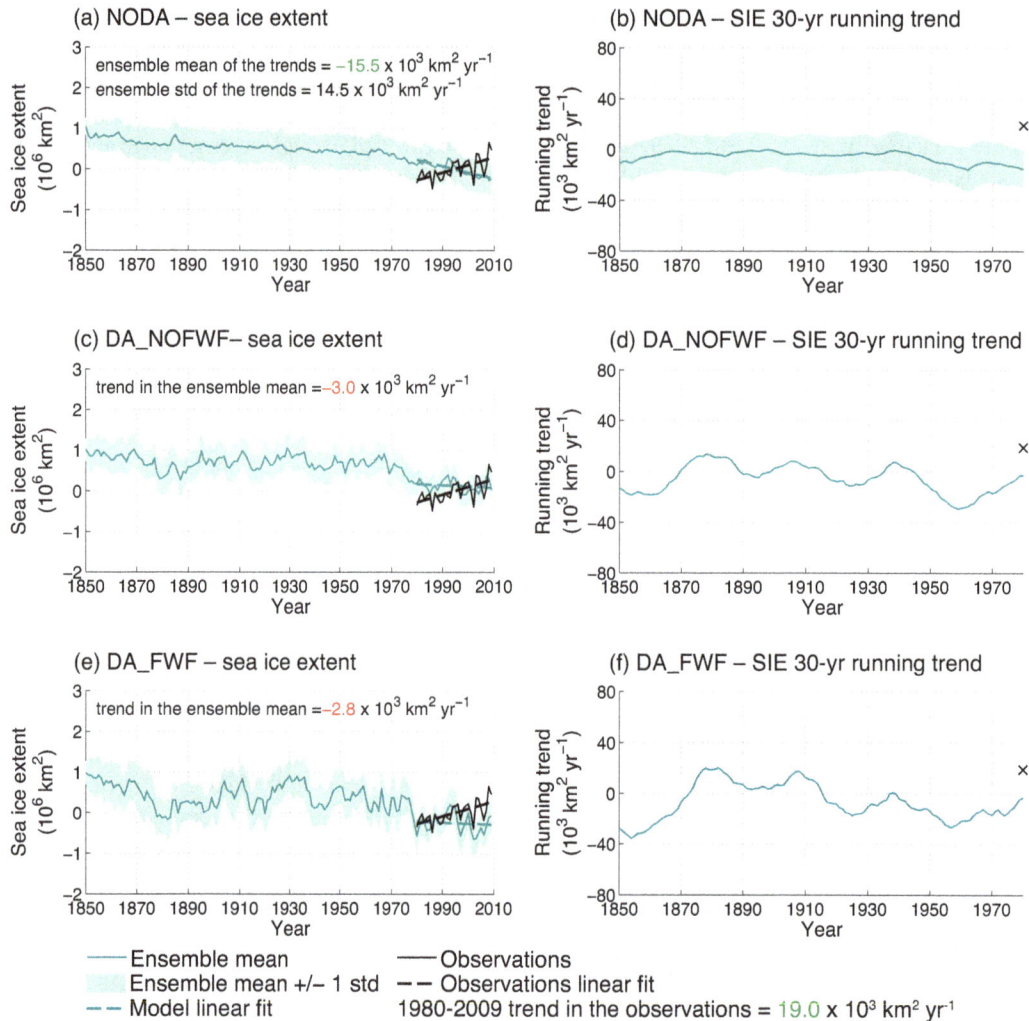

Figure 2. (**a, c, e**) Yearly mean sea ice extent anomalies with regard to 1980–2009 and (**b, d, f**) 30-year running trend in sea ice extent. Results are from (**a, b**) the simulation without data assimilation (NODA), (**c, d**) the model simulation that assimilates anomalies of surface air temperature (DA_NOFWF) and (**e, f**) the model simulation that assimilates anomalies of surface air temperature and that is forced by an additional autoregressive freshwater flux following Eq. (1) (DA_FWF). The model ensemble mean is shown as the dark green line surrounded by one standard deviation shown as the light green shading. Observations (Comiso, 1999) are shown as the black line (cross) in (**a, c, e**) (in **b, d, f**). The green (black) dashed line shows the linear fit of the model simulation (observations) in (**a, c, e**). The values of the trend indicated in (**a, c, e**) correspond to the ensemble mean of the trends, computed over the period 1980–2009, along with the ensemble standard deviation for NODA. Trends that are (non-)significant at the 99 % level are shown in green (red).

sen seas is, however, overestimated by the model, leading to the decrease of the overall extent.

In the simulation DA_NOFWF, the ocean heat content in both the upper and interior ocean is lower than the ones obtained in the simulation NODA until about 1980 (Fig. 4a and b). This arises from the lower surface air temperature in DA_NOFWF compared to NODA (Fig. 5a and b) that cools down the whole system. The correlation between the upper and interior ocean heat contents equals 0.34 over the period 1980–2009 (Table 2) and is thus lower than for the ensemble mean in NODA. This could be due to the interannual variability captured thanks to the data assimilation that miti-

gates the global warming signal (see below). The ocean salt content is larger in DA_NOFWF than in NODA until 1980, likely because of the weakening of the hydrological cycle associated with the lower simulated temperature. Indeed, in DA_NOFWF, the freshwater input associated with precipitation integrated over the area south of 60° S equals 363 mSv on average between 1850 and 1980, against 368 mSv in NODA over the same period. From 1980 onwards, the ocean heat content, in both the upper and middle layer, increases and the salt content decreases in response to the external forcing, as in NODA. Nevertheless, as the ocean heat content is still slightly lower in the simulation DA_NOFWF than in the

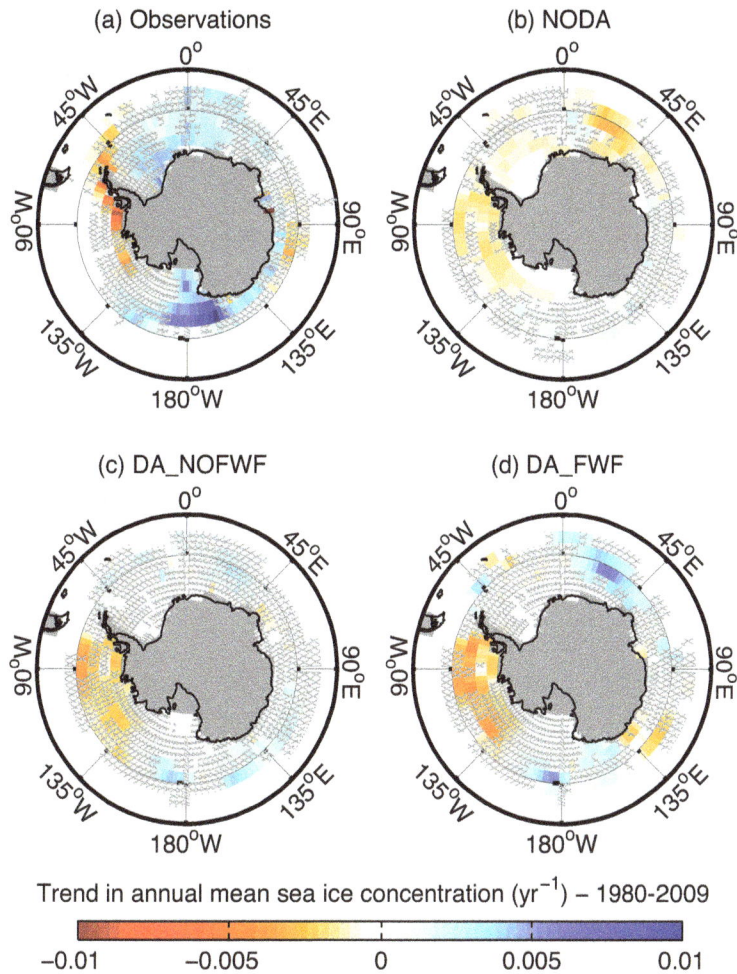

Figure 3. Trend in yearly mean sea ice concentration between 1980 and 2009, shown for (**a**) the observations (Comiso, 1999), (**b**) the model simulation without data assimilation (NODA), (**c**) the model simulation that assimilates anomalies of surface air temperature (DA_NOFWF) and (**d**) the model simulation that assimilates anomalies of surface air temperature and that is forced by an additional autoregressive freshwater flux following Eq. (1) (DA_FWF). Hatched areas highlight the grid cells where the trend is not significant at the 99 % level. The shaded grey areas correspond to the land mask of the ocean model.

Table 2. Correlation between the ocean heat content in the first 100 m below the surface and the ocean heat content between -500 and -100 m (second column) and correlation between the ocean heat content and the ocean salt content in the first 100 m below the surface (third column), for the different simulations summarised in Table 1. The correlation is computed over the period 1980 and 2009, from the ensemble mean of the variables. For the simulation NODA, the correlation computed for each member of the simulation and averaged over the ensemble is given in brackets.

Simulation	Correlation between the upper and interior ocean heat content	Correlation between the upper ocean heat and salt contents
NODA	0.89 (0.03)	-0.94 (-0.02)
DA_NOFWF	0.34	-0.28
DA_FWF	-0.24	0.35
HINDCAST_1	0.86	-0.94
HINDCAST_2.1	0.07	-0.03
HINDCAST_2.2	-0.44	0.44
HINDCAST_2.3	-0.32	0.27

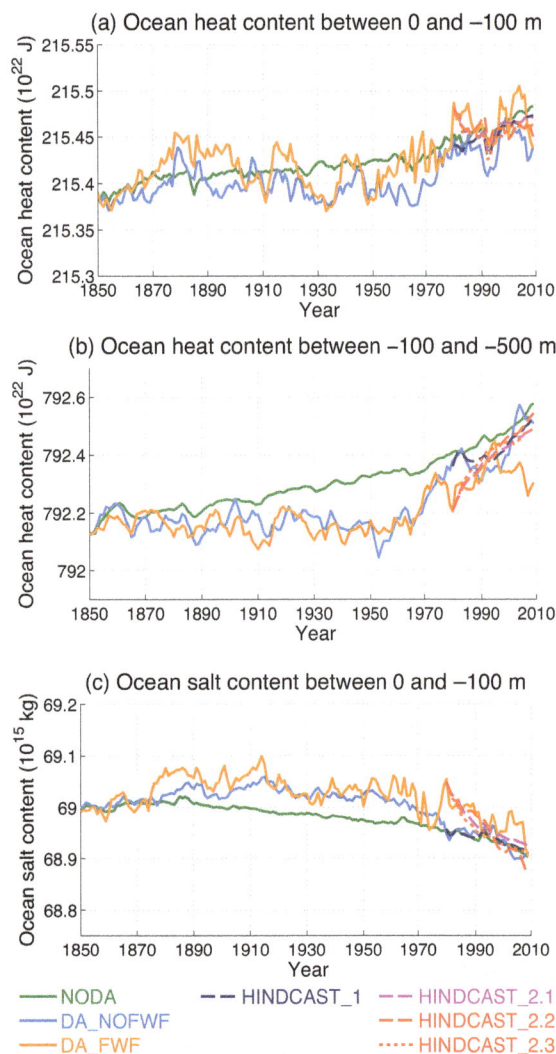

Figure 4. Ensemble mean of yearly mean (a) ocean heat content in the first 100 m below the surface, (b) ocean heat content between −100 and −500 m and (c) ocean salt content in the first 100 m below the surface, for the simulations summarised in Table 1. The ocean heat and salt contents are computed south of 60° S. The ocean heat content is computed against absolute zero.

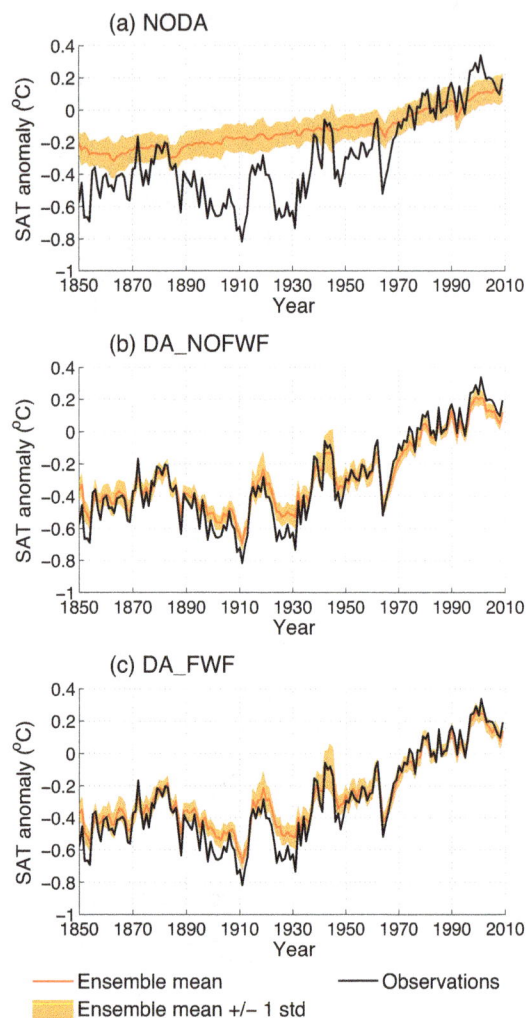

Figure 5. Yearly mean surface air temperature anomalies with regard to 1961–1990, averaged over the area south of 30° S, from (a) the model simulation without data assimilation (NODA), (b) the model simulation that assimilates anomalies of surface air temperature (DA_NOFWF) and (c) the model simulation that assimilates anomalies of surface air temperature and that is forced by an additional autoregressive freshwater flux following Eq. (1) (DA_FWF). The model ensemble mean is shown as the orange line, surrounded by one standard deviation shown as the light orange shading. Observations (Brohan et al., 2006) are shown as the black line.

simulation NODA, the quantity of energy available to melt the sea ice at the surface is also lower. This can explain why the absolute value of the trend in sea ice extent between 1980 and 2009 is smaller in DA_NOFWF than in NODA.

Including a freshwater flux following the autoregressive process defined in Eq. (1) in the simulation DA_FWF increases the variance of the ensemble of particles. This also slightly enhances the variability of the ensemble mean sea ice extent at interannual and multi-decadal timescales (Fig. 2e, f). Over the period 1850–2009, the values of the 30-year trend in sea ice extent, computed from the ensemble mean, lie between -35.2×10^3 km^2 yr^{-1} and 20.3×10^3 km^2 yr^{-1} (Fig. 2f). Over the period 1980–2009, the trend in sea ice

extent in DA_FWF equals -2.8×10^3 km^2 yr^{-1} and is thus slightly less negative than in the simulation DA_NOFWF. The spatial distribution of the trends in sea ice concentration in DA_FWF is also in good agreement with the observations (Fig. 3d). The decrease in sea ice concentration occurring in the Bellingshausen and Amundsen seas is less widespread than in DA_NOFWF but it is still overestimated. The increase in the eastern Weddell and Ross seas is better represented than in DA_NOFWF as well.

The additional freshwater flux in DA_FWF also induces a higher variability of the heat and salt contents in the up-

Diagnosed freshwater flux

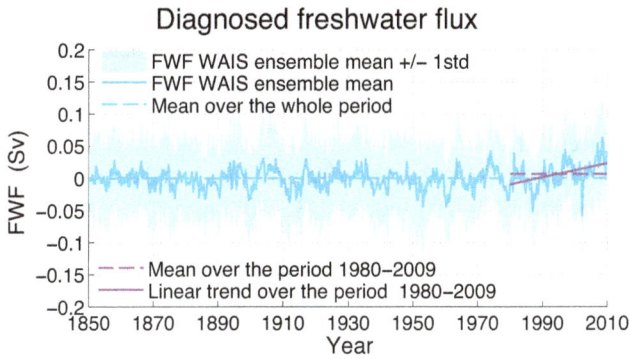

Figure 6. Freshwater flux from the model simulation with data assimilation and additional autoregressive freshwater flux following Eq. (1) (DA_FWF). The ensemble mean is shown as the blue solid line, surrounded by one standard deviation shown as the light blue shading. The dashed blue (purple) line shows the mean over the period 1850–2009 (1980–2009). The linear fit between 1980 and 2009 is shown as the solid purple line.

per ocean compared to the simulation DA_NOFWF (Fig. 4a, c). The correlation between the upper and interior ocean heat contents has a negative value of −0.24 over the period 1980–2009 (see Table 2), which means that when the ocean surface is colder, the intermediate layer is warmer and vice versa. This indicates that, in this experiment, the heat content in the water column is strongly influenced by vertical mixing. The amplitude of this mixing depends on the difference in density between the surface and the deeper layers, which is in turn determined by the difference in temperature and salinity. In the simulation DA_FWF, the correlation between the ocean salt and heat contents in the first 100 m reaches a value of 0.35, while it is negative for the ensemble mean in NODA and in DA_NOFWF (see Table 2). This confirms that, during periods of increase in salt content in the upper layer, the vertical mixing in the ocean is enhanced, allowing positive heat anomalies to be transported from the interior to the upper ocean. The heat content in the first 100 m increases while the one between −100 and −500 m decreases. In contrast, when the salt content in the upper layer decreases, the ocean becomes more stratified, preventing the heat exchange between the surface and the interior ocean. The heat is trapped in the interior ocean that gets warmer, and the upper ocean cools down. This process appears more important in DA_FWF than for the individual members of NODA (see Table 2) because of the effect of the additional freshwater flux on the stratification. Keep in mind that correlation between the heat content in the upper and intermediate layers is very high in the ensemble mean of NODA because of the contribution of the forcing.

Because of the additional freshwater flux that tends to stabilise the water column during some periods and to destabilise it in others (Fig. 6), the general behaviour of the ocean in the simulation DA_FWF differs from the simula-

tion NODA and DA_NOFWF. While the latter simulations appear mainly driven by the external forcing, the interaction between the different layers in the ocean seems to be dominant in DA_FWF. In the simulation DA_FWF, the ocean heat and salt contents of the surface layer are particularly large in 1980 while the heat content between 100 and 500 m is low. This implies that the heat storage at depth in 1980 is much lower in DA_FWF than in NODA. Note that the heat content of the top 500 m in DA_FWF is also lower than in NODA. After 1980, the salt content in DA_FWF decreases until 2009 (Fig. 4c). This is associated with a decrease (increase) in the upper (interior) ocean heat content until the early 1990s, suggesting a reduction of the vertical ocean heat flux. This is likely responsible for the weaker decrease in sea ice extent between 1980 and 2009 in DA_FWF (Fig. 2e). In DA_FWF, the additional freshwater flux is the main cause of the variability of the stratification. Additionally, internal processes can be responsible for such changes in vertical exchanges, as discussed in detail in Goosse and Zunz (2014), also leading to a negative correlation between the heat content in surface and intermediate layers. This explains why the correlation between those two variables is lower for the ensemble mean of DA_NOFWF than in NODA. It is also much lower in individual simulations of NODA (0.03 on average, Table 2) than in the ensemble mean (0.89, Table 2), the ensemble mean amplifying the contribution of the response to the forcing associated with high positive value.

The additional freshwater flux also weakens the link between the sea ice and the surface air temperature because of the larger role of the changes in oceanic stratification. The correlation between the sea ice and the surface air temperature remains negative in the presence of an additional freshwater flux, i.e. a warmer ocean surface is still associated with a smaller sea ice extent. Nevertheless, the correlation between the ensemble mean of the averaged sea surface temperature and the ensemble mean of the sea ice extent over the period 1850–2009 is smaller in absolute value in the simulation with data assimilation including an additional freshwater flux (−0.78 in DA_FWF) compared to the simulations without any additional freshwater flux (−0.97 in NODA and −0.86 in DA_NOFWF). Keep in mind that the reconstruction of the surface air temperature provided by both DA_NOFWF and DA_FWF is based on the assimilation of surface air temperature data. As expected, the surface air temperature simulated in DA_NOFWF is thus very similar to the one in DA_FWF, both simulations achieving a clear model bias reduction. This bias reduction is, however, obtained differently in the two simulations DA_NOFWF and DA_FWF. For instance, the sea ice simulated in DA_NOFWF, in particular the trend in sea ice extent between 1980 and 2009, differs from the one in DA_FWF. These differences in the simulated sea ice extent are consistent with the modification of the link between the surface air temperature and the sea ice extent induced by the additional freshwater flux.

3.2   Hindcast simulations

In this section, we focus on simulations that are initialised on 1 January 1980 with a state that has been extracted from the data assimilation simulations discussed in Sect. 3.1. After the initialisation, the hindcast simulation is driven by external forcing but no further observations are taken into account. The discussion of analyses here aims at answering two questions. (1) Can the information contained in the initial state persist long enough to impact the simulated trend in sea ice extent? (2) How does an additional freshwater flux impact the sea ice in hindcast simulations? Including an additional freshwater flux appears indeed to be relevant to improve the efficiency of data assimilation (see Sect. 3.1). The results of HINDCAST_1, initialised from DA_NOFWF, and HIND-CAST_2.1, initialised from DA_FWF, bring answers to the first question, these hindcasts including no additional freshwater flux. The second question is specifically addressed in the analyses of HINDCAST_2.2 and HINDCAST_2.3, initialised from a state provided by the simulation DA_FWF, a freshwater perturbation being applied during these two hindcasts. Given that it is not clear whether it is the mean value of the additional freshwater flux or its variations that matters, two configurations for the additional freshwater flux have been tested. In HINDCAST_2.2, the additional freshwater flux corresponds to the one that has been diagnosed from DA_FWF, shown on Fig. 6, and evolves in time. In contrast, in HINDCAST_2.3, the freshwater flux is constant in time and equals 0.01 Sv, the average freshwater flux diagnosed in DA_FWF between 1980 and 2009.

In HINDCAST_1, the sea ice extent is high at the beginning of the simulation and decreases between 1980 and 2009 (Fig. 7a). The ensemble mean of the trends equals -14.2×10^3 km^2 yr^{-1}, with an ensemble standard deviation of 13.2×10^3 km^2 yr^{-1}. This provides a 95 % range that does not encompass the observed trend of 19.0×10^3 km^2 yr^{-1}. In this hindcast, the trend in sea ice concentration is negative over a large area in the Bellingshausen and Amundsen seas and slightly positive elsewhere (Fig. 8b). This pattern thus roughly fits the observed one (Fig. 8a) but the decrease obtained in the western part of the Southern Ocean covers too large an area and the increase in the Weddell and Ross seas is too weak. The regional distribution of the trend in sea ice concentration in HINDCAST_1 (Fig. 8b) is thus very similar to the one in DA_NOFWF, i.e. the simulation that provided the initial state for HINDCAST_1. This suggests that the information provided at the initialisation can slightly impact the solution of the hindcast over multi-decadal timescales. The too large decrease in sea ice concentration occurring in the Bellingshausen and Amundsen seas already noticed in DA_NOFWF is however amplified in HINDCAST_1, leading to an overall decrease in sea ice extent similar to the mean of NODA. The ocean heat and salt contents in HIND-CAST_1 follow roughly the evolution of these variables for the ensemble mean in NODA (Fig. 4). The correlation be-

tween the upper and interior ocean heat content equals 0.86 and the correlation between the upper ocean heat and salt content equals -0.94 (see Table 2). This points out the role played by the external forcing in this hindcast, as discussed in Sect. 3.1.

In HINDCAST_2.1, the ensemble mean of the trends over the period 1980–2009 equals 1.3×10^3 km^2 yr^{-1}, with an ensemble standard deviation of 14.5×10^3 km^2 yr^{-1} (Fig. 7b). The observed trend is thus included in the 95 % range of the ensemble. The spatial distribution of the trends in sea ice concentration in HINDCAST_2.1 is also in acceptable agreement with the observations (Fig. 8a, c). Given that no additional freshwater flux is applied in this hindcast, the positive trend in its sea ice extent likely arises from the state used to initialise this simulation. This initial state is characterised by relatively large heat and salt contents in the upper ocean (Fig. 4a, c) and a small heat content in the interior ocean (Fig. 4b). This situation corresponds to a weakly stratified ocean column in 1980 that stabilises during the following years in HINDCAST_2.1, leading to a cooling of the ocean surface that in turn favours the production of sea ice.

HINDCAST_2.2 provides an ensemble mean of the trends over the period 1980–2009 equal to 13.0×10^3 km^2 yr^{-1}, with an ensemble standard deviation of 12.4×10^3 km^2 yr^{-1} (Fig. 7c). This value of the trend is thus closer to the observation of 19.0×10^3 km^2 yr^{-1} (corresponding to version 2 of the Bootstrap algorithm) than the one provided by HIND-CAST_2.1. Nevertheless, in realistic conditions, this would require obtaining information on the mass balance of the ice sheets spanning the period of the prevision itself. The spatial distribution of the trends in sea ice concentration in HIND-CAST_2.2 is very similar to the one in HINDCAST_2.1 (Fig. 8c, d). In HINDCAST_2.3, a constant additional freshwater flux equal to 0.01, corresponding to the average over the period 1980–2009 of the freshwater flux diagnosed from DA_FWF_1, is applied. This also provides trends in sea ice extent and concentration over the period 1980–2009 that are compatible with the observations (Figs. 7d and 8a, e). For both HINDCAST_2.2 and HINDCAST_2.3, no clear change in the ocean heat and salt contents is noticed compared to HINDCAST_2.1 (Fig. 4). Nevertheless, the additional freshwater flux results in a slightly higher increase in sea ice extent compared to HINDCAST_2.1.

The results of our hindcast simulations demonstrate that the state used to initialise these simulations plays a fundamental role in determining the trends in sea ice extent and concentration over the 3 decades following the initialisation, in agreement with the idealised experiments presented in Zunz et al. (2014). In our simulations, the additional freshwater flux improves the reconstruction of the evolution of the system in the simulation with data assimilation and thus helps to provide an adequate initial state for the hindcasts. An appropriate freshwater input during the last 30 years may further improve the agreement with observations derived from both version 1 and version 2 of the Bootstrap algorithm

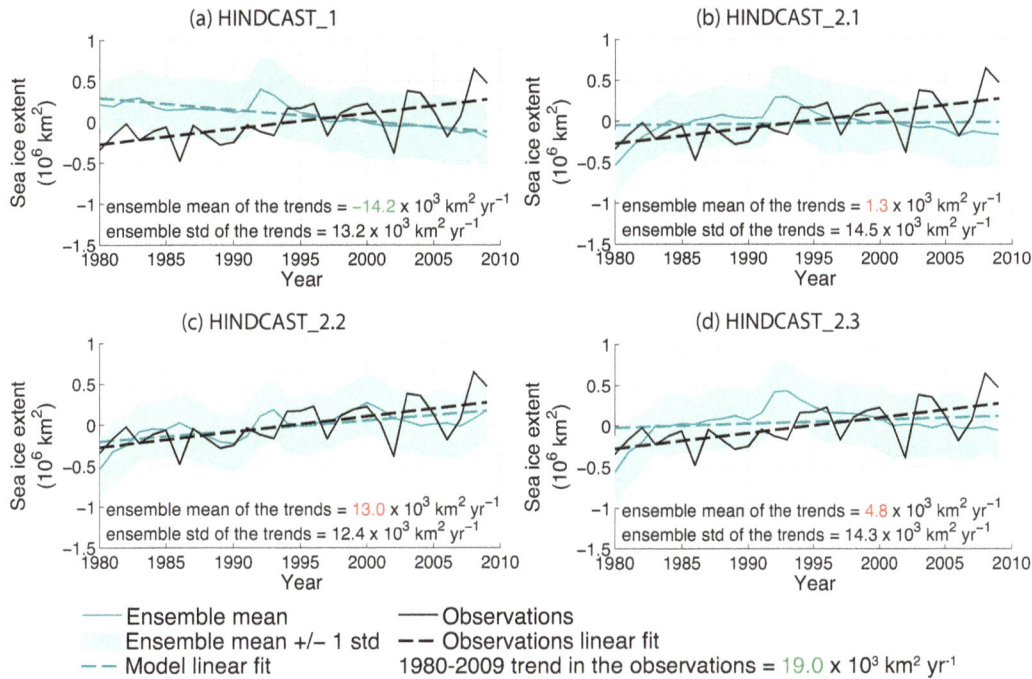

Figure 7. Yearly mean sea ice extent anomalies with regard to 1980–2009, for the four hindcast simulations initialised on 1 January 1980 through data assimilation (see Table 1 for details). The model ensemble mean is shown as the dark green line, surrounded by one standard deviation shown as the light green shading. Observations (Comiso, 1999) are shown as the black line. The green (black) dashed line shows the linear fit of the model simulation (observations). The values of the trend indicated in each panel correspond to the ensemble mean of the trends, computed over the period 1980–2009, along with the ensemble standard deviation. Trends that are (non-)significant at the 99 % level are shown in green (red).

(Eisenman et al., 2014), as shown by the results of HIND-CAST_2.2 and HINDCAST_2.3.

As mentioned in Sect. 2.3, another formulation of the additional freshwater flux that allows stronger variations has also been tested. The results of this additional simulation are not discussed in detail here for brevity's sake (for details, see Supplement Sect. S1). In the corresponding simulation with data assimilation, the additional freshwater flux seems to contribute to a reduction of the model biases. Nevertheless, the state associated with such a strongly varying additional freshwater flux is characterised by an enhanced interannual and multi-decadal variability of the sea ice extent as well as the ocean heat and salt contents that may be unrealistic (Figs. S2 and S4). In addition, the strongly varying additional freshwater flux applied during this simulation with data assimilation induces a shift of the system compared to the solution of the model in the absence of any additional freshwater flux. The hindcasts initialised in January 1980 from a state extracted from this simulation provide trends in sea ice extent and concentration, as HINDCAST_2.1, HIND-CAST_2.2 and HINDCAST_2.3, that agree relatively well with the observations. Nevertheless, since the initial state used in these hindcasts is shifted, it is essential to apply a constant additional freshwater flux of adequate magnitude during the hindcast simulation in order to ensure the con-

sistency of the experimental design and to prevent a drift of the model (for details see Sect. S1).

A change in the freshwater input from one period to the other (for instance between the 30 years preceding and following 1980), in the absence of an adequate initialisation of the simulation, is not sufficient to account for the observed positive trend in sea ice extent between 1980–2009. This conclusion is supported by the results of an additional simulation, initialised in January 1960 from a state extracted from NODA. This simulation is driven by external forcing and receives an additional freshwater input, following the spatial distribution displayed in Fig. 1, equal to −0.03 Sv between January 1960 and December 1979 and abruptly increased to −0.01 Sv in January 1980, i.e. a larger shift than in any of our simulations with data assimilation or hindcasts. The additional freshwater flux then remains constant until the end of the simulation in December 2009. In this simulation, the sea ice extent decreases between 1960 and 1980 in response to the external radiative forcing and the negative freshwater perturbation (see Sect. S2). The sea ice extent then rapidly increases after the abrupt change in the additional freshwater input in January 1980 but decreases again after a few years, in contrast to observations.

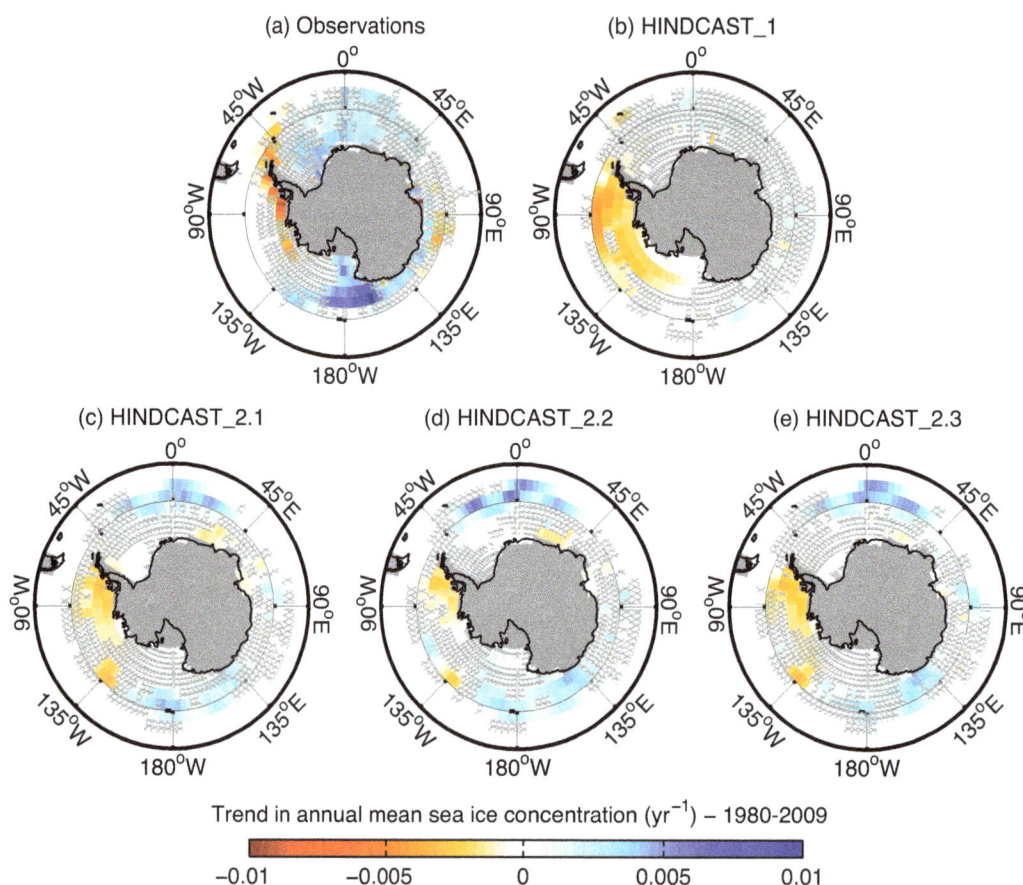

Figure 8. Trend in yearly mean sea ice concentration between 1980 and 2009, for (**a**) the observations (Comiso, 1999) and (**b, c, d, e**) the four hindcast simulations initialised on 1 January 1980 through data assimilation (see Table 1 for details). Hatched areas highlight the grid cells where the trend is not significant at the 99 % level. The shaded grey areas correspond to the land mask of the ocean model.

4 Summary and conclusions

The trend in sea ice extent derived from satellite observations is subject to uncertainties (e.g. Eisenman et al., 2014) but even the lowest estimate of this trend indicates a slight increase in Antarctic sea ice extent that is not reproduced in our simulation driven by external forcing only. Assimilating anomalies of the surface air temperature through the nudging proposal particle filter induces an increase in the trend in simulated sea ice extent over recent decades in the Southern Ocean, compared to the case where no observation is taken into account. This leads to a better agreement with satellite data than in the simulation without data assimilation. Further improvement is achieved if an additional autoregressive freshwater flux is included during the data assimilation. This freshwater flux induces a larger spread of the ensemble and thus allows a better efficiency of the particle filtering. The additional freshwater input may also compensate for model deficiencies that affect the representation of the freshwater cycle (in particular the variability of the meltwater input), the ocean dynamics, the internal variability, etc. Overall, in combination with the data assimilation, the additional fresh-

water input leads to simulated trends in sea ice extent and concentration between 1980 and 2009 that reproduce reasonably well the observations. The freshwater flux thus appears to play an important role on the simulated evolution of the sea ice, as already pointed out in previous studies (e.g. Hellmer, 2004; Swingedouw et al., 2008; Bintanja et al., 2013).

Hindcasts initialised from those simulations with data assimilation identify several factors that can help increase the model skill for predictions of trends in Antarctic sea ice extent and concentration for coming decades. Specifically, we highlight two findings.

1. Initialising a hindcast simulation with a state extracted from a simulation that has assimilated observations through a nudging proposal particle filter has a significant impact on the simulated trends in sea ice extent and concentration over the period 1980–2009. This indicates that the information contained in the initial state influences the results of the simulation over multi-decadal timescales, confirming the results of Zunz et al. (2014). As a consequence, an initial condition that adequately represents the observed state is required in order

to perform skillful predictions for the trend in sea ice extent over the next decades. Nevertheless, the conclusions drawn from our hindcast simulations have to be considered cautiously since they are based on the analyses of the only 30-year period for which we have relevant observations. Similar analyses could be performed for periods starting before 1980, using the reconstruction of the sea ice provided by the simulation with data assimilation as target for the hindcast instead of actual observations. However, this approach would be nearly equivalent to a perfect model study, as proposed in Zunz et al. (2014).

2. In hindcast simulations, the additional freshwater input may help to correctly reproduce the observed positive trend in sea ice extent. Nevertheless, this additional freshwater flux is not the dominant element in our experimental design, in agreement with the results of Swart and Fyfe (2013). Indeed, an abrupt increase in the additional freshwater flux at the beginning of the hindcast simulation, without an adequate initialisation of the simulation, does not provide a long-term increase in sea ice extent such as the one derived from the observations over the last 30 years (Fig. S7).

Our results suggest that the increase in ice extent and the surface cooling between 1980 and 2009 are not due to the greenhouse gas forcing or to a particular large melting of the ice sheet during this period. The evolution of the variables at the surface of the ocean seems rather influenced by the state of the ocean in the 1970s, characterised by a warm and salty surface layer, a cold intermediate layer and strong vertical mixing. This state of the system is consistent with the results of de Lavergne et al. (2014). It then evolves towards a fresher and cooler upper ocean that allows a greater production of sea ice after 1980. In our experiments, this state in the late 1970s is reached thanks to variations in the freshwater input to the Southern Ocean. This flux is very likely playing a role but we could not determine if it is amplified or not by our experimental design that allows variations of this flux only and not of other forcings or model parameters. Overall, the results that have been discussed here are rather encouraging and open perspectives to perform predictions of the sea ice in the Southern Ocean over the next decades.

Acknowledgements. The authors warmly thank Antoine Barthélemy for his careful reading and helpful comments on the paper. V. Zunz is Research Fellow with the Fonds pour la formation à la Recherche dans l'Industrie et dans l'Agronomie (FRIA-Belgium). H. Goosse is Senior Research Associate with the Fonds National de la Recherche Scientifique (F. R. S. – FNRS-Belgium). This work is supported by the Belgian Federal Science Policy (Research Programme on Science for a Sustainable Development). Computational resources have been provided by the supercomputing facilities of the Université catholique de Louvain (CISM/UCL) and the Consortium des Equipements de Calcul Intensif en Fédération Wallonie Bruxelles (CECI) funded by the Fond de la Recherche Scientifique de Belgique (FRS-FNRS).

Edited by: H. Eicken

References

Bintanja, R., van Oldenborgh, G. J., Drijfhout, S. S., Wouters, B., and Katsman, C. A.: Important role for ocean warming and increased ice-shelf melt in Antarctic sea-ice expansion, Nat. Geosci., 6, 376–379, 2013.

Bitz, C. M. and Polvani, L. M.: Antarctic climate response to stratospheric ozone depletion in a fine resolution ocean climate model, Geophys. Res. Lett., 39, L20705, doi:10.1029/2012GL053393, 2012.

Bitz, C. M., Gent, P. R., Woodgate, R. A., Holland, M. M., and Lindsay, R.: The Influence of Sea Ice on Ocean Heat Uptake in Response to Increasing CO_2, J. Climate, 19, 2437–2450, 2006.

Brohan, P., Kennedy, J. J., Harris, I., Tett, S. F. B., and Jones, P. D.: Uncertainty estimates in regional and global observed temperature changes: A new data set from 1850, J. Geophys. Res., 111, D12106, doi:10.1029/2005JD006548, 2006.

Brovkin, V., Bendtsen, J., Claussen, M., Ganopolski, A., Kubatzki, C., Petoukhov, V., and Andreev, A.: Carbon cycle, vegetation, and climate dynamics in the Holocene: Experiments with the CLIMBER-2 model, Global Biogeochem. Cy., 16, 86-1–86-20, doi:10.1029/2001GB001662, 2002.

Comiso, J.: Bootstrap Sea Ice Concentrations from Nimbus-7 SMMR and DMSP SSM/I-SSMIS, Version 2, January 1980 to December 2009, Boulder, Colorado USA: NASA DAAC at the National Snow and Ice Data Center, 1999 (updated daily).

de Lavergne, C., Palter, J. B., Galbraith, E. D., Bernardello, R., and Marinov, I.: Cessation of deep convection in the open Southern Ocean under anthropogenic climate change, Nat. Clim. Change, 4, 278–282, 2014.

Dubinkina, S. and Goosse, H.: An assessment of particle filtering methods and nudging for climate state reconstructions, Clim. Past, 9, 1141–1152, doi:10.5194/cp-9-1141-2013, 2013.

Dubinkina, S., Goosse, H., Sallaz-Damaz, Y., Crespin, E., and Crucifix, M.: Testing a particle filter to reconstruct climate changes over the past centuries, Int. J. Bifurc. Chaos, 21, 3611–3618, doi:10.1142/S0218127411030763, 2011.

Dunstone, N. J. and Smith, D. M.: Impact of atmosphere and subsurface ocean data on decadal climate prediction, Geophys. Res. Lett., 37, L02709, doi:10.1029/2009GL041609, 2010.

Eisenman, I., Meier, W. N., and Norris, J. R.: A spurious jump in the satellite record: has Antarctic sea ice expansion been overestimated?, The Cryosphere, 8, 1289–1296, doi:10.5194/tc-8-1289-2014, 2014.

Ferreira, D., Marshall, J., Bitz, C. M., Solomon, S., and Plumb, A.: Antarctic Ocean and Sea Ice Response to Ozone Depletion: A Two-Time-Scale Problem, J. Climate, 28, 1206–1226, 2015.

Fyfe, J. C., Gillett, N. P., and Marshall, G. J.: Human influence on extratropical Southern Hemisphere summer precipitation, Geophys. Res. Lett., 39, L23711, doi:10.1029/2012GL054199, 2012.

Goosse, H. and Fichefet, T.: Importance of ice-ocean interactions for the global ocean circulation: A model study, J. Geophys. Res.-Oceans, 104, 23337–23355, 1999.

Goosse, H. and Zunz, V.: Decadal trends in the Antarctic sea ice extent ultimately controlled by ice-ocean feedback, The Cryosphere, 8, 453–470, doi:10.5194/tc-8-453-2014, 2014.

Goosse, H., Lefebvre, W., de Montety, A., Crespin, E., and Orsi, A.: Consistent past half-century trends in the atmosphere, the sea ice and the ocean at high southern latitudes, Clim. Dynam., 33, 999–1016, 2009.

Goosse, H., Brovkin, V., Fichefet, T., Haarsma, R., Huybrechts, P., Jongma, J., Mouchet, A., Selten, F., Barriat, P.-Y., Campin, J.-M., Deleersnijder, E., Driesschaert, E., Goelzer, H., Janssens, I., Loutre, M.-F., Morales Maqueda, M. A., Opsteegh, T., Mathieu, P.-P., Munhoven, G., Pettersson, E. J., Renssen, H., Roche, D. M., Schaeffer, M., Tartinville, B., Timmermann, A., and Weber, S. L.: Description of the Earth system model of intermediate complexity LOVECLIM version 1.2, Geosci. Model Dev., 3, 603–633, doi:10.5194/gmd-3-603-2010, 2010.

Hellmer, H. H.: Impact of Antarctic ice shelf basal melting on sea ice and deep ocean properties, Geophys. Res. Lett., 31, L10307, doi:10.1029/2004GL019506, 2004.

Holland, M. M., Blanchard-Wrigglesworth, E., Kay, J., and Vavrus, S.: Initial-value predictability of Antarctic sea ice in the Community Climate System Model 3, Geophys. Res. Lett., 40, 2121–2124, doi:10.1002/grl.50410, 2013.

Holland, P. R. and Kwok, R.: Wind-driven trends in Antarctic sea-ice drift, Nat. Geosci., 5, 872–875, 2012.

Kalnay, E.: Atmospheric Modeling, Data Assimilation and Predictability, Cambridge University Press, Cambridge, 4 Edn., 2007.

Keenlyside, N., Latif, M., Jungclaus, J. H., Kornbueh, L., and Roeckner, E.: Advancing decadal-scale climate prediction in the North Atlantic sector, Nature, 453, 84–88, doi:10.1038/nature06921, 2008.

Kirkman, C. H. and Bitz, C. M.: The Effect of the Sea Ice Freshwater Flux on Southern Ocean Temperatures in CCSM3: Deep-Ocean Warming and Delayed Surface Warming, J. Climate, 24, 2224–2237, 2010.

Kröger, J., Müller, W., and von Storch, J.-S.: Impact of different ocean reanalyses on decadal climate prediction, Clim. Dynam., 39, 795–810, 2012.

Landrum, L., Holland, M. M., Schneider, D. P., and Hunke, E.: Antarctic Sea Ice Climatology, Variability, and Late Twentieth-Century Change in CCSM4, J. Climate, 25, 4817–4838, 2012.

Lefebvre, W. and Goosse, H.: An analysis of the atmospheric processes driving the large-scale winter sea ice variability in the Southern Ocean, J. Geophys. Res., 113, C02004, doi:10.1029/2006JC004032, 2008.

Liu, J. and Curry, J. A.: Accelerated warming of the Southern Ocean and its impacts on the hydrological cycle and sea ice, P. Natl. Acad. Sci., 107, 14987–14992, 2010.

Mahlstein, I., Gent, P. R., and Solomon, S.: Historical Antarctic mean sea ice area, sea ice trends, and winds in CMIP5 simulations, J. Geophys. Res.-Atmos., 118, 1–6, doi:10.1002/jgrd.50443, 2013.

Matei, D., Pohlmann, H., Jungclaus, J., Müller, W., Haak, H., and Marotzke, J.: Two Tales of Initializing Decadal Climate Prediction Experiments with the ECHAM5/MPI-OM Model, J. Climate, 25, 8502–8523, 2012.

Mathiot, P., Goosse, H., Crosta, X., Stenni, B., Braida, M., Renssen, H., Van Meerbeeck, C. J., Masson-Delmotte, V., Mairesse, A., and Dubinkina, S.: Using data assimilation to investigate the causes of Southern Hemisphere high latitude cooling from 10 to 8 ka BP, Clim. Past, 9, 887–901, doi:10.5194/cp-9-887-2013, 2013.

Opsteegh, J. D., Haarsma, R., Selten, F., and Kattenberg, A.: EC-BILT: a dynamic alternative to mixed boundary conditions in ocean models, Tellus A, 50, 348–367, 1998.

Pohlmann, H., Jungclaus, J. H., Köhl, A., Stammer, D., and Marotzke, J.: Initializing Decadal Climate Predictions with the GECCO Oceanic Synthesis: Effects on the North Atlantic, J. Climate, 22, 3926–3938, 2009.

Polvani, L. M. and Smith, K. L.: Can natural variability explain observed Antarctic sea ice trends? New modeling evidence from CMIP5, Geophys. Res. Lett., 40, 3195–3199, 2013.

Pritchard, H. D., Ligtenberg, S. R. M., Fricker, H. A., Vaughan, D. G., van den Broeke, M. R., and Padman, L.: Antarctic ice-sheet loss driven by basal melting of ice shelves, Nature, 484, 502–505, 2012.

Rignot, E., Bamber, J. L., van den Broeke, M. R., Davis, C., Li, Y., van de Berg, W. J., and van Meijgaard, E.: Recent Antarctic ice mass loss from radar interferometry and regional climate modelling, Nat. Geosci., 1, 106–110, 2008.

Santer, B. D., Wigley, T. M. L., Boyle, J. S., Gaffen, D. J., Hnilo, J. J., Nychka, D., Parker, D. E., and Taylor, K. E.: Statistical significance of trends and trend differences in layer-average atmospheric temperature time series, J. Geophys. Res., 105, 7337–7356, doi:10.1029/1999JD901105, 2000.

Servonnat, J., Mignot, J., Guilyardi, E., Swingedouw, D., Séférian, R., and Labetoulle, S.: Reconstructing the subsurface ocean decadal variability using surface nudging in a perfect model framework, Clim. Dynam., 44, 315–338, doi:10.1007/s00382-014-2184-7, 2014.

Shepherd, A., Ivins, E. R., A, G., Barletta, V. R., Bentley, M. J., Bettadpur, S., Briggs, K. H., Bromwich, D. H., Forsberg, R., Galin, N., Horwath, M., Jacobs, S., Joughin, I., King, M. A., Lenaerts, J. T. M., Li, J., Ligtenberg, S. R. M., Luckman, A., Luthcke, S. B., McMillan, M., Meister, R., Milne, G., Mouginot, J., Muir, A., Nicolas, J. P., Paden, J., Payne, A. J., Pritchard, H., Rignot, E., Rott, H., Sørensen, L. S., Scambos, T. A., Scheuchl, B., Schrama, E. J. O., Smith, B., Sundal, A. V., van Angelen, J. H., van de Berg, W. J., van den Broeke, M. R., Vaughan, D. G., Velicogna, I., Wahr, J., Whitehouse, P. L., Wingham, D. J., Yi, D., Young, D., and Zwally, H. J.: A Reconciled Estimate of Ice-Sheet Mass Balance, Science, 338, 1183–1189, doi:10.1126/science.1228102, 2012.

Sigmond, M. and Fyfe, J. C.: Has the ozone hole contributed to increased Antarctic sea ice extent?, Geophys. Res. Lett., 37, L18502, doi:10.1029/2010GL044301, 2010.

Sigmond, M. and Fyfe, J. C.: The Antarctic Sea Ice Response to the Ozone Hole in Climate Models, J. Climate, 27, 1336–1342, doi:10.1175/JCLI-D-13-00590.1, 2013.

Silva, T. A. M., Bigg, G. R., and Nicholls, K. W.: Contribution of giant icebergs to the Southern Ocean freshwater flux, J. Geophys. Res.-Oceans, 111, C03004, doi:10.1029/2004JC002843, 2006.

Smith, D. M., Eade, R., Dunstone, N. J., Fereday, D., Murphy, J. M., Pohlmann, H., and Scaife, A. A.: Skilful multi-year predictions of Atlantic hurricane frequency, Nat. Geosci., 3, 846–849, doi:10.1038/NGEO1004, 2010.

Smith, K. L., Polvani, L. M., and Marsh, D. R.: Mitigation of 21st century Antarctic sea ice loss by stratospheric ozone recovery, Geophys. Res. Lett., 39, L20701, doi:10.1029/2012GL053325, 2012.

Solomon, S.: Stratospheric ozone depletion: A review of concepts and history, Rev. Geophys., 37, 275–316, 1999.

Stammerjohn, S. E., Martinson, D. G., Smith, R. C., Yuan, X., and Rind, D.: Trends in Antarctic annual sea ice retreat and advance and their relation to El Niño Southern Oscillation and Southern Annular Mode variability, J. Geophys. Res., 113, C03S90, doi:10.1029/2007JC004269, 2008.

Stroeve, J. C., Kattsov, V., Barrett, A., Serreze, M., Pavlova, T., Holland, M., and Meier, W. N.: Trends in Arctic sea ice extent from CMIP5, CMIP3 and observations, Geophys. Res. Lett., 39, L16502, doi:10.1029/2012GL052676, 2012.

Swart, N. C. and Fyfe, J. C.: The influence of recent Antarctic ice sheet retreat on simulated sea ice area trends, Geophys. Res. Lett., 40, 4328–4332, 2013.

Swingedouw, D., Fichefet, T., Huybrechts, P., Goosse, H., Driesschaert, E., and Loutre, M.-F.: Antarctic ice-sheet melting provides negative feedbacks on future climate warming, Geophys. Res. Lett., 35, L17705, doi:10.1029/2008GL034410, 2008.

Swingedouw, D., Mignot, J., Labetoulle, S., Guilyardi, E., and Madec, G.: Initialisation and predictability of the AMOC over the last 50 years in a climate model, Clim. Dynam., 40, 2381–2399, doi:10.1007/s00382-012-1516-8, 2012.

Talagrand, O.: Assimilation of Observations, an Introduction, J. Meteorol. Soc. Jpn. Ser. II, 75, 191–209, 1997.

Taylor, K. E., Stouffer, R. J., and Meehl, G. A.: An Overview of CMIP5 and the Experiment Design, B. Am. Meteorol. Soc., 93, 485–498, 2011.

Turner, J. and Overland, J.: Contrasting climate change in the two polar regions, Polar Res., 28, 146–164, doi:10.1111/j.1751-8369.2009.00128.x, 2009.

Turner, J., Bracegirdle, T. J., Phillips, T., Marshall, G. J., and Hosking, J. S.: An Initial Assessment of Antarctic Sea Ice Extent in the CMIP5 Models, J. Climate, 26, 1473–1484, doi:10.1175/JCLI-D-12-00068.1, 2013.

van Leeuwen, P. J.: Particle Filtering in Geophysical Systems, Mon. Weather Rev., 137, 4089–4114, doi:10.1175/2009MWR2835.1, 2009.

Vaughan, D. G., Comiso, J. C., Allison, I., Carrasco, J., Kwok, R., Mote, P., Murray, T., Paul, F., Ren, J., Rignot, E., Solomina, O., Steffen, K., and Zhang, T.: Observations: Cryosphere, in: Climate Change 2013: The Physical Science Basis. Contribution of Working Group I to the Fifth Assessment Report of the Intergovernmental Panel on Climate Change, edited by: Stocker, T. F., Qin, D., Plattner, G.-K., Tignor, M., Allen, S. K., Boschung, J., Nauels, A., Xia, Y., Bex, V., and Midgley, P. M., Cambridge University Press, Cambridge, United Kingdom and New York, NY, USA, 2013.

Velicogna, I.: Increasing rates of ice mass loss from the Greenland and Antarctic ice sheets revealed by GRACE, Geophys. Res. Lett., 36, L19503, doi:10.1029/2009GL040222, 2009.

Zhang, J.: Increasing Antarctic Sea Ice under Warming Atmospheric and Oceanic Conditions, J. Climate, 20, 2515–2529, 2007.

Zunz, V., Goosse, H., and Massonnet, F.: How does internal variability influence the ability of CMIP5 models to reproduce the recent trend in Southern Ocean sea ice extent?, The Cryosphere, 7, 451–468, doi:10.5194/tc-7-451-2013, 2013.

Zunz, V., Goosse, H., and Dubinkina, S.: Impact of the initialisation on the predictability of the Southern Ocean sea ice at interannual to multi-decadal timescales, Clim. Dynam., 1–20, doi:10.1007/s00382-014-2344-9, 2014.

9

Comparing C- and L-band SAR images for sea ice motion estimation

J. Lehtiranta, S. Siiriä, and J. Karvonen

Finnish Meteorological Institute, Marine Research Programme, Helsinki, PB 503, 00101 Finland

Correspondence to: J. Lehtiranta (jonni.lehtiranta@fmi.fi)

Abstract. Pairs of consecutive C-band synthetic-aperture radar (SAR) images are routinely used for sea ice motion estimation. The L-band radar has a fundamentally different character, as its longer wavelength penetrates deeper into sea ice. L-band SAR provides information on the seasonal sea ice inner structure in addition to the surface roughness that dominates C-band images. This is especially useful in the Baltic Sea, which lacks multiyear ice and icebergs, known to be confusing targets for L-band sea ice classification. In this work, L-band SAR images are investigated for sea ice motion estimation using the well-established maximal cross-correlation (MCC) approach. This work provides the first comparison of L-band and C-band SAR images for the purpose of motion estimation. The cross-correlation calculations are hardware accelerated using new OpenCL-based source code, which is made available through the author's web site. It is found that L-band images are preferable for motion estimation over C-band images. It is also shown that motion estimation is possible between a C-band and an L-band image using the maximal cross-correlation technique.

1 Introduction

The Baltic Sea gets an ice cover every winter, covering 45 % of its area on an average year. In the northern Bay of Bothnia, the typical duration of ice cover is from late October to late May, and the greatest level ice thickness ranges from 50 to 110 cm. The bay has an average depth of 41 m and typically has large areas of landfast ice on the eastern and northeastern coasts (Myrberg et al., 2006). Observations of the Baltic sea ice are for winter navigation safety. Work has been done to calculate sea ice motion from two consecutive satellite im-

ages using different optical flow estimation algorithms (e.g., Fily and Rothrock, 1987; Vesecky et al., 1988; Liu et al., 1997; Karvonen et al., 2007; Thomas et al., 2011), and this approach has provided acceptable results using the C-band synthetic aperture radar, which is regarded as a good compromise for sea ice remote sensing (Dierking and Busche, 2006). This work will compare C-band (38–75 mm wavelength) with L-band (150–300 mm wavelength) for sea ice motion estimation.

Motion estimation from consecutive satellite images has its limitations. Only an average velocity can be determined, and that only if the ice surface remains mostly unchanged. Weather conditions can change ice surface properties enough to make feature detection impossible. Generally the method only works for image pairs typically less than 3 days apart, naturally depending on the rate of the ice drift and deformation. Previous work has also concentrated on sequential images from a single instrument, which places a limitation on the availability of suitable image pairs. A satellite might fly over the area of interest only once per day or less. For longer time intervals, velocities due to short-duration events such as storms are lost.

If observations from multiple satellites are used, image pairs mere hours apart are easier to find, but the benefit comes with the added difficulty of comparing images of fundamentally different character. To improve the situation, this work will examine the idea of calculating sea ice motion using two pictures from different instruments, namely EnviSAT ASAR (56.2 mm wavelength), RadarSAT-2 SAR (55.5 mm wavelength) and ALOS PALSAR (236 mm wavelength).

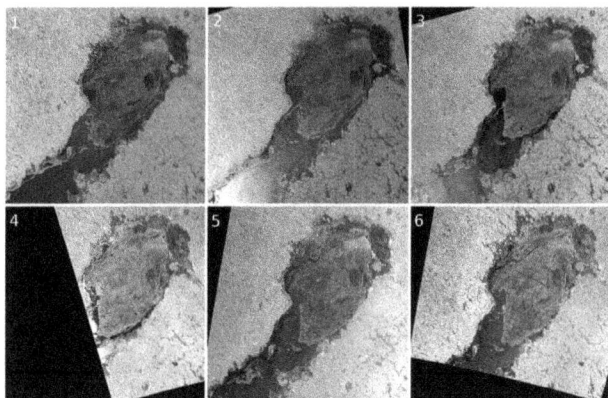

Figure 1. Satellite images used in this work, normalized for viewing. Details given in Table 1. ©MDA, ESA and JAXA.

Table 1. List of satellite images used in this work.

#	tag	satellite	time (UTC)	t	band
1	R1	RadarSAT	16 Mar 2009, 04:59	t_0	C
2	E1	EnviSAT	16 Mar 2009, 19:54	$t_0 + 14{:}55$	C
3	R2	RadarSAT	17 Mar 2009, 16:00	$t_0 + 35{:}01$	C
4	A1	ALOS	17 Mar 2009, 20:12	$t_0 + 39{:}13$	L
5	E2	EnviSAT	18 Mar 2009, 09:04	$t_0 + 51{:}05$	C
6	A2	ALOS	18 Mar 2009, 09:36	$t_0 + 51{:}37$	L

2 Data and methods

For this work, a set of synthetic-aperture radar (SAR) images from March 2009 were used (see Fig.). C-band images were available from both EnviSAT ASAR and RadarSAT 2, while L-band images were available from ALOS PALSAR. A set of six images were chosen for the time period between 16 and 18 March. These days were chosen because there were a relatively large number of images available, including two L-band images. Additionally, two of the images were of different frequency bands and almost simultaneous, with only 32 m between them. This is desirable for comparing frequency bands, and a unique occurrence in the set of images that were available. The images were resampled to 100 m pixel size, approximately corresponding to the nominal resolution of the employed ScanSAR capturing mode.

Lots of changes including compaction and lead opening were present during this period. Landfast ice and open water areas were seen in visual inspection, as well as different types of drift ice. As the ice cover in other parts of the Baltic was sparse, only the seas north of 63° N latitude were considered.

2.1 Weather and ice conditions during the experiment period

For the Baltic Sea, the winter 2008–2009 was milder and shorter than average. Freezing commenced in the Bay of Bothnia in the second half of November, but the ice cover ex-

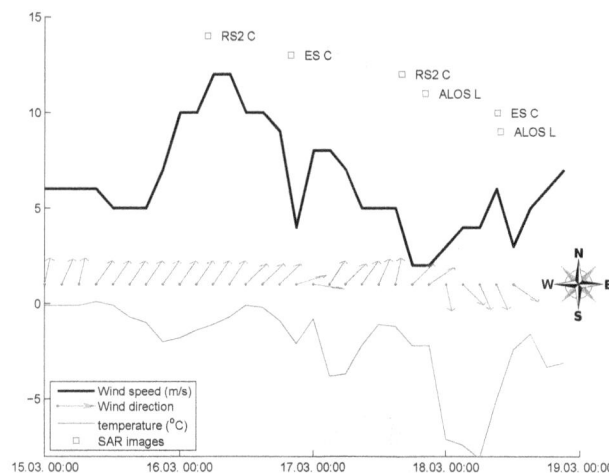

Figure 2. Wind and air temperature recorded by the Kemi 1 lighthouse weather station (65.385° N, 25.096° E) during the experiment period. Timing of SAR images is also marked, red for C-band and blue for L-band images.

tended across the Bay of Bothnia only in the end of January. February was a normal winter month, and the maximum ice cover, 110 000 km^2, was recorded on 20 February. Much of this ice was thin, and after a cold period, warmer southwesterly winds pushed ice northwards during March. On March 16, only the Bay of Bothnia and northern Gulf of Finland had a significant ice cover (The Baltic Sea Portal, 2009).

Figure 2 summarizes the weather conditions recorded by a weather station at the Kemi 1 lighthouse (located at 65.385° N, 25.096° E) during the acquisition of the satellite images. During 16 and 17 March, strong southwesterly winds were pushing the ice pack towards the north. Eventually the wind turned north. On 18 much of the ice had returned southwards and new leads had formed. The temperature remained at or below the freezing point. It is assumed that no significant melting took place during the experiment and that melting did not affect the motion estimation results. Formation of new ice, however, needs to be taken into account.

As reported in ice charts, most of the drift ice in the Bay of Bothnia is deformed, mostly by ridging but also rafting. Not much level ice remains, the well-defined areas being west of the island of Hailuoto and southwest from Tornio. There is no new ice to be found, but large sections of landfast ice lie around the coastline. Reported level ice thicknesses range from 10 to 50 cm in the drift ice and up to 70 cm in landfast ice. Six icebreakers were on duty assisting ships.

2.2 The motion estimation approach

For this work, a straightforward block cross-correlation program was written in the general purpose C++ programming language. The code works directly in the spatial domain, to allow normalized cross-correlation, more flexibility in fine-tuning the computational parameters (Emery et al., 1991) and

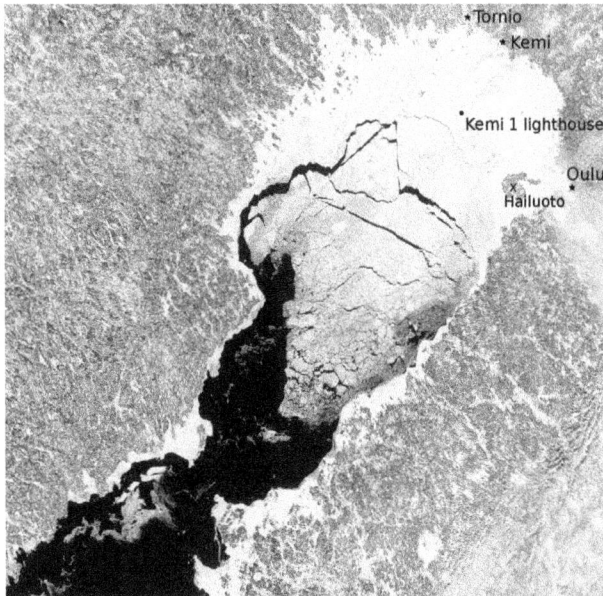

Figure 3. True colour satellite image of the Bay of Bothnia, 18 March 2009, 10:05 UTC. Image captured by the MODIS instrument on board the Terra satellite, courtesy of NASA.

easy parallelization. Critical parts of the algorithm were programmed in OpenCL C, which is a portable language for writing code that can be run in a parallel fashion on a variety of devices (Stone et al., 2010). The cross-correlation code was run on a Graphics Processing Unit (GPU) produced by NVIDIA. This approach cut down the calculation time significantly. The OpenCL cross-correlation program can process one pair of images in roughly 20 s, as opposed to 20 min for a single-core program running on the CPU. This source code is available through the author's website at http://jonni.lehtiranta.net/.

The motion vectors were calculated using a multi-resolution approach. This is usually done to limit the area that has to be processed, but because of the GPU approach, only 48 kB of fast local memory was available. The size of the search domain was limited to 96×96 pixels. First, motion vectors were calculated in a coarse resolution (1/8 of the original or 800 m pixel^{-1}, which allows almost 40 km displacements), and median-filtered result vectors were used as initial guesses for the high-resolution matching step. Finally, the high-resolution result was median-filtered to remove problematic values. For this work, the median filtering radius was chosen to be 3 (as in Karvonen et al., 2007).

For the image windows that were cross-correlated with the search domain, a size of 16×16 pixels was chosen. There is a tradeoff involved in choosing this window size, as it has to be large enough to contain a discernible pattern and at the same time small enough to retain its structure in the time interval separating the pair of images. The chosen size is at the small end of practical options. It was chosen to minimize

errors due to deformations, and to concentrate on errors due to lack of discernible patterns within these windows. This way the error fractions are maximally useful for comparing C-band images to L-band images.

The method consists of the following steps:

1. re-projecting and cropping satellite images using the GDAL toolset,

2. loading the GeoTIFF images, translating 16-bit greyscale values to floating-point numbers,

3. generating a resolution pyramid for both images, using a 2-D low-pass filter and decimating for every level,

4. running normalized cross-correlation for coarse-resolution image windows,

5. median-filtering the coarse result to produce the average motion field and first guess for next step,

6. running normalized cross-correlation for the finest-resolution image windows,

7. saving this result and a median-filtered version (radius 3) of it in an ASCII text file.

The results were analysed and plotted using the Matlab and Octave programs.

2.3 Performance metrics for motion estimation

For this study, no ground truth data was available for comparison. It was necessary to define some performance metric that could be calculated from the results alone. In this work, the cross-correlation method was not tuned for the image types, and especially between C- and L-band images, low cross-correlation coefficients were expected. Instead of the cross-correlation coefficient itself, we consider the ratio of the two highest peaks. While a high peak-to-peak ratio is not conclusive evidence of correctness, it is assumed to be a necessary requirement. A motion vector is rejected if the margin between two highest cross-correlation peaks is less than 15 %, and otherwise accepted in a "peak margin" sense.

Additionally, each motion result is evaluated against the expectation of uniformity, flagging as errors all vectors that differ significantly from the median-filtered vector field. It is assumed that the median filtering succeeds at removing spurious values and retains real stepwise changes in the ice motion field (Astola et al., 1990), so that the median-filtered motion field represents the real average motion. Even when this is not the case, unrealistic vectors will not match it so these cases cannot produce false successes. A motion vector is rejected if it differs from the median of its neighbourhood by more than 500 m. Otherwise it is considered acceptable in a "regularity" sense.

Both criteria are arbitrary. However, they appear to be sensible choices for this study.

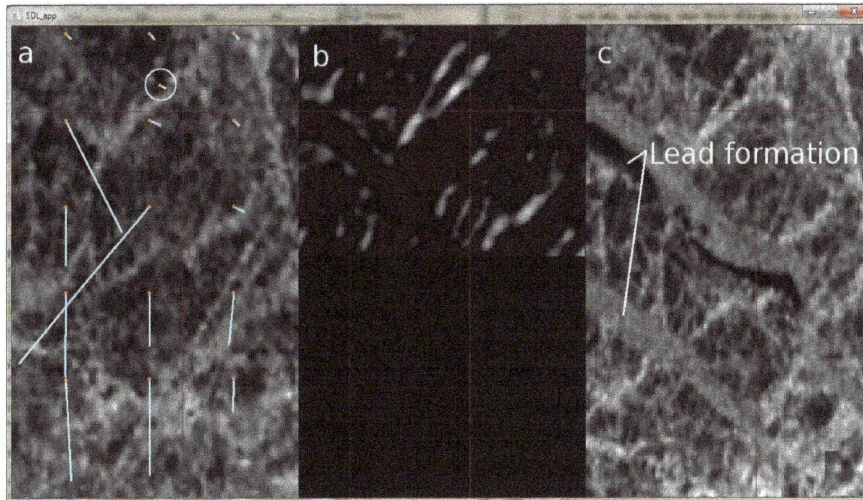

Figure 4. Screenshot of the motion estimation program written for this work. (**a**) Zoom-in of the first image with some detected motion vectors. (**b**) The cross-correlation result for the circled vector. White represents maximum cross-correlation, black represents zero correlation and the area left outside of the calculation. Red represents negative cross-correlation. (**c**) Aligned zoom-in of the second image of the pair. Notice the newly formed NW–SE aligned leads. The thin red lines are rulers that highlight the mouse cursor's location.

2.4 Satellite image processing

Algorithms used for operational satellite image analysis are often tuned to the specific instruments. As the objective of this study is to compare different instruments, no instrument-specific tuning was done. The images still need georectification, and typically a landmask is used.

For this work, SAR images are rectified to the Mercator projection with a reference latitude of $61°40'$. This projection was chosen as it matches the one used in both the nautical charts for this area and previous ice motion estimation work for the Baltic Sea (Karvonen, 2012). There still remains a slight error after this projection step. It could be corrected by matching static features between the images.

An incidence angle correction was not performed. It was deemed unnecessary, as the method calculates normalized cross-correlations for small image windows. No speckle filtering was applied.

2.5 Masking land points

For sea ice motion estimation in the narrow basins of the Baltic Sea, land points are sometimes masked out before analysis (Karvonen, 2012). In this work, motion detection was performed using unmasked images. Result vectors for land and sea areas were then analysed separately. As a drawback, image windows that include the coastline generate two valid cross-correlation peaks. Land points and shallow areas were distinguished by topographical data produced by the Leibniz Institute for Baltic Sea Research (Seifert et al., 2001).

The satellite images were found to suffer from a spatially varying registration error. This was corrected using the finest-

resolution motion estimates for land points. These were interpolated in order to generate a seamless estimate for the image registration error. This registration error field was finally substracted from the motion results recorded for the drift ice.

3 Visual comparison between L- and C-band images

The PALSAR L-band images have been compared to RADARSAT-1 SAR by the Canadian Ice Service. They report that the L-band images contain a far superior amount of ridge information compared to C-band. Large ridges are clearly defined, and detail remains well into the spring melt season. It is also reported that PALSAR allows clearer delineation between ice floes. PALSAR also allows thin ice to be easily distinguished from thick ice, while C-band images could confuse rough thin ice with thicker ice types (Arkett et al., 2008).

As images 5 and 6 (see Table 1 and Fig.) are separated by only 32 min, they are assumed to represent the same ice situation in C- and L-bands. No ice-related change can be distinguished visually, so all differences are taken to result from differences between the imaging instruments. As a general difference, the L-band image (f) has more contrast within the sea area. The coastline is also more easy to distinguish, while in the C-band image, the coastline disappears in some, especially northern, locations. Below, specific differences in these two images are evaluated in detail.

To summarize, ice types in the drift ice region appear similarly in images of both frequency bands. Sometimes the C-band image is better at distinguishing the edge of an ice floe, and sometimes the L-band shows features not visible in the C-band image (see east edge of Fig. 9), but for most features,

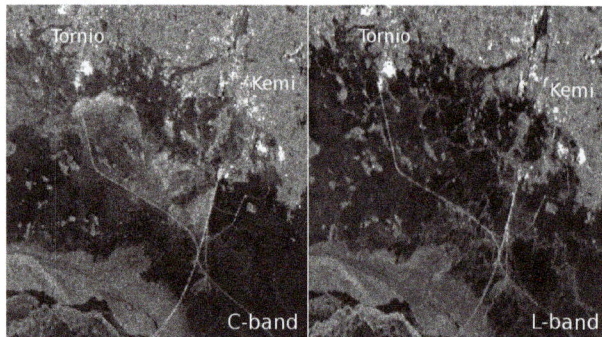

Figure 5. Detail of landfast ice in northern Bay of Bothnia on 18 March 2009. White tracks are shipping lanes to Tornio and Kemi, which appear very bright in SAR images.

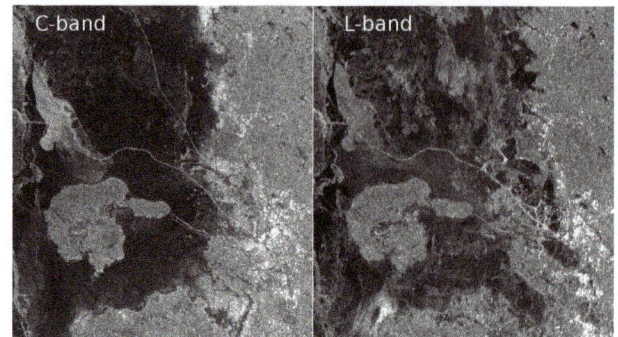

Figure 6. Detail of landfast ice in northern Bay of Bothnia around Hailuoto, offshore from Oulu, on 18 March 2009.

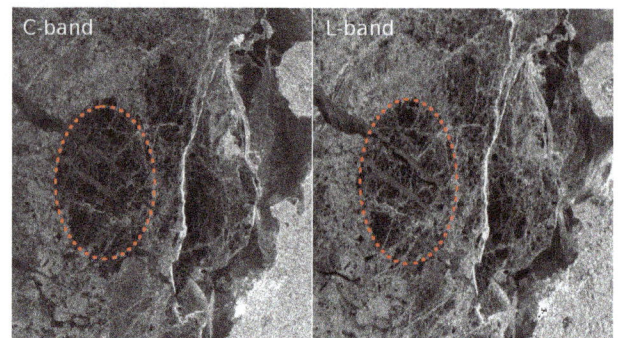

Figure 7. Elliptic dark area classified as level ice near Raahe on 18 March 2009.

the L-band image simply seems to provide stronger contrast. On the other hand, many features in landfast ice appear differently in C- and L-band images. Perhaps a long, relatively peaceful evolution of an ice surface produces surface roughness in length scales comparable to the radar wavelengths.

3.1   Landfast ice

Landfast ice is immobile and non-dynamic by definition. It is assumed that no recent deformation took place in the landfast zone. Discernible features are assumed to be either old deformations or weather-related. As can be seen in Fig. 5, the archipelago looks more homogenous and dark in the L-band image. Conversely, the C-band image shows a large hazy feature, conspicuously framed by the shipping lanes.

The linear or web-like features visible in the L-band image but missing from the C-band image are probably due to the greater volume scattering in L-band. The surface scattering is weaker and less extended, perhaps due to snowfall or melt–freeze events.

Features missing from the L-band image but visible on the C-band image, on the other hand, are probably caused by surface roughness smaller than the L-band wavelength (23.6 cm). The shipping lanes that constrict the bright haze in the C-band image, provide a hint of its formation. This was possibly mobile broken slush, which froze to form a rough surface on the northern side of the shipping lanes.

Near the southwest corner, there is a brighter gray band without clear features. This is the shear zone at the landfast ice boundary, experiencing deformation by external forces but still attached to the landfast ice, islands, or the shallow sea floor. The dark feature under it is open water or thin ice in a lead, and we also see some drift ice in the corner of the image. These features look similar in both images.

In Fig. 6, the L-band image has ill-defined bright features in the landfast ice zone while the C-band shows little scattering. To know the evolutionary history of these features, one would need to track their formation from the beginning of the freezing period. Here, too, early-season deformations could

be masked by smoothing surface processes. The bright feature north of Hailuoto island, which appears similar in both images, is probably a field of broken ice, often called a rubble field, analogous to a very wide pressure ridge.

Comparing these images, it can be concluded that landfast ice can be a tricky substance for matching windows of SAR images of different bands. Some features will appear similar but at different intensities, and some areas will look completely different.

3.2   Level ice

Some ice classified as level ice can be seen in the southwest corner of Fig. 6, southwest from Tornio in Fig. , and in the dark ovals in Fig. 7. These areas show up as relatively dark areas, presumably because of relatively low specular reflection, in SAR images of both wavelengths. In general, C-band shows these features darker than L-band, as L-band will cause more scattering from beneath the level surface (Dierking and Busche, 2006). In some areas, level ice is relatively featureless and in others rather detailed. Some of the areas look identical in C- and L-bands, others show more contrast in L-band. However, based on visual inspection, correlating image windows in level ice seems feasible. This analysis is limited by the small amount of level ice.

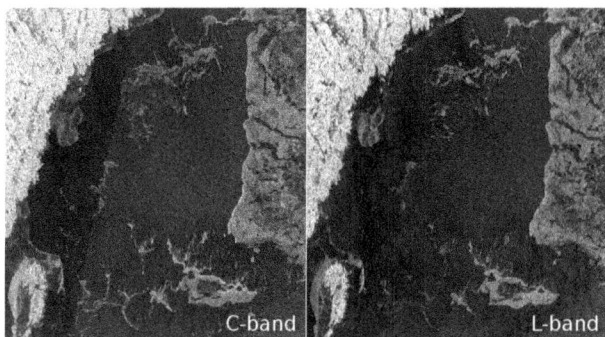

Figure 8. Open ice between the Swedish coast and the compact ice pack in North Kvarken on 18 March 2009.

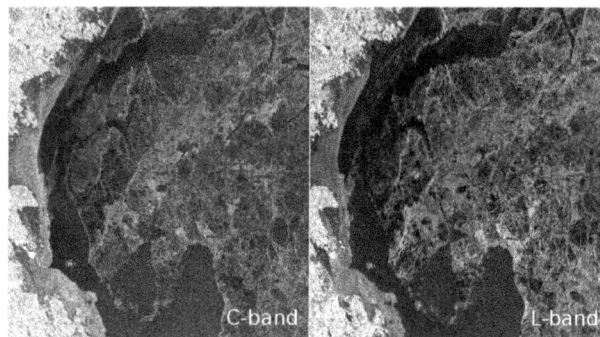

Figure 9. Southern tip of the compact drift ice on the Bay of Bothnia on the 18 March 2009. Encircled the area of faint, barely distinguishable ice floes.

3.3 Open ice

Sea areas with less than 60 % ice cover are classified as open ice. In open ice, separate ice floes drift freely among waves. Using both frequency bands, ice forms similar gray curls, visible in Fig. 8, that should allow motion detection using cross-correlation to work well. Most notable visible differences are dark lines in the open water in the L-band image, and slightly better contrast in the C-band image. However, these formations appear fragile and susceptible to changes, which makes tracking them rather demanding.

3.4 Compact drift ice

Drift ice, classified in Finnish ice maps as consolidated, compact or very close ice, often covers the central Bay of Bothnia during winters. It is a mobile continuum, it deforms readily and it transmits compressive forces over large distances.

In Fig. 9, separate but closely packed floes of compact drift ice can be seen, sometimes separated by leads or other open water features. Many distinct ice floes are recognizable in both images, but the fainter floes near the eastern edge are not visible in the L-band image despite standing out very clearly in the true-colour Fig. 3. The L-band image seems less able

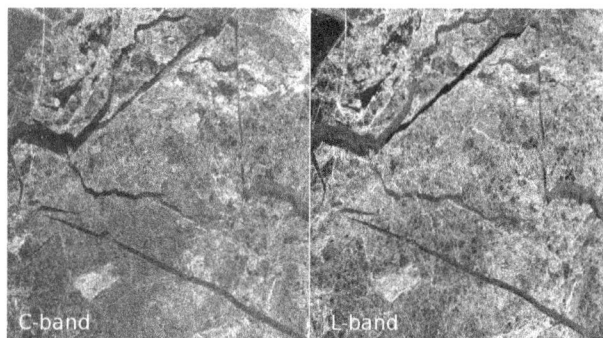

Figure 10. Drift ice on the western Bay of Bothnia, 18 March 2009.

Figure 11. Leads in drift ice, Bay of Bothnia, 18 March 2009.

to distinguish the edge between a lead and a smooth ice floe. Occasionally there is texture not present in the C-band image, such as the bright features in the southeast corner. However, the edge of open water is well visible and similar in both frequency bands, and most ice floes are similar enough for motion estimation.

In Fig. 10, a compact and mostly continuous ice pack is seen in both C- and L-band. Both images reveal the same features, though L-band in better contrast.

It is evident from Figs. 10 and 11 that sometimes leads appear very dark in L-band images. In general however, leads are visible in both kinds of images, and should pose no special problem for motion estimation in a mixed-frequency image pair.

4 Results and discussion

4.1 Motion estimates

To summarize, the motion estimates calculated for image pairs covering the same time interval are similar in all cases. For a C–C or L–L band image pair, the matching is better and motion results may be found for a larger area than in a mixed pair. Based on the metrics defined in Sect. 2.3, an L–L image pair is superior for motion estimates compared to C–C pairs,

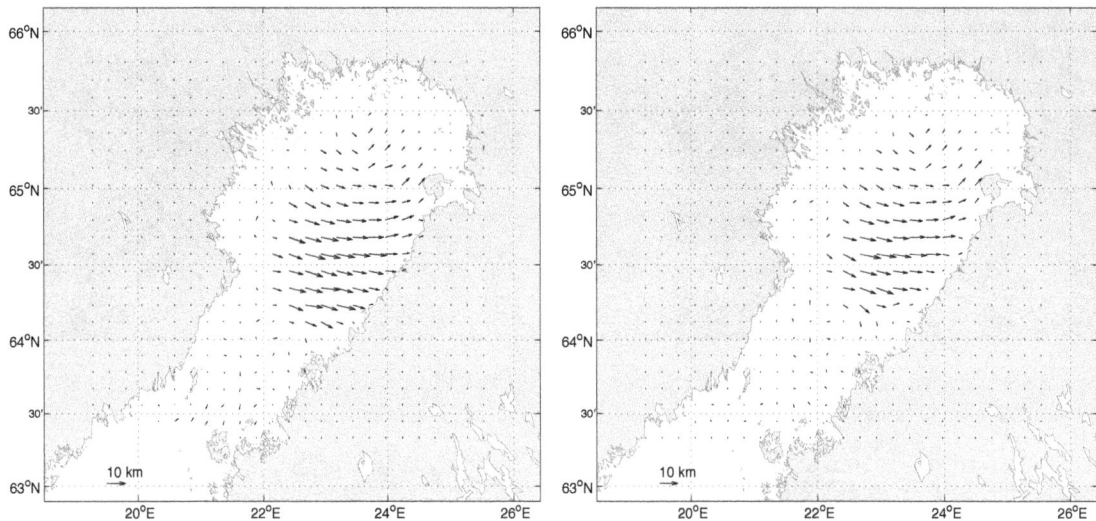

Figure 12. (**a**) Motion vectors from combining images 1 and 6, of C- and L-band, respectively. (**b**) Motion vectors from combining images 1 and 5, both C-band.

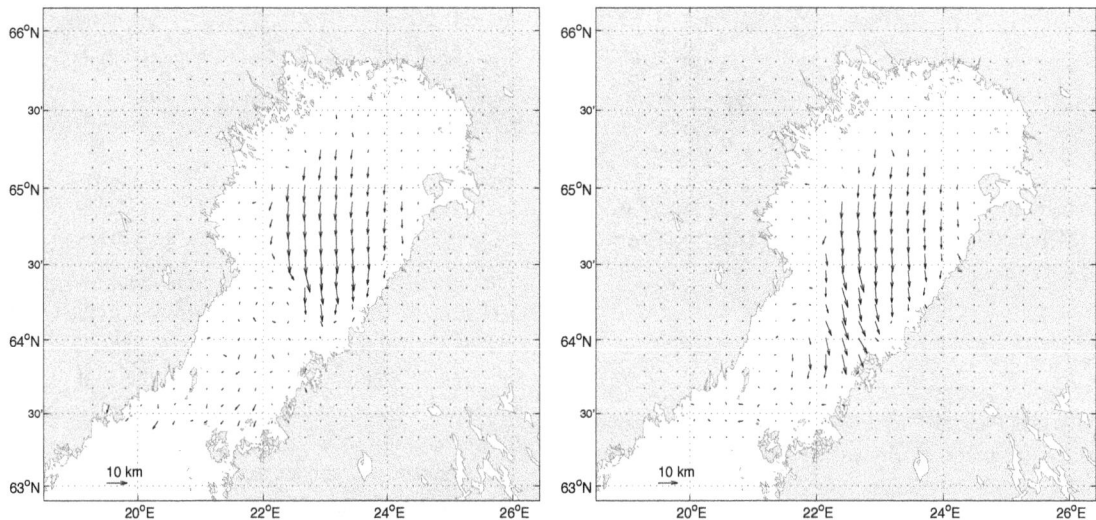

Figure 13. (**a**) Motion vectors from combining images 2 and 6, of C- and L-band, respectively. (**b**) Motion vectors from combining images 2 and 5, both C-band.

while mixed pairs are still feasible despite them presenting the most problematic case.

The average motion for the whole experiment period is shown in Fig. 12. Both a C–C pair and a mixed L–C pair produce an acceptable result for most of the drift ice. The motion fields are almost identical, and the average eastward motion is well supported by the southwesterly winds that turned north towards the end of the period. It is notable though, that neither image pair produces motion for the southern tip of the drift ice area. This is probably because the ice edge changed shape completely, and the numerous ice floes were too small to be distinguished. These two parallel estimates correspond to the R1-A2 and R1-E2 rows in Table 2. Of the

motion vectors in the R1–A2 image pair, 17.6 % had an acceptable cross-correlation peak margin, and 14.0 % of the vectors were close to the local median. For the concurrent image pair R1–E2, both C-band, an additional 2 % of the motion vectors passed both criteria.

In Fig. 13, we see an average southward movement for the latter 36 h of the experiment. This is in line with the prevailing winds as well, as the northward transport of ice had stopped before the winds turned north. This time, for the C-band pair, the southern ice edge is also successful but Fig. 13a shows no motion where Fig. 13b finds realistic vectors. These two parallel estimates correspond to the E1-A2 and E1-E2 rows in Table 2. Again, the C-band pair produces

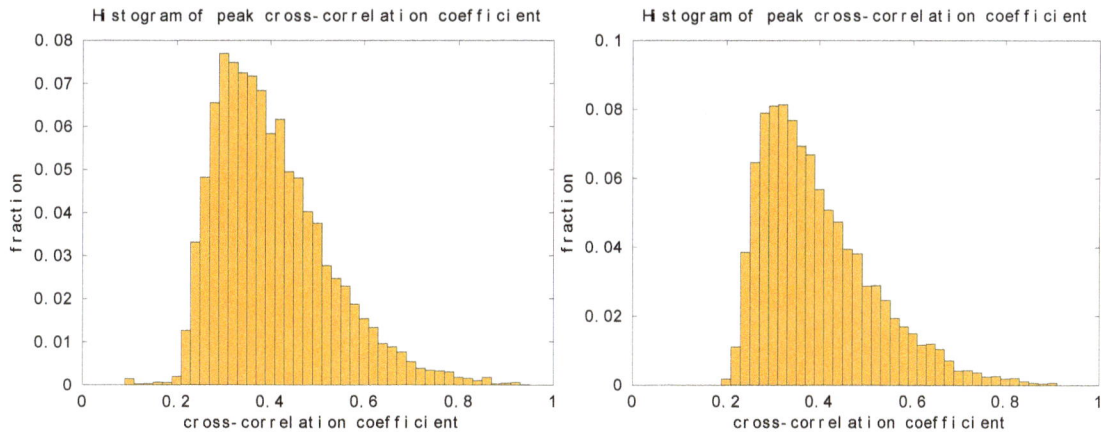

Figure 14. Maximum cross-correlation for matched windows in the R2–A2 image pair (C–L, left) and the R2–E2 image pair (C–C, right)

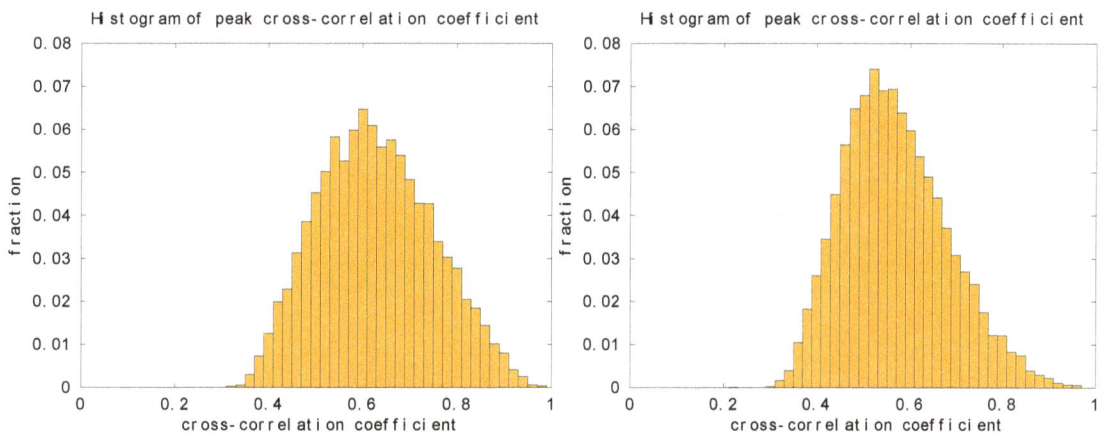

Figure 15. Maximum cross-correlation coefficient histogram for the A1–A2 image pair (L–L), left, and the A1–E2 image pair (L–C), right.

more acceptable vectors, some of which must be located in the southern ice edge, less deformed during the shorter time span covered by these image pairs.

The four latter motion estimates, represented on the two bottom rows of Table 2.3, appear very much like Fig. 13b. This is because each of these image pairs cover the whole period of northerly winds.

Comparing the performance of parallel image pairs, some observations were made. As expected, the motion estimation algorithm works better for shorter timescales, as less deformation has had time to occur. For all image pairs, large-scale motion estimation was successful. All motion estimates contained a large number of spurious vectors too, but a radius 3 median filtering was found to produce a realistic and smooth motion field. Due to the median filtering, the algorithm works even if only 10–20 % of motion vectors are correct. This success rate is thus found sufficient for detecting the large-scale motion. However, as evident in Fig. 13, a mixed image pair can fail in details in some sub-regions.

Same-band image pairs (C–C, L–L) are found better than mixed-band (C–L) pairs. Further, the L-band is found more

suitable for motion estimation in this data set than C-band. Unfortunately, it seems that a large peak margin in cross-correlation is not sufficient as an indicator of correctness. Many motion vectors were found to be nonsensical even when they were produced by a unique cross-correlation peak. This can happen, for example, when the ice surface pattern is lost between images. Upon closer investigation, it was found that a motion estimate using the highest peak is often correct even if the second-highest peak is just barely lower.

4.2 Statistical performance of image pairs

Overall, both C- and L-band image pairs and mixed image pairs show similar statistical properties in the motion results.

For most image windows, the highest found normalized cross-correlation coefficient was between 0.2 and 0.6. The best matches had a cross-correlation coefficient up to 0.95. As can be seen in Fig. 14, for C-band pairs the worst match is around 0.2. This is closer to 0.4 in the L-band pair of Fig. 15, which has overall higher correlation coefficients.

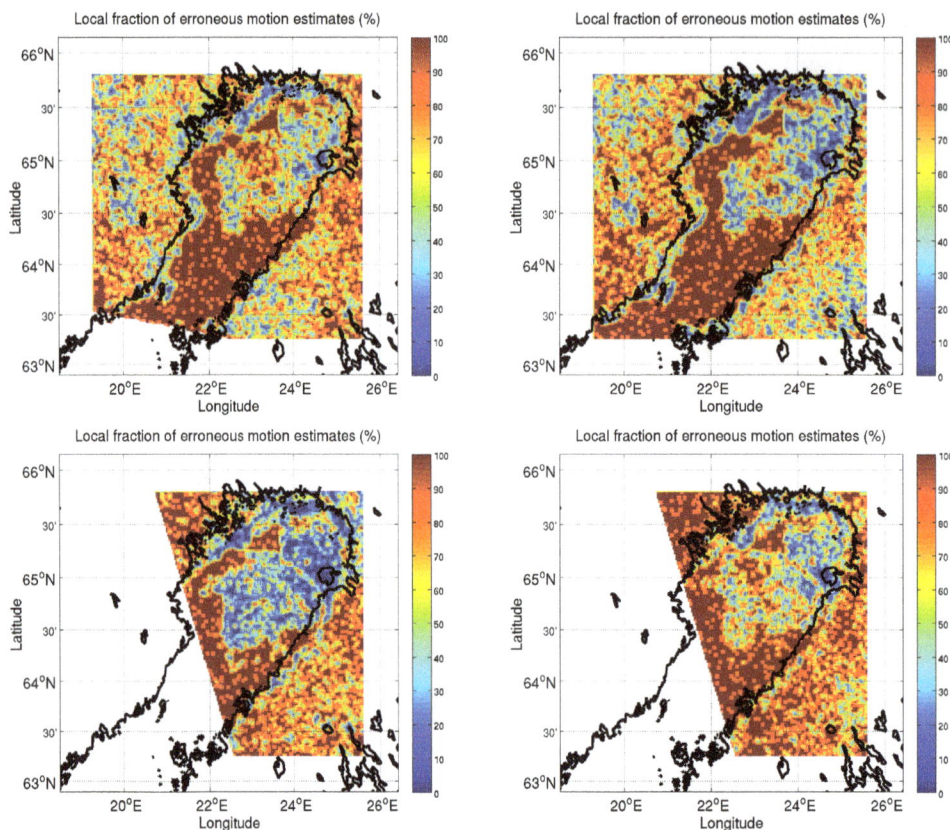

Figure 16. Geographical distributions of errors, (**a**) pair R2–A2 (C–L), (**b**) R2–E2 (C–C), (**c**) A1–A2 (L–L) and (**d**) A1–E2 (L–C)

Table 2. Performance values for parallel image pairs, as the percentage of motion vectors that are accepted based on the peak margin -criterion (pm-good) and regularity-criterion (reg-good), both defined in Sect. 2.3.

Image pair	pm-good	reg-good
R1–A2 (C–L)	17.6 %	14.0 %
E1–A2 (C–L)	20.1 %	14.2 %
R2–A2 (C–L)	24.7 %	15.8 %
A1–A2 (L–L)	45.6 %	28.4 %
R1–E2 (C–C)	19.6 %	16.2 %
E1–E2 (C–C)	22.7 %	16.7 %
R2–E2 (C–C)	27.9 %	18.6 %
A1–E2 (L–C)	30.7 %	18.7 %

The ice conditions and their change are the most important factors of success. This is evident from Fig. 15b. The A1–E2 image pair boasts large cross-correlation coefficients despite mixing two different wavelengths.

The histograms for motion estimation error magnitude, as estimated by the difference in metres between each motion vector and the local median, are all rather similar. The histograms of error show a strong peak for no or very small error and a distribution characteristic of this problem. This distribution roughly corresponds to the idealized theoretical distribution of the distance of a random point. This distribution arises from the fact that the search window is square and it allows at most 40 pixels of movement in each dimension. It is concluded that there are no systematic errors in the motion estimation algorithm.

Considering the margin between the two highest correlation peaks, it was found that a C–C pair is better than a mixed C–L pair at finding unique peaks. The difference is small though, and very often the highest cross-correlation peak stands only slightly above the second contender. This was expected, as the maximal cross-correlation (MCC) method is known to often produce multiple cross-correlation peaks for noisy signals. To improve performance, the algorithm should consider multiple cross-correlation peaks, not just the highest one.

4.3 Geographical distribution of errors

The geographical distribution of errors was calculated for the test cases with smallest time difference in order to evaluate problems stemming from local effects and not changes that occur over longer time intervals. Figures 16a and 16b correspond to the same time interval and show that a C–C pair is stronger than a C–L pair in all localities, but the mixed-band pair also succeeds to some extent everywhere the C–C pair does. Figure 16c and d correspond to another time interval and shows that an L–L pair is much better than a mixed pair, again without any clear difference in the areas of successful motion estimation.

To summarize, all image combinations have trouble with the northwesterly lead opening near the northeast edge of landfast ice, and all combinations behave better in the central ice pack. It is clear that a single-frequency pair is desirable, but also that for most regions, a mixed-frequency pair performs reasonably well. No image pair finds more than an occasional good motion vector in open ice of less than 30 % coverage. It seems that the C-band is better than L-band for matching image patterns on land. While this is of no concern for perfectly georeferenced images, this might mean that georectifying L–L image pairs might be more problematic.

5 Conclusions

We show that it is possible to calculate sea ice motion using an L-band SAR image together with a C-band image. The program written for this purpose works and produces convincing results, so the chosen algorithm of maximal cross-correlation suits this purpose.

L-band images are fundamentally different than C-band images as the ratio of surface and volume scattering is different and some C-band scatterers are invisible to L-band radar. This difference manifests itself primarily in landfast ice, possibly because long periods of thermodynamical changes create different surface features near the length scales of the employed wavelengths. Fortunately, the motion estimation largely succeeds for landfast ice, and most features in drift ice appear much easier targets for motion detection.

The different frequency bands complement each other when plentiful data is available, but they are somewhat poorer for backup purposes as each band has distinct strengths and weaknesses. On C-band, ice floe edges appear in a more reliable manner, while the L-band distinguishes the coastline better and generally shows more features and better contrast.

For motion estimation, a pair of two L-band SAR images is found to be desirable among the compared options. A pair of two C-band images also performs well, and a mixed pair performs adequately. The introduction of L-band SAR instruments can thus present both more reliable motion estimates by using L–L pairs and better time resolution, albeit at a cost of increased uncertainty, by using mixed L–C pairs.

This work provides a new tool for motion estimation. It also provides insights into the usage of L-band SAR images, both alone and in combination with C-band images. Thus, it is good preparation for the future launch of the ALOS-2 satellite and for the handling of its L-band images, and utilizing the GPGPU computational framework was both a strength in this work and a valuable lesson for the future.

Acknowledgements. This study was supported by the Finnish Meteorological Institute and the Polar View project. The authors wish to thank Eero Rinne, Lars Kaleschke and Lang Wenhui for their detailed and insightful comments on the original manuscript.

Edited by: L. Kaleschke

References

Arkett, M., Flett, D., De Abreu, R., Clemente-Colon, P., Woods, J., and Melchior, B.: Evaluating ALOS-PALSAR for Ice Monitoring-What Can L-band do for the North American Ice Service?, in: Geoscience and Remote Sensing Symposium, IGARSS 2008, IEEE International, 5, V–188, 2008.

Astola, J., Haavisto, P., and Neuvo, Y.: Vector median filters, P. IEEE, 78, 678–689, 1990.

The Baltic Sea Portal: Ice winter 2008–2009, available at: http://www.itameriportaali.fi/en/tietoa/jaa/jaatalvi/en_GB/2009/ (last access: 18 December 2014), 2009.

Dierking, W. and Busche, T.: Sea Ice Monitoring by L-Band SAR: An Assessment Based on Literature and Comparisons of JERS-1 and ERS-1 Imagery, IEEE T. Geosci. Remote, 44, 957–970, 2006.

Emery, W., Fowler, C., Hawkins, J., and Preller, R.: Fram Strait satellite image-derived ice motions, J. Geophys. Res., 96, 4751–4768, 1991.

Fily, M. and Rothrock, D.: Sea ice tracking by nested correlations, IEEE T. Geosci. Remote, GE-25, 570–580, 1987.

Karvonen, J.: Operational SAR-based sea ice drift monitoring over the Baltic Sea, Ocean Sci., 8, 473–483, doi:10.5194/os-8-473-2012, 2012.

Karvonen, J., Similä, M., and Lehtiranta, J.: Sar-based estimation of the baltic sea ice motion, in: Geoscience and Remote Sensing Symposium, IGARSS 2007, IEEE International, 2605–2608, 2007.

Liu, A. K., Martin, S., and Kwok, R.: Tracking of Ice Edges and Ice Floes by Wavelet Analysis of SAR Images, J. Atmos. Ocean. Tech., 14, 1187–1198, 1997.

Myrberg, K., Leppäranta, M., and Kuosa, H.: Itämeren fysiikka, tila ja tulevaisuus, Palmenia, Helsinki University Press, 2006.

Seifert, T., Tauber, F., and Kayser, B.: A high resolution spherical grid topography of the Baltic Sea, 2nd Edn., Baltic Sea Science Congress, Stockholm, 25–29 November 2001, Poster# 147, available at: www.io-warnemuende.de/iowtopo (last access: 18 December 2014), 2001.

Stone, J. E., Gohara, D., and Shi, G.: OpenCL: A parallel programming standard for heterogeneous computing systems, Comput. Sci. Eng., 12, 66–72, 2010.

Thomas, M., Kambhamettu, C., and Geiger, C.: Motion Tracking of Discontinuous Sea Ice, IEEE T. Geosci. Remote, 49, 5064–5079, 2011.

Vesecky, J. F., Samadani, R., Smith, M. P., Daida, J. M., and Bracewell, R. N.: Observation of sea-ice dynamics using synthetic aperture radar images: Automated analysis, IEEE T. Geosci. Remote, 26, 38–48, 1988.

10

Response of ice cover on shallow lakes of the North Slope of Alaska to contemporary climate conditions (1950–2011): radar remote-sensing and numerical modeling data analysis

C. M. Surdu[1], C. R. Duguay[1], L. C. Brown[2], and D. Fernández Prieto[3]

[1]Department of Geography & Environmental Management and Interdisciplinary Centre on Climate Change (IC[3]), University of Waterloo, Waterloo, Canada
[2]Climate Research Division, Environment Canada, Toronto, Canada
[3]EO Science, Applications and Future Technologies Department, European Space Agency (ESA), ESA-ESRIN, Frascati, Italy

Correspondence to: C. M. Surdu (csurdu@uwaterloo.ca)

Abstract. Air temperature and winter precipitation changes over the last five decades have impacted the timing, duration, and thickness of the ice cover on Arctic lakes as shown by recent studies. In the case of shallow tundra lakes, many of which are less than 3 m deep, warmer climate conditions could result in thinner ice covers and consequently, in a smaller fraction of lakes freezing to their bed in winter. However, these changes have not yet been comprehensively documented. The analysis of a 20 yr time series of European remote sensing satellite ERS-1/2 synthetic aperture radar (SAR) data and a numerical lake ice model were employed to determine the response of ice cover (thickness, freezing to the bed, and phenology) on shallow lakes of the North Slope of Alaska (NSA) to climate conditions over the last six decades. Given the large area covered by these lakes, changes in the regional climate and weather are related to regime shifts in the ice cover of the lakes. Analysis of available SAR data from 1991 to 2011, from a sub-region of the NSA near Barrow, shows a reduction in the fraction of lakes that freeze to the bed in late winter. This finding is in good agreement with the decrease in ice thickness simulated with the Canadian Lake Ice Model (CLIMo), a lower fraction of lakes frozen to the bed corresponding to a thinner ice cover. Observed changes of the ice cover show a trend toward increasing floating ice fractions from 1991 to 2011, with the greatest change occurring in April, when the grounded ice fraction declined by 22 % ($\alpha = 0.01$). Model results indicate a trend toward thinner ice covers by 18–22 cm (no-snow and

53 % snow depth scenarios, $\alpha = 0.01$) during the 1991–2011 period and by 21–38 cm ($\alpha = 0.001$) from 1950 to 2011. The longer trend analysis (1950–2011) also shows a decrease in the ice cover duration by ~ 24 days consequent to later freeze-up dates by 5.9 days ($\alpha = 0.1$) and earlier break-up dates by 17.7–18.6 days ($\alpha = 0.001$).

1 Introduction

Lake ice cover has been shown to be a robust indicator of climate variability and change. Previous studies have identified lake ice as a highly sensitive cryospheric component to climate conditions (Schindler et al., 1990; Robertson et al., 1992; Heron and Woo, 1994; Vavrus et al., 1996; Walsh et al., 1998; Magnuson et al., 2000; Hodgkins et al., 2002; Assel et al., 2003; Bonsal et al., 2006; Duguay et al., 2006). Climate-driven changes have significantly impacted high-latitude environments over recent decades, changes that are predicted to continue or even accelerate in the near future as projected by global climate models (Overland et al., 2011; Dufresne et al., 2013; Koenigk et al., 2013). With projected amplified warming of polar regions, the ice cover of shallow Arctic lakes is expected to continue reducing in both thickness and duration. Although the response of lakes may be heterogeneous depending on latitude, lake depth and size, water composition and water dynamics, the majority of lakes demonstrate an overall strong response to surface air temperatures (Palecki and Barry, 1986). Persistent warmer air temperatures (Serreze et al., 2000; Trenberth et al., 2007) and increased snowfall observed in the Arctic over the last decades (Jones et al., 2011; Arp et al., 2012), associated with amplified reduction of sea-ice concentrations, thickness and extent (Serreze et al., 2007; Comiso et al., 2008; Walsh et al., 2011), have accelerated during recent years (Walsh et al., 2011). These changes in the Arctic climate system have likely had an impact on ice phenology of lakes in coastal regions adjacent to the Arctic Ocean.

Changes in lake ice cover could in turn have an important feedback effect on energy exchanges between the lake surface and the atmosphere, and on water levels and therefore on lake water balance, water properties and quality. As a result, water resources, food supply, aquatic habitat, and underlying permafrost conditions will undergo changes at various spatial and temporal extents. Through their heat and water budgets, lakes play an important role in the local and regional climate of high-latitude regions. Longer open-water seasons lead to increased exposure to solar radiation that, through evaporation, results in extended latent heat release from lakes to the atmosphere, the amount of latent heat being twice that released by the adjacent tundra (Mendez et al., 1998). In permafrost areas such as the North Slope of Alaska (NSA), changes in lake water balance, dynamics or temperature can also disturb the underlying permafrost layer, resulting in thaw (Romanovsky et al., 2010) with consequent talik formation and lateral lake water drainage, and also in carbon dioxide and methane release to the atmosphere (Walter et al., 2006). The presence of liquid water underneath ice extends fresh water availability for residential and industrial use throughout the winter. The changing ice cover of high-latitude lakes not only alters the physical and thermal properties of lakes but also affects the chemical properties and the dependent biota; warming lake water temperatures may lead to extinction, blooming or migration of various biological species. However, the magnitude to which changes within the Arctic lakes affect the dependent ecosystem is complex but yet poorly understood and remains to be further investigated.

In response to warmer climatic conditions and to changes in snow cover in recent decades, break-up dates in particular have been occurring earlier in many parts of the Northern Hemisphere (Magnuson et al., 2000; Duguay et al., 2006). The presence of trends in lake ice duration may be occasionally masked by the seasonal, annual or decadal variability that is influenced by the intensity and duration of a climatic episode. Under warmer climate conditions, shallow tundra lakes, many of which are less than 3 m deep, are expected to develop thinner ice covers, likely resulting in a smaller fraction of lakes that freeze to their bed in winter, earlier ice-off dates, and overall shorter ice seasons. Shallow lakes of the Alaskan Arctic Coastal Plain (ACP), Arctic Siberia (Grosswald et al., 1999; Smith et al., 2005; Sobiech and Dierking, 2012), the Hudson Bay Lowlands (Duguay et al., 1999, 2003;

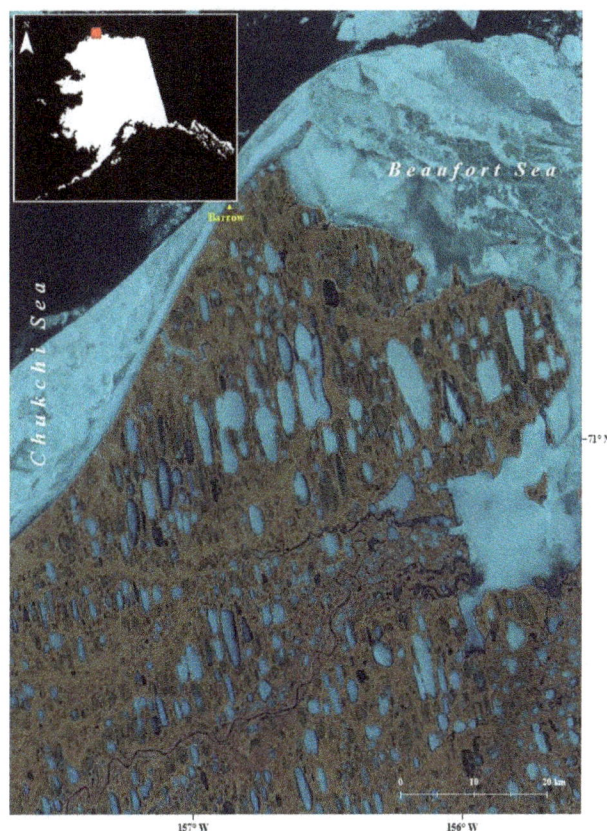

Fig. 1. Sub-region of the Alaskan Arctic Coastal Plain, near Barrow (71°17′ N, 156°46′ W). The satellite view of Barrow was provided by NASA, Landsat program 2011, Landsat TM L1T scene (ID: LT50790102011170GLC00). Publisher: USGS. Acquisition date – 9 June 2011.

Duguay and Lafleur, 2003; Brown and Duguay, 2011a) and other similar regions in the Arctic, have likely been experiencing changes in seasonal ice thickness and phenology (e.g., freeze-up, break-up, and ice cover duration) over the last decades but few studies have documented these changes.

Past changes in lake ice cover have mostly been identified only through non-spatially representative point in situ measurements, which have been almost unavailable over the last two decades following the decline of the global terrestrial monitoring network for fresh-water ice (Lenormand et al., 2002; Prowse et al., 2011). Recent studies have demonstrated that satellite remote sensing provides a viable alternative to detecting and monitoring changes of the ice cover on high-latitude lakes (Latifovic and Pouliot, 2007; Arp et al., 2012; Duguay et al., 2012). Previous remote-sensing investigations indicate that optical sensors are not the ideal tool for comprehensive monitoring of lakes since they are limited by the presence of cloud cover and extended polar darkness (Hinkel et al., 2012), and in most cases, by moderate spatial resolution (i.e., 100–1000 m). Instead, with fewer restrictions (i.e., allowing imaging under cloudy and darkness

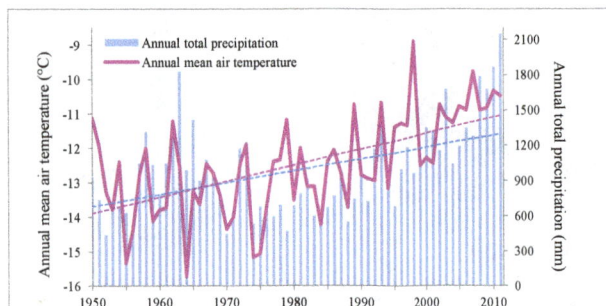

Fig. 2. 1950–2011 annual mean air temperature and total precipitation (rain and snowfall) as recorded at the National Weather Service station, Barrow, AK. The dashed lines indicate the trend for annual mean air temperature (2.9 °C increase in 62 yr, $\alpha = 0.001$) and for annual total precipitation (639 mm increase in 62 yr, $\alpha = 0.001$).

conditions), spaceborne synthetic aperture radar (SAR) has been shown to be the most efficient tool for detecting changes in Arctic lake ice (Jeffries et al., 1994, 2005; Morris et al., 1995; Duguay et al., 2002; Cook and Bradley, 2010; Arp et al., 2012; Jones et al., 2013). Recent attempts to identify the response of shallow lakes of the NSA to contemporary climate conditions exist and changes in the grounded ice fraction has been noticed. However, the short period covered by these studies using satellite observations, 2003–2011 (Arp et al., 2012) and 2008–2011 (Engram et al., 2013) precludes identification of a trend.

The climate trajectory in the Barrow region has taken an abrupt turn during the first decade of the 21st century, with mean air temperatures increasing by 1.7 °C (Wendler et al., 2012), a change that has been shown to impact the lake ice regimes in this coastal region. Consequent to warmer air temperatures and increased precipitation (Callaghan et al., 2011) during recent decades (Fig. 2), the ice regimes of these shallow lakes are expected to develop thinner ice covers, earlier melt, and shorter ice seasons. Transition toward higher floating ice fractions, with more lakes maintaining liquid water underneath the ice cover and fewer lakes freezing to the bottom by the end of winter is also likely to occur.

The objectives of this study are to identify changes in the maximum ice thickness by Arctic shallow lakes as derived from both SAR satellite observations and numerical modeling scenarios during recent decades, to report observed and simulated changes in the ice phenology of these lakes from 1950 to 2011, and to determine potential ice regime trends during recent decades. To achieve these goals, this study (1) analyzes and reports monthly changes in the fraction of lakes that froze to the bed in winter between 1991 and 2011, (2) evaluates the rate of change of late winter maximum ice thickness during the past two decades, (3) presents the identified changes in lake ice thickness and duration as derived from a numerical lake ice model (1950–2011), and (4) cor-

relates SAR-detected changes within the lake ice cover with model-simulation results (1991–2011).

2   Background

Airborne X-band and C-band SAR images acquired over shallow lake regions have been shown to be useful for determining the presence of floating or grounded ice (Weeks et al., 1978; Mellor, 1982) and timing of lakes that freeze to their bed in winter (Elachi et al., 1976). The first analysis of ERS-1 SAR data over lakes on the NSA was performed by Jeffries et al. (1994) during the ice season of 1991/1992. The study shows that monitoring the evolution of radar return, also referred to as radar backscatter intensity (σ°), is an efficient tool in detecting ice onset and melt, as well as floating or grounded ice. The low temporal resolution of the ERS sensors, a repeat cycle of 35 days, would not be suitable for the determination of freeze-up and break-up dates. However, considering that this study focuses on the determination of grounded and floating ice during the ice growth season, this is not a significant issue since high-temporal resolution is not necessary (i.e., ERS observations still allow for the monitoring of monthly changes in the fraction of floating and grounded ice, and appropriate comparison between years). In order to complement these observations and be able to simulate freeze-up and break-up dates with a daily temporal resolution, CLIMo was employed.

Similar work by Jeffries et al. (1996) used SAR coupled with a numerical lake ice model to determine the timing of maximum ice thickness and the number of lakes on the NSA that freeze to their bottom, and estimate the depth of these lakes. These results are summarized in the study area section of this paper. Likewise, Duguay et al. (1999) and Duguay and Lafleur (2003) evaluated the presence of floating and grounded lake ice, and the timing of maximum ice thickness, with ERS-1 SAR observations of the Hudson Bay Lowlands, near Churchill, Manitoba. Methods developed in these earlier investigations have been recently applied to map fish overwintering habitat in channels of the Sagavanirktok River, Alaska (Brown et al., 2010), to estimate methane sources (Walter et al., 2008) or to determine winter water availability in Alaska (White et al., 2008). Providing that discrimination between floating and grounded ice is facilitated by the high contrast displayed in SAR images, C-band SAR has been shown to be the most useful frequency for distinguishing between the two different ice cover conditions (Engram et al., 2013).

Generally, low radar returns (-17 to -12 dB) in C-band ERS-1/2, VV polarized SAR imagery indicate the presence of a thin, relatively uniform ice cover at the beginning of the ice season, associated with specular reflection off the ice–water interface (Duguay et al., 2002). Low radar backscatter also attests the existence of grounded ice later into the growing season, explained by the low dielectric contrast at the ice–lake-bottom interface and the absorption of the radar signal into the substrate (Jeffries et al., 1994). Steady increase in radar backscatter persists during the ice season, from November to April, as ice continues to grow and remains afloat. Maximum returns (-11 to -2 dB) are associated with the presence of floating ice. Higher backscatter from floating ice is a combination of high difference in dielectric properties between the ice and the underlying liquid water (Weeks et al., 1978), and the presence of air inclusions in the ice layer. The higher radar return could also be explained by the presence of smaller tubular bubbles formed during freeze-up or of larger ebullition spherical bubbles, resulting in a double-bounce effect (Mellor et al., 1982). Ice decay at the end of the season is characterized by low radar returns from the melting ice and snow, and/or ponding water that reflect the radar signal in a direction away from the sensor (Duguay et al., 2002).

3　Study area

The study focuses on a region that encompasses 402 lakes, near Barrow ($71°31'$ N, $156°45'$ W) on the NSA (Fig. 1), an area that is dominated by the ubiquitous presence of shallow thermokarst lakes, lakes that are reported to cover up to 40 % of the coastal plain (Sellmann et al., 1975; Hinkel et al., 2005). The area is dominated by the polar marine climate, with cold air temperatures and high winds. The mean annual air temperature (1921–2011) recorded at Barrow is $-12\,°C$ and the mean annual precipitation fall is 845 mm (106 mm liquid precipitation and 739 mm snowfall). The east, east-northeast prevailing wind has a mean annual speed of $19.1\,\mathrm{km\,h^{-1}}$ (National Climate Data Center, 2012). Summer air temperatures are usually highest in July, with a mean air temperature of $4.4\,°C$, and lowest in February, with a mean of $-26.6\,°C$.

The area of lakes investigated in this study ranges from 0.1 to $58\,\mathrm{km^2}$. Despite the unknown bathymetry for the majority of these lakes, using a numerical ice-growth modeling approach, Jeffries et al. (1996) determined that 23 % of the lakes may be deeper than 2.2 m, 10 % with depths ranging from 1.5 m to 2.2 m, 60 % between 1.4 m and 1.5 m, and 7 % less than 1.4 m. A considerable number of lakes on the Alaskan ACP freeze to their bed each ice season (Mellor, 1982), and are ice free for only eight to ten weeks per year (Jeffries et al., 1996). Ice formation, mostly a function of lake morphometry and air temperature, commences in mid-September (Jeffries et al., 1994; Liston et al., 2002; Jones et al., 2009) or early October (Hinkel et al., 2003) and attains a maximum growth rate in November (Jones et al., 2009) that is followed by a slower growth rate until early March or later (Jeffries et al., 1996), when many shallow lakes freeze to the bottom. Depending on lake water depth, the timing of maximum ice thickness varies and can occur any time between late April (Jeffries et al., 1996) and May (Jones et al., 2009). Changes in air temperature, snowfall timing and snow

depth prior to and during all months of freeze-up (ice-on) and break-up (ice-off) affect the timing of these ice events. However, maximum ice thickness is primarily driven by changes in the April air temperature and it happens earlier by six days following higher air temperatures and is delayed by seven days if lower April air temperatures occur. Changes in April snow depth do not affect the timing of maximum ice thickness (Morris et al., 2005).

April air temperature (monthly mean of the 2 m air temperature) was shown to also strongly affect the ice decay of central Alaskan lakes, with a $\pm 1\,°C$ change in air temperature resulting in an advance or delay of break-up dates by ± 1.86 days (Jeffries and Morris, 2007). Ice break-up of lakes on the NSA is driven by changes in air temperatures and the presence of an insulating snow cover, and may commence as early as April and last until June (Hinkel et al., 2003) or even July (Hinkel et al., 2012), when lakes become completely ice free. Field measurements indicate that snow is still present on lakes during the month of April (Jeffries et al., 1994; Sturm and Liston, 2003) hence ice break-up commences after the disappearance of the snow cover on top of lakes in late April or May.

4　Data and methods

4.1　SAR-image processing

A time series of 79 SAR images from 1991 to 2011, between December and early May, standard low resolution (100 m pixel size and 240 m spatial resolution) ERS-1/2 (C-band, 5.3 GHz), VV polarized (vertical transmit and vertical receive), ascending and descending passes, was radiometrically calibrated and geocoded with the MapReady software (v2.3.17) provided by the Alaska Satellite Facility (ASF). Following calibration and geocoding, each individual image was segmented in order to derive the ice cover fractions for both floating and grounded (bedfast) ice.

In order to ensure that the images used in the analysis were not affected by possible melt at the end of the ice season, daily air temperatures recorded at the Barrow meteorological station were also taken into consideration in order to confirm that air temperature values prior to and at the time of SAR acquisitions were below $0\,°C$. Optimum radar images – not affected by possible melt at the end of the ice season – are acquired during April (Mellor, 1982), also coinciding with the approximate timing of maximum ice thickness in this study area. As SAR imagery was not consistently available on the same date during the 20 yr of study, assessment of differences in the grounded ice fraction between images acquired a few weeks apart was also performed. April to early May imagery was selected to derive the fraction of lakes frozen to the bed since images acquired later in the season may be affected by the presence of wet snow or ponding water on the ice surface and therefore result in erroneous results (Hall et al.,

1994). Evaluation of differences between the ascending and descending pass acquisitions at a two-day interval showed a difference of 1.5 % to 2 % in the fraction of grounded ice. The higher fraction of grounded ice was consistently noticed in the descending pass images, acquired two days after the ascending pass. The differences in the grounded ice fraction observed in the overlapping ascending and descending images are attributed to the right-looking ERS geometry. The SAR looking geometry of ERS – from the east in ascending mode and from the west in descending mode – limits the identification of the exact same ground features in overlapping images due to the angle of illumination. Issues such as foreshortening and layover are known to result in possible deformations in area where the topographic slope is greater than 10°. Considering that the study area is a coastal plain, such deformations are likely minimum and the difference in the grounded ice fraction is associated with the illumination differences.

ERS imagery (December to March) was not available on a monthly basis during the 20 yr period. However, SAR acquisitions during April were available for each year included in the study, except for 1996, 2002 and 2004 when images acquired on 3 May, 2 May, and 6 May were used to obtain the late-winter grounded ice fractions.

In order to map lake areas frozen to the bed and those with ice afloat in the Barrow region of the ACP, image segmentation was performed for the 79 ERS-1/2 acquisitions (Fig. 3 shows segmentation results of late winter images). The automated segmentation combines gradual increased edge penalty and region growing techniques, both incorporated in the Iterative Region Growing with Semantics (IRGS) algorithm and implemented in the MAp-Guided Ice Classification System (MAGIC) software (Clausi et al., 2010). The method, proved to be robust, has been fully validated and is being successfully used by Environment Canada's Canadian Ice Service (CIS) for sea-ice classification. The statistical and spatial characteristics of pixels in SAR images have been effectively modeled with IRGS and successfully used in a recent study to map and monitor ice cover on large northern lakes (Ochilov et al., 2010). Given that different ice types are present on lakes, a three-cluster segmentation (two floating ice classes and one grounded ice class) was used. In order to further verify the performance of the three-cluster segmentation, a five-cluster segmentation was at times performed. Following the input of each individual SAR image and the corresponding vector file of lakes included in the study area, automated image segmentation is performed with IRGS, and the output is a file that includes fractions for all three classes that were initially selected by the operator. In order to determine the total fraction of grounded and floating ice, visual assessment of each segmentation result against the original SAR image is performed, and all resulting ice classes are merged into two classes (grounded and floating ice) by a human operator. Once merging was completed, a two-class map was generated for each date of SAR imagery included

in the analysis. Low-resolution (100 m pixels) images were segmented with IRGS and further classified as floating and grounded ice. The current study extends the use of IRGS in documenting and analyzing changes in ice cover on shallow lakes.

4.2   Lake-ice modeling

The lake ice model CLIMo was used to derive lake ice thickness, freeze-up and break-up dates. CLIMo was forced using data obtained from the online archives of the National Climate Data Center (mean daily 2 m screen air temperature, relative humidity, wind speed, cloud cover fraction, snow depth) for the Barrow meteorological station (1950–2011). As meteorological data was not available for all years prior to 1950, the model was forced with available data from 1950 onward. In order to capture the typical observed variability in snow depth on Arctic coastal lakes, simulations were performed with two scenarios: one that assumed that no snow was present on the ice surface (0 %) and a second one that assumed a 53 % snow cover depth, calculated as a fraction from the total snow depth measured over land. The mean lake depth specified in the lake model simulations was 3 m. This model has been extensively tested over various lake regions, including the NSA (Duguay et al., 2003). CLIMo results presented in the study were in good agreement with both ERS-1 SAR observations and in situ measurements during the winter of 1991–1992. For example, the simulated ice-on date was 19 September, while satellite observations indicated that freeze-up occurred between 11–20 September 1991 (Jeffries et al., 1994). Similarly, the latest ice-off date simulated with CLIMo in a no-snow scenario was 14 July 1992 and the SAR-derived one was 15 July 1992 (Zhang and Jeffries, 2000), only one day apart. Maximum ice thickness simulations (165–221 cm) in snow depth scenarios ranging from zero to 100 % displayed differences of 5–6 cm when compared to field measurements (159–216 cm) during 19–29 April (Jeffries et al., 1994). More recently, the model has been further evaluated for a shallow lake near Churchill, Manitoba (Brown and Duguay, 2011a), and at the pan-Arctic scale for lakes of various depths (Brown and Duguay, 2011b). Analysis of model performance at the pan-Arctic scale showed that the average absolute error for determining ice-on and ice-off dates was less than one week when compared to field observations on 15 lakes in northern Canada. The mean maximum ice thickness difference between simulations and in situ measurements for three sites was 12 cm (6.5 %).

Wind redistributes snow, resulting in a thinner and denser snow layer over lakes than over land (Sturm and Liston, 2003), with a reported average fraction of 52 % between the snow depth measured over lake ice and the snow depth measured over land at the Barrow weather station (Zhang and Jeffries, 2000). Considering the wide fluctuations in snow cover fraction associated with its redistribution during the

Floating ice
Grounded ice

62%	46%	42%	51%
20 April 1992	21 April 1993	29 April 1994	14 April 1995
53%	42%	39%	34%
3 May 1996	19 April 1997	23 April 1998	24 April 1999
51%	48%	57%	40%
22 April 2000	28 April 2001	2 May 2002	17 April 2003
45%	51%	39%	41%
17 April 2004	21 April 2005	22 April 2006	26 April 2007
31%	29%	28%	26%
5 April 2008	25 April 2009	29 April 2010	16 April 2011

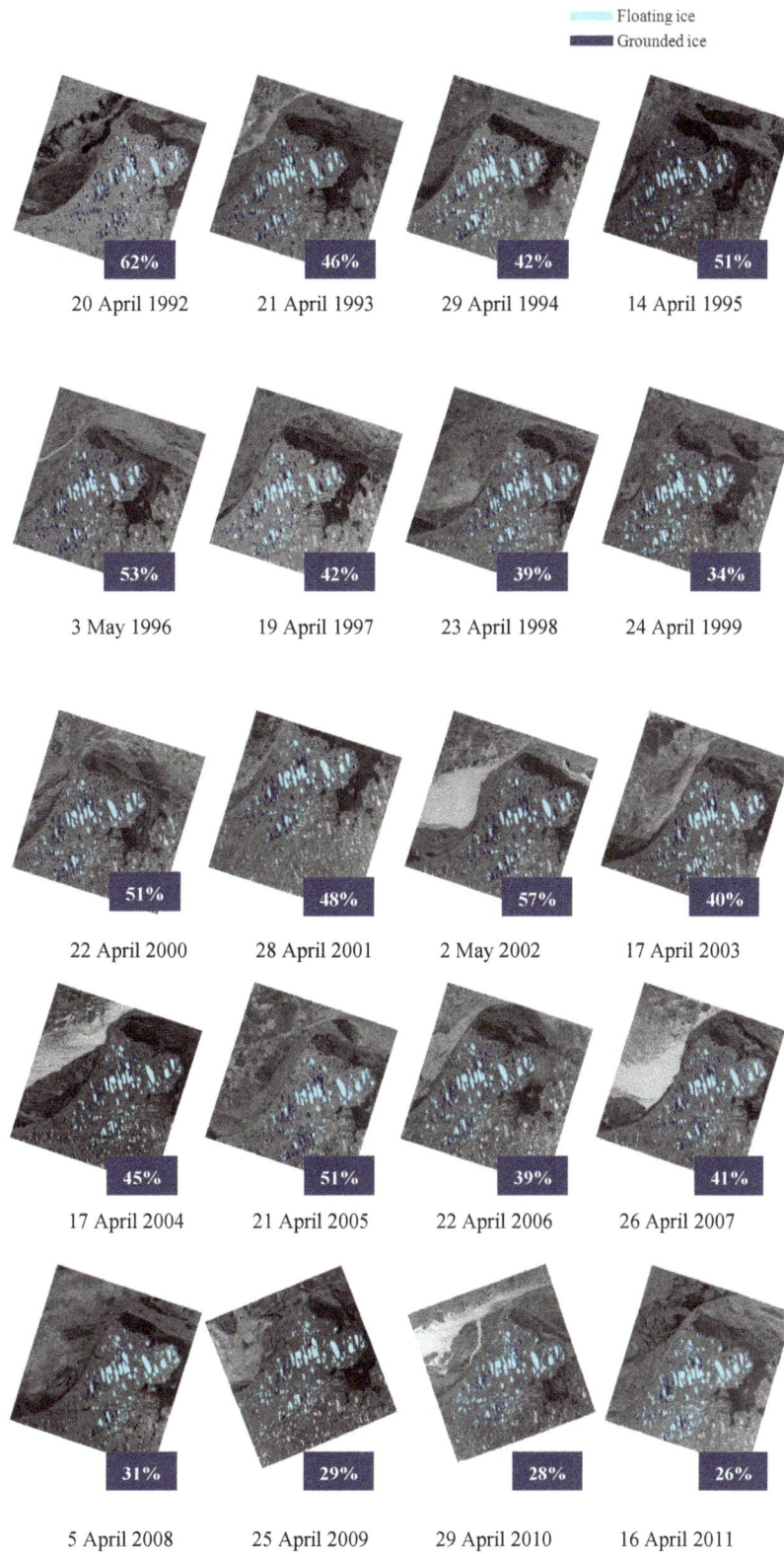

Fig. 3. Image segmentation results of ERS-1/2 SAR images acquired near the time of maximum ice thickness for lakes near Barrow, from 1992 to 2011. The fraction of grounded ice for each date is also shown. Data source: Alaska Satellite Facility. All SAR images are copyright ESA (1992–2011).

Table 1. Dates of ERS-1/2 acquisitions used for image segmentation in order to determine the monthly fraction of grounded ice (1991–2011).

	Day and month of SAR acquisition					
Year	Dec	Jan	Feb	Mar	Apr	May
1991	30					
1992	16	20	13	17	20	
1993	25	25	10	17	21	
1994	10	21	17	16	29	
1995	15	16	22	11	14	
1996		04	23			3
1997	20		8		19	
1998	24	24	12	19	23	
1999		28			24	
2000	28	13	17	23	8	
2001	13			24	28	
2002		17				2
2003	18		6		17	
2004	18	17	21	11		6
2005	22	22	10	17	21	
2006				16	22	
2007	22			22	26	
2008	27	12		22	10	
2009	31	15	19	26	9	
2010		16	20	25	29	
2011		20		17, 26	16	

winter season and accounting for wide variations observed in snow density (198–390 kg m^{-3}) in this area (Sturm and Liston, 2003), model simulations for the snow scenario were performed with a 53 % snow depth fraction and a fixed snow density of 335 kg m^{-3}. The calculated snow depth fraction over lakes and model input for snow density was based on available field measurements in the Barrow region from 1991 to 2006.

5 Results

5.1 SAR-data analysis

A 20 yr time series of ERS-1/2 SAR images (1991–2011), with acquisition dates between mid-December and early May, was analyzed (Table 1). The results show not only an expected inter-annual variability but also a gradual transition toward higher floating ice fractions, particularly noticed during recent years. The observed fraction of grounded ice, calculated as a monthly mean, gradually increased during the winter, from a December mean of 15 % to a mean of 43 % in April, when ice is most likely to grow to its maximum thickness. Assessment of grounded ice fractions during the winter seasons (1991–2011) with available ERS imagery indicates a gradual trend toward lower fractions of grounded ice in all months of observations in the image time series. The

greatest change was observed to occur in April, with maximum deviation values ($\pm$15–18 %) from the monthly mean of 43 % calculated from all years (1992–2011) and a standard deviation of 9.83. The highest positive deviation was observed in 1992 (more grounded ice) and the highest negative value in 2011 (less grounded ice). The transition toward lower fraction of grounded ice during late winter (April) correlates well with the trend toward thinner ice covers as indicated by model simulations during the same period ($r = 0.75$, $p < 0.001$; Fig. 4).

The trend accentuated from 2006 onward (as observed in Fig. 5), observation also reported in a similar study from an adjacent region of the NSA (Arp et al., 2012). Trend detection was performed using the Mann–Kendall test, a method often used for detecting the presence of linear trends in long-term lake ice observations (Futter, 2003; Duguay et al., 2006). Trend magnitude (slope) was estimated with Sen's method (Sen, 1968; Duguay et al., 2006; Noguchi et al, 2011). Statistical analysis of SAR data over the 20 yr period, indicates a decrease of 22 % in the fraction of lakes that freeze to the bed in April (1.1 % yr^{-1}, $\alpha = 0.01$). The maximum number of lakes froze to their beds in 1992 when the fraction of grounded ice was 62 %, as opposed to 2011 when a minimum fraction of 26 % of bedfast ice was noticed. The 2011 lowest April fraction of grounded ice was also observed with Envisat Advanced Synthetic Aperture Radar (ASAR) imagery from 2003 to 2011 (Arp et al., 2012).

5.2 Model results

The performance of CLIMo vs. field-measured ice-on, ice-off dates, and thickness and against ice-on and ice-off dates from satellite observations was previously shown to agree well (Duguay et al., 2003). To further demonstrate the good agreement of CLIMo with in situ measurements, model results were statistically compared to observed mean ice thickness during several years with available field data. Accurate ice thickness simulations are dependent on using representative snow cover depths and densities over lakes for model runs. Data on ice thickness, snow cover depth, and (rarely) snow density over lakes is available for a selection of lakes near Barrow for the 1978/79 (Imikpuk Lake, Ikroavik Lake and West Twin Lake; Mellor, 1982) and 1991/1992 (Ikroavik Lake, Emaiksoun Lake and Emikpuk Lake; Jeffries et al., 1994) ice seasons. The mean bias error (MBE) with both snow scenarios (0 % and 53 %) compared to in situ measurements was +8 cm (4 %), indicating that CLIMo generates reliable ice thickness simulations for lakes in the Barrow region.

The longer historical-trend analysis (1950–2011) of maximum ice thickness derived from CLIMo simulations indicates the development of thinner ice covers on the Alaskan shallow lakes. Model runs with two different snow scenarios (0 % and 53 % snow cover depth), forced with data from the Barrow meteorological station, show a significant

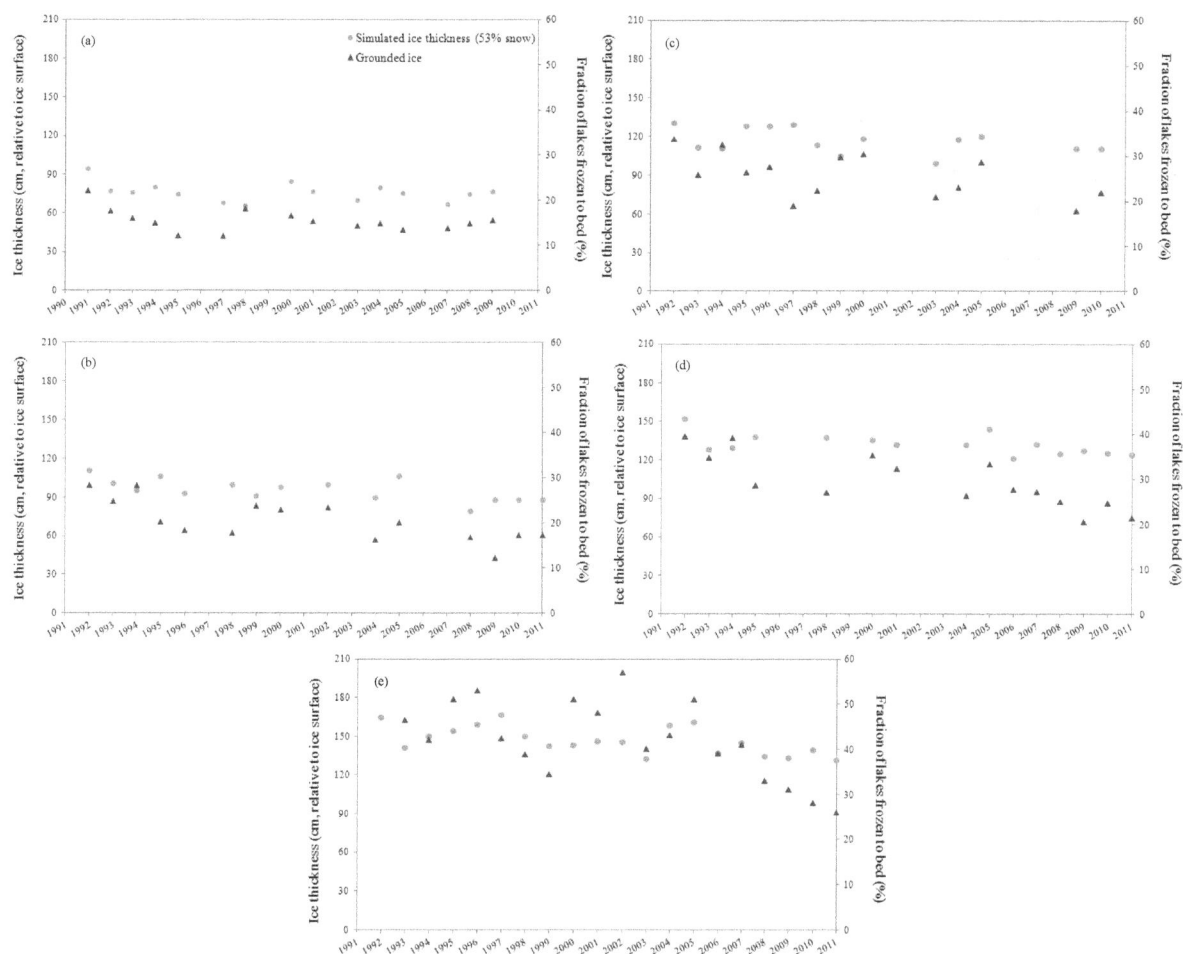

Fig. 4. Monthly fractions of lakes frozen to bed as derived from analysis of available ERS images and simulated ice thickness on day of ERS acquisition (1991 to 2011) – **(a)** December; **(b)** January; **(c)** February; **(d)** March; **(e)** April.

decline ($\alpha = 0.001$) in the maximum ice thickness of a total of 21 cm (no-snow cover) and 38 cm (53 % snow cover depth) for the period 1950–2011. Simulated maximum ice thickness with 0 % and 53 % snow cover depth, respectively, ranged from 196 cm and 140 cm in 2011 to 238 cm and 209 cm ($\alpha = 0.001$) in 1976. The ice thickness simulated with CLIMo using a snow depth of 53 % correlates well with the SAR-derived ice cover fractions for lakes frozen to the bed vs. lakes with floating ice ($r = 0.75$, $p < 0.001$), a thinner ice cover corresponding to a lower fraction of lakes frozen to the bed and thicker ice indicating a higher grounded ice fraction (Fig. 6). For the overlapping years with the ERS-1/2 SAR images (1991–2011), model simulations with no-snow and 53 % snow depth scenarios show a decline in the maximum ice thickness by 18–22 cm ($\alpha = 0.01$).

Additionally, CLIMo simulations indicate that, in response to warmer climatic conditions as reflected by the increase in annual mean air temperature and total precipitation during recent decades, the duration of the ice cover has reduced, with later freeze-up dates by 5.9 days and break-up

dates occurring earlier in the season by 18.6 days with 0 % snow cover and by 17.7 days with the 53 % snow cover depth scenario from 1950 to 2011 (Fig. 7). During the 62 yr period, CLIMo indicates a decrease in the duration of the ice seasons by a total of 24.8 days in the absence of a snow cover and by 23.6 days with 53 % snow cover depth. Statistically, freeze-up and break-up, and the duration of the ice season trends, with both snow cover depth scenarios, are equally significant at the $\alpha = 0.001$ level. For the period of the ERS imagery analysis (1991–2011), model simulations indicate later ice-on dates by 14.5 days ($\alpha = 0.05$), earlier ice-off dates by 5.3 days ($\alpha > 0.1$) with the no-snow scenario and no change in the ice-off dates ($\alpha > 0.1$) with the 53 % snow depth scenario.

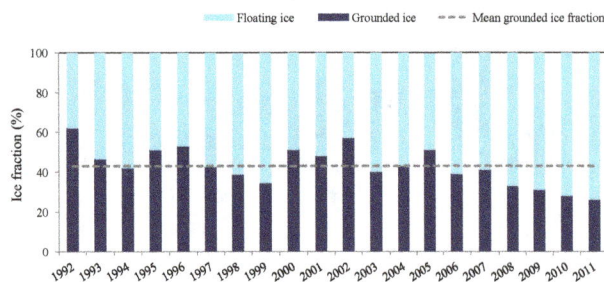

Fig. 5. Late winter (April/May) floating and grounded ice fractions from 1992 to 2011 resulting from segmentation of ERS-1/2 images.

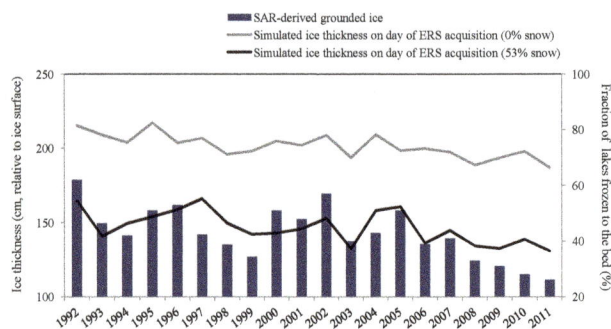

Fig. 6. SAR-derived fraction of grounded ice and simulated ice thickness from CLIMo on day of ERS acquisitions from 1992 to 2011.

6 Discussion

6.1 Ice cover changes: 1950–2011

Analysis of ice-thickness trends from CLIMo simulations during the 1950–2011 period indicates a trend toward thinner ice covers for the Alaskan lakes under study, a trend that is more evident with 53 % snow cover depth conditions and that indicates a decrease of a total of 38 cm in ice thickness, at a rate of $0.6 \, \text{cm} \, \text{yr}^{-1}$.

Albeit inter-decadal and inter-annual variability is noted, trend analysis of ice phenology from 1950 to 2011 indicates a slight change in freeze-up dates, ice onset occurring later in the season by 5.9 days and a significant advancement of ice melt by 17.7 to 18.6 days (0 % and 53 % snow depth scenarios). These results are supported by similar findings that show a significant trend toward later ice-on and earlier ice-off dates, and overall shorter ice seasons of lakes across the Northern Hemisphere, a trend accentuated during recent decades (Magnuson et al., 2000; Duguay et al., 2006; Benson et al., 2011). Shorter ice seasons have been mainly attributed to the advance of break-up days earlier in the spring (Bonsal et al., 2006), earlier ice-off being associated with higher spring air temperatures and earlier snow melt onset (Duguay et al., 2006). CLIMo simulations for the 1950–2011 period suggest that the length of the ice season has reduced by 23.6–24.8 days (53 % and 0 % snow cover depth scenarios). In a 53 % snow cover depth scenario, the shortest ice seasons were identified to have been occurring in recent years, with 1998 being the year when lakes were ice free for a total of 101 days, followed by 2006 (ice free for 98 days) and 2009 (ice free for 97 days). Using the same climate scenario (53 % snow depth), model simulations show that 1955, 1960 and 1965 (51, 52, and 59 days respectively) were the years when lakes had the most extended ice coverage. To support the strong correlation previously shown to exist between ice-off dates and ice cover duration (Duguay et al., 2006), shorter ice seasons occurred in all years of early ice-off dates in both snow cover depth scenarios but not all years with reduced ice duration had later ice-on dates.

6.2 Ice cover changes: 1991–2011

6.2.1 SAR-observed changes

During the 20 yr period of SAR analysis, a specific temporal pattern in the evolution of the grounded ice fraction for individual lakes on a yearly basis was not observed. Considering that the fraction of grounded ice is strongly dependent on climate conditions, the inter-annual variability of air temperatures and that of the snow cover impacts is being reflected by the variations in the yearly bedfast ice fraction. For example, the climatic conditions of a cold winter (1991/1992) and a warm winter (2010/2011) season, differed largely. The ice-growing season (October to April) of 1991/1992 was characterized by lower mean air temperatures ($-22\,^\circ\text{C}$) and reduced total snowfall (561 mm) as opposed to the higher winter mean air temperatures ($-17.7\,^\circ\text{C}$) and greater amount of total snowfall (1199 mm) of 2010/2011. The observed fraction of lakes frozen to the bed was greater during the colder ice season, with values ranging from a minimum of 18 % in December to a maximum of 62 % in April. Noteworthy is also the fact that the fraction of grounded ice observed in April 1992 (62 %) was the highest among all years of available SAR data. In addition to inter-annual variability of the grounded ice fraction, inter-lake differences were observed during the 20 yr of available SAR data. For instance, the shallower West Twin Lake (71°16′ N, 156°29′ W; 1.2 m maximum depth), located close to the Beaufort Sea, developed bedfast ice during all years of observations, whereas the deeper Ikroavik Lake (71°13′ N, 156°37′ W; 2.1 m maximum depth), further from the Arctic coast, maintained a floating ice cover throughout all winter seasons.

Ice regimes of shallow coastal lakes on the NSA correlate with the distance from the coast, with lakes closer to the coast in this study area preserving their ice cover later into the season (Hinkel et al., 2012). The discrepancies in ice regimes of these lakes may therefore be attributed to temperature gradients or snow-cover redistribution within the area and that are associated with the distance from the coast. Ice regimes

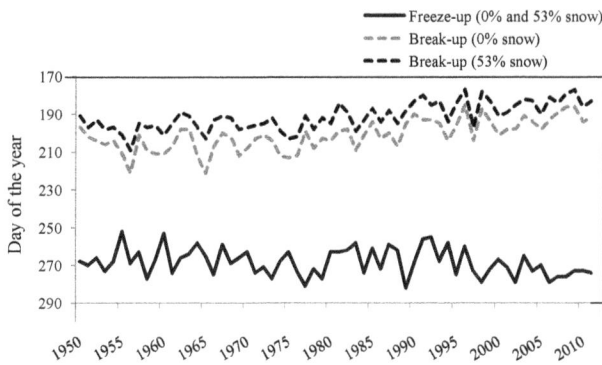

Fig. 7. CLIMo-simulated freeze-up and break-up dates from 1950 to 2011.

are also related to lake depth, with deeper lakes maintaining liquid water underneath the ice (Jeffries et al., 1996).

6.2.2   Simulated changes

Ice growth and downward thickening is strongly influenced by snow cover over lake ice (Vavrus et al., 1996; Ménard et al., 2002; Gao and Stefan, 2004), which influences the vertical conductive heat flow from the ice to the atmosphere through heat loss at the ice–snow interface (Jeffries et al., 1999). A thinner and denser snowpack provides less insulation and allows higher rates of heat flow to the atmosphere (Sturm and Liston, 2003). Thus, the accentuated reduction of ice thickness and number of lakes that freeze to the bottom from 2006 onwards may be associated with deeper snow resulting from increased winter precipitation (Callaghan et al., 2011), a consequence of higher air temperatures (Schindler and Smol, 2006; Kaufman et al., 2009; Walsh et al., 2011). From 1991 to 2011, simulated ice thickness declined by 18–22 cm (no-snow and 53 % snow depth scenarios). Consequently, thinner ice covers are linked to the noticeable trend toward fewer lakes freezing to the bed during the past 20 yr, as observed from the analysis of SAR data (Fig. 6). Since ice cover conditions are better captured assuming a 53 % snow depth atop lakes vs. no-snow cover when compared to available field measurements, analysis of overlapping CLIMo and SAR results indicate a closer agreement between the ice thickness and the fraction of grounded ice during the former scenario, with slight discrepancies in 1994, 1997, 2001 and 2010. The disagreement between model simulations and SAR observations is possibly associated with the timing of snow accumulation during the winter season, greater snow depth at the time of ice formation leading to thinner ice.

From 2006 onwards, model simulations indicate smaller changes in ice thickness whereas satellite observations indicate a significant reduction of the grounded ice fraction. One hypothesis to consider when explaining the considerable SAR-derived changes in grounded ice and the small changes in simulated ice thickness from 2006 onwards is the yearly

variation in ice thickness. Minimum changes in ice thickness should occur, once a threshold is reached, thinner ice covers are reflected by the higher fractions of floating ice. Increasing lake water levels at the time of freeze-up or greater lake depths are aspects to also consider in explaining the differences between the changes in grounded ice and those in ice thickness. Since calculated precipitation (P) minus evaporation (E) ($P - E$) values do not indicate higher water levels from 2006 to 2011, changes in lake depth may explain the minor ice thickness changes. However, considering the lack of data on lake bathymetry, this hypothesis needs to be further investigated.

6.3   Teleconnections and lake ice regimes

Air temperature changes at high latitudes are often related to the decreasing extent of sea ice and that allows more heat to be absorbed by the Arctic Ocean, heat that is further released in the atmosphere. Additionally, changes in the air temperature are also associated with large-scale atmospheric and oceanic circulations. Previous analysis of teleconnection patterns that affect the Northern Hemisphere climate and weather reveals that lake ice conditions are partly driven by these large-scale circulations and exhibit stronger correlations with spring (January–April) climatic indices (Bonsal et al., 2006). For western Canada, the strongest correlation between teleconnections and ice phenology was associated with the Pacific–North American (PNA), a positive phase of the PNA being highly correlated with earlier break-up dates ($r = -0.74$) and vice versa (Bonsal et. al., 2006). To articulate this relationship, one third of the variability in Northern Hemisphere winter temperatures variability of previous decades can be explained by the positive phase of the North Atlantic Oscillation (NAO; Hurell, 1996). Likewise, a shift of the Pacific Decadal Oscillation (PDO) in 1976 toward a positive phase contributed to increased northward advection of warm air (Morris et al., 2005), which is highly noticeable in Alaska, being well reflected by the recent increased warming of the area (Hartmann and Wendler, 2005). The PDO shift is also reflected in the CLIMo simulations that indicate a transition toward thinner ice covers from 1976 onward. Given that El Niño years have been associated with up to ten days shorter ice seasons for lakes in western Canada (Bonsal et al., 2006), the fact that the longest open-water season occurred during an extreme El Niño year (1998) is associated with considerable warmer air temperatures recorded that year at the Barrow meteorological station.

6.4   Water levels and lake ice regimes

In response to warmer climate conditions during recent decades, the spatial distribution and the surface area of Arctic lakes has been noted to change. In ice-rich permafrost areas such as the NSA, changes in air temperature alter the frozen ground layer that, by thawing, may result in the appearance of new water bodies or increasing surface areas of the existing lakes as a result of thermally induced lateral expansion (Jones et al., 2011). Alternatively, as many lakes in this area have low water volumes, lake water levels will rapidly respond to changes in the water budget, expressed as total precipitation (P) minus evaporation (E), $P - E$. Calculated water balance during the ice-free season also includes spring snow-water equivalent (SWE). Thus, a negative water balance ($P < E$) results in reduced lake levels and/or intermittent lake disappearance (Smol and Douglas, 2007) with lakes disappearing during dry seasons and refilling during wetter seasons. Hinkel et al. (2007) showed that from 1975 to 2000, over 25 % of lakes on the western ACP experienced shoreline retreat through lateral drainage, the lake area change not being strongly supported by climate conditions during the period of analysis. This may be seen in the case of Sikulik Lake ($71°18'$ N, $156°40'$ W) that appears to have experienced fluctuating water levels during the 20 yr period as indicated by the differences in radar returns from this lake in late winter (Fig. 8). As radar returns for this lake were similar to those of the adjacent land, the assumption made was that the lake drained in most years and that it may have filled in 2000 and 2002, both years with a positive water balance. Lake water levels are greatly controlled by precipitation and evaporation rates during the summer season when lakes are exposed to energy exchanges with the atmosphere and variations in the lake water balance can be explained by fluctuations in the $P - E$ (Bowling et al., 2003). In the case of Arctic lakes, the overall lake water balance is generally negative, as the high evaporation during summer is not compensated by higher amounts of precipitation (Rovansek et al., 1996). Positive values of the $P - E$ index are associated with lower annual mean air temperatures and wetter conditions during the ice-free season while lower water levels are recorded during warm and dry years (Labrecque et al., 2009). Extreme $P - E$ values (i.e., 1993, 2010, 2011 – warm years, and 1995, 2005 – cold years) correlate well with the grounded ice fraction, lower grounded ice fraction being strongly related to the positive $P - E$ values, and higher grounded ice values matching well those of negative $P - E$. During the years of extreme $P - E$ values, grounded ice fractions also correlate with mean air temperatures ($r = 0.68$, $p < 0.0010$). Additional periodic recharge through ground water, spring snowmelt and river inflows, or lateral drainage and ice melt within the underlying permafrost (Young and Woo, 2000) can occur and consequently affect lake water balance, and thus explaining the discrepancies between lake water levels and ice conditions. These are tentative explanations and the relationship between water balance and grounded ice fraction needs to be further investigated.

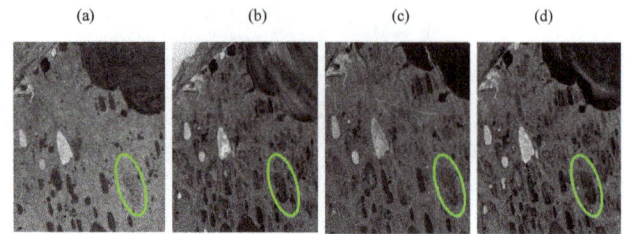

Fig. 8. Late winter (April/May) differences in radar returns from Sikulik Lake – **(a)** 1992; **(b)** 2000; **(c)** 2001; **(d)** 2002.

7   Summary and conclusions

This study aimed to detect changes in ice thickness, the fraction of lakes/lake areas freezing to the bed vs. those developing a floating ice cover and phenology (freeze-up and break-up dates, and ice cover duration) of Arctic shallow lakes near Barrow, in a sub-region of the NSA. The methods employed were image segmentation of ERS-SAR images acquired over the region from 1991 to 2011 and simulations with two different snow depth scenarios (0 % and 53 %) of a numerical lake model forced with climate data from the Barrow meteorological station between 1950–2011.

A trend toward an increasing number of lakes that maintain liquid water underneath the floating ice atop in all months of available SAR imagery (December to early May), and thinner ice covers during the winter months was identified from ERS analysis and CLIMo simulations. Statistical analysis showed that in the case of thermokarst lakes near Barrow, the fraction of bedfast ice as extracted from the analysis of ERS-1/2 SAR data (1991–2011) correlates well with the thickness of the ice layer simulated with CLIMo. The most significant decrease in grounded ice was noticed to occur in late winter; grounded ice that considerably declined since 2006 and reached its lowest in 2011. Model outputs indicate thinner ice covers by 18 to 22 cm (1991–2011) and by 21 to 38 cm (1950–2011), and extended duration of the ice season – a function of later ice-on and earlier ice-off dates – by a total of 23.6 days (1950–2011) with a 53 % snow depth scenario.

SAR data provides the opportunity to effectively monitor Arctic lakes and assess the degree of changes in winter lake ice growth in response to climate conditions. Low-resolution ERS imagery allows an adequate detection of the rate at which lakes freeze to their bed for the duration of the ice season and of the grounded ice fraction at the end of winter, thus providing a valuable data set. The use of satellite sensors that provide higher temporal coverage, such as ASAR Wide Swath (2002–2012), would further improve the investigation of ice regimes of high-latitude lakes. Future satellite missions of the European Space Agency (Sentinel-1), the National Aeronautics and Space Administration (Surface Water and Ocean Topography – SWOT) and the Canadian Space Agency (the RADARSAT constellation) are planned for launch in 2013, 2019 and 2018, respectively. These missions will not only continue the C-band SAR operational applications (Sentinel-1 and the RADARSAT constellation) and enable accurate monitoring of water levels (SWOT) but also provide increased temporal resolution, thus ensuring frequent, long-term SAR acquisitions for the Arctic regions.

Acknowledgements. This research was supported by European Space Agency (ESA-ESRIN) contract no. 4000101296/10/I-LG (Support to Science Element, North Hydrology Project) and a Discovery Grant from the Natural Sciences and Engineering Research Council of Canada (NSERC) to C. Duguay. The ERS-1/2 SAR images are copyright ESA 1991–2011 and were provided by the Alaska Satellite Facility (ASF).

Edited by: D. Hall

References

Arp, C. D., Jones, B. M., Lu, Z., and Whitman, M. S.: Shifting balance of thermokarst lake ice regimes across the Arctic Coastal Plain of northern Alaska, Geophys. Res. Lett., 39, L16503, doi:10.1029/2012GL052518, 2012.

Assel, R. A., Cronk, K., and Norton, D.: Recent trends in Laurentian Great Lakes ice cover, Clim. Change, 57, 185–204, 2003.

Benson, B. J., Magnuson, J. J., Jensen, O. P., Card, V. M., Hodgkins, G., Korhonen, J., and Granin, N. G.: Extreme events, trends, and variability in Northern Hemisphere lake-ice phenology (1855–2005), Clim. Change, 12, 1–25, 2011.

Bonsal, B. R., Prowse, T. D., Duguay, C. R., and Lacroix, M. P.: Impacts of large-scale teleconnections on freshwater-ice break/freeze-up dates over Canada, J. Hydrol., 330, 340–353, 2006.

Bowling, L. C., Kane, D. L., Gieck, R. E., Hinzman, L. D., and Lettenmaier, D. P.: The role of surface storage in a low-gradient Arctic watershed, Water Resour. Res., 39, 1087, doi:10.1029/2002WR001466, 2003.

Brown, L. C. and Duguay, C. R.: A comparison of simulated and measured lake ice thickness using a shallow water ice profiler, Hydrol. Process., 25, 2932–2941, 2011a.

Brown, L. C. and Duguay, C. R.: The fate of lake ice in the North American Arctic, The Cryosphere, 5, 869–892, doi:10.5194/tc-5-869-2011, 2011b.

Brown, R. S., Duguay, C. R., Mueller, R. P., Moulton, L. L., Doucette, P. J., and Tagestad, J. D.: Use of synthetic aperture radar to identify and characterize overwintering areas of fish in ice-covered arctic rivers: a demonstration with broad whitefish and their habitats in the Sagavanirktok River, Alaska, T. Am. Fish. Soc., 139, 1711–1722, 2010.

Callaghan, T. V., Johansson, M., Brown, R. D., Groisman, P. Y., Labba, N., Radionov, V., Bradley, R. S., Blangy, S., Bulygina, O. N., Christensen, T. R., Colman, J. E., Essery, R. L. H., Forbes, B. C., Forchhammer, M. C., Golubev, V. N., Honrath, R. E., Juday, G. P., Meshcherskaya, A. V., Phoenix, G. K., Pomeroy, J., Rautio, A., Robinson, D. A., Schmidt, N. M., Serreze, M. C., Shevchenko, V. P., Shiklomanov, A. I., Shmakin, A. B., Sköld, P., Sturm, M., Woo, M.-K., and Wood, E. F.: Multiple effects of changes in arctic snow cover, Ambio, 40, 32–45, 2011.

Clausi, A., Qin, A. K., Chowdhury, M. S., Yu, P., and Maillard, P.: MAGIC: Map-Guided Ice Classification System, Can. J. Remote Sens., 36, S13–S25, 2010.

Comiso, J. C., Parkinson, C. L., Gersten, R., and Stock, L.: Accelerated decline in the Arctic sea ice cover, Geophys. Res. Lett., 35, doi:10.1029/2007GL031972, 2008.

Cook, T. L. and Bradley, R. S.: An analysis of past and future changes in the ice cover of two High-Arctic lakes based on synthetic aperture radar (SAR) and Landsat imagery, Arct. Antarct. Alp. Res., 42, 9–18, 2010.

Dufresne, J.-L., Foujols, M.-A., Denvil, S., Caubel, A., Marti, O., Aumont, O, Balkanski, Y., Bekki, S., Bellenger, H., Benshila, R., Bony, S., Bopp, L., Braconnot, P., Brockmann, P., Cadule, P., Cheruy, F., Codron, F., Cozic, A., Cugnet, D., de Noblet, N., Duvel, J.-P., Ethé, C., Fairhead, L., Fichefet, T., Flavoni, S., Friedlingstein, P., Grandpeix, J.-Y., Guez, L., Guilyardi, E., Hauglustaine, D., Hourdin, F., Idelkadi, A., Ghattas, J., Joussaume, S., Kageyama, M., Krinner, G., Labetoulle, S., Lahellec, A., Lefebvre, M.-P., Lefevre, F., Levy, C., Li, Z.X., Lloyd, J., Lott, F., Madec, G., Mancip, M., Marchand, M., Masson, S., Meurdesoif, Y., Mignot, J., Musat, I., Parouty, S., Polcher, J., Rio, C., Schulz, M., Swingedouw, D., Szopa, S., Talandier, C., Terray, P., and Viovy, N.: Climate change projections using the IPSL-CM5 Earth System Model: from CMIP3 to CMIP5, Clim. Dynam., 40, 2123–2165, 2013.

Duguay, C. R. and Lafleur, P. M.: Determining depth and ice thickness of shallow subarctic lakes using spaceborne optical and SAR data, Int. J. Remote Sens., 24, 475–489, 2003.

Duguay, C. R., Rouse, W. R., Lafleur, P. M., Boudreau, D. L., Crevier, Y., and Pultz, T. J.: Analysis of multi-temporal ERS-1 SAR data of subarctic tundra and forest in the northern Hudson Bay Lowland and implications for climate studies, Can. J. Remote Sens., 25, 21–33, 1999.

Duguay, C. R., Pultz, T. J., Lafleur, P. M., and Drai, D.: RADARSAT backscatter characteristics of ice growing on shallow sub-Arctic lakes, Churchill, Manitoba, Canada, Hydrol. Process., 16, 1631–1644, 2002.

Duguay, C. R., Flato, G. M., Jeffries, M. O., Ménard, P., Morris, K., and Rouse, W. R.: Ice-cover variability on shallow lakes at high latitudes: model simulations and observations, Hydrol. Process., 17, 3465–3483, 2003.

Duguay, C. R., Prowse, T. D., Bonsal, B. R., Brown, R. D., Lacroix, M. P., and Ménard, P.: Recent trends in Canadian lake ice cover, Hydrol. Process., 20, 781–801, 2006.

Duguay, C., Brown, L., Kang, K.-K., and Kheyrollah Pour, H.: The Arctic lake ice, in "State of the climate in 2011", B. Am. Meteorol. Soc., 93, S152–S154, 2012.

Elachi, C. M., Bryan, M. L., and Weeks, W. F.: Imaging radar observations of frozen Arctic lakes, Remote Sens. Environ., 5, 169–175, 1976.

Engram, M., Anthony, K. W., Meyer, F. J., and Grosse, G.: Characterization of L-band synthetic aperture radar (SAR) backscatter from floating and grounded thermokarst lake ice in Arctic Alaska, The Cryosphere, 7, 1741–1752, doi:10.5194/tc-7-1741-2013, 2013.

Futter, M. N.: Patterns and trends in southern Ontario lake ice phenology, Environ. Monit. Assess., 88, 431–444, 2003.

Gao, S. and Stefan, H. G.: Potential climate change effects on ice covers of five freshwater lakes, J. Hydrol. Eng., 9, 226–234, 2004.

Grosswald, M. G., Hughes, T. J., and Lasca, N. P.: Oriented lake-and-ridge assemblages of the Arctic coastal plains: glacial landforms modified by thermokarst and solifluction, Polar Rec., 35, 215–230, 1999.

Hall, D. K., Fagre, D. B., Klasner, F., Linebaugh, G., and Liston, G. E.: Analysis of ERS 1 synthetic aperture radar data of frozen lakes in northern Montana and implications for climate studies, J. Geophys. Res., 99, 22473–22482, 1994.

Hartmann, B. and Wendler, G.: The significance of the 1976 Pacific climate shift in the climatology of Alaska, J. Climate, 18, 4824–4839, 2005.

Heron, R. and Woo, M. K.: Decay of a High Arctic lake-ice cover: Observations and modelling, J. Glaciol., 40, 283–292, 1994.

Hinkel, K. M., Nelson, F. E., Klene, A. F., and, Bell, J. H.: The urban heat island in winter at Barrow, Alaska, Int. J. Climatol., 23, 1889–1905, 2003.

Hinkel, K. M., Frohn, R. C., Nelson, F. E., Eisner, W. R., and Beck, R. A.: Morphometric and spatial analysis of thaw lakes and drained thaw lake basins in the western Arctic Coastal Plain, Alaska, Permafrost Periglac., 16, 327–341, 2005.

Hinkel, K. M., Jones, B. M., Eisner, W. R., Cuomo, C. J., Beck, R. A., and Frohn, R. C.: Methods to assess natural and anthropogenic thaw lake drainage on the western Arctic Coastal Plain of northern Alaska, J. Geophys. Res., 112, F02S16, doi:10.1029/2006JF000584, 2007.

Hinkel, K. M., Zheng, L., Yongwei, S., and Evan, A.: Regional lake ice meltout patterns near Barrow, Alaska, Polar Geogr., 35, 1–18, 2012.

Hodgkins, G. A., James, I. C., and Huntington, T. G.: Historical changes in lake ice-out dates as indicators of climate change in New England, 1850–2000, Int. J. Climatol., 22, 1819–1827, 2002.

Hurrell, J. W.: Influence of variations in extratropical wintertime teleconnections on Northern Hemisphere temperature, Geophys. Res. Lett., 23, 665–668, 1996.

Jeffries, M. O. and Morris, K.: Some aspects of ice phenology on ponds in central Alaska, USA, Ann. Glaciol., 46, 397–403, 2007.

Jeffries, M. O., Morris, K., Weeks, W. F., and Wakabayashi, H.: Structural and stratigraphic features and ERS 1 synthetic aperture radar backscatter characteristics of ice growing on shallow lakes in NW Alaska, winter 1991–1992, J. Geophys. Res., 99, 22459–22471, 1994.

Jeffries, M. O., Morris, K., and Liston, G. E.: A method to determine lake depth and water availability on the North Slope of Alaska with spaceborne imaging radar and numerical ice growth modelling, Arctic, 49, 367–374, 1996.

Jeffries, M. O., Zhang, T., Frey, K., and Kozlenko, N.: Estimating late winter heat flow to the atmosphere from the lake-dominated Alaskan North Slope, J. Glaciol., 45, 315–324, 1999.

Jeffries, M. O., Morris, K., and Kozlenko, N.: Ice characteristics and processes, and remote sensing of frozen rivers and lakes, in: Remote Sensing in Northern Hydrology: Measuring Environmental Change, edited by: Duguay, C. R. and Pietroniro, A., Washington, American Geophysical Union, 63–90, 2005.

Jones, B. M., Arp, C. D., Hinkel, K. M., Beck, R. A., Schmutz, J. A., and Winston, B.: Arctic lake physical processes and regimes with implications for winter water availability and management in the National Petroleum Reserve Alaska, Environ. Manage., 43, 1071–1084, 2009.

Jones, B. M., Grosse, G., Arp, C. D., Jones, M. C., Walter, A. K. M., and Romanovsky, V. E.: Modern thermokarst lake dynamics in the continuous permafrost zone, northern Seward Peninsula, J. Geophys. Res., 16, G00M03, doi:10.1029/2011JG001666, 2011.

Jones, B. M., Gusmeroli, M. A., Arp, C. D., Strozzi, T., Grosse, G., Gaglioti, B. V., and Whitman, B. S.: Classification of freshwater ice conditions on the Alaskan Arctic Coastal Plain using ground penetrating radar and TSX satellite data, Int. J. Remote Sens., 34, 8253–8265, 2013.

Kaufman, D. S., Schneider, D. P., McKay, N. P., Ammann, C. M., Bradley, R. S., Briffa, K. R., and Thomas, E.: Recent warming reverses long-term Arctic cooling, Science, 325, 1236–1239, 2009.

Koenigk, T., Brodeau, L., Graversen, R. G., Karlsson, J., Svensson, G., Tjernström, M., and Willén, K.: Arctic climate change in 21st century CMIP5 simulations with EC-Earth, Clim. Dynam., 40, 2719–2743, 2013.

Labrecque, S., Lacelle, D., Duguay, C. R., Lauriol, B., and Hawkings, J.: Contemporary (1951–2001) evolution of lakes in the Old Crow Basin, northern Yukon, Canada: Remote sensing, numerical modeling, and stable isotope analysis, Arctic, 62, 225–238, 2009.

Latifovic, R. and Pouliot, D.: Analysis of climate change impacts on lake ice phenology in Canada using the historical satellite data record, Remote Sens. Environ., 106, 492–507, 2007.

Lenormand, F., Duguay, C. R., and Gauthier, R.: Development of a historical ice database for the study of climate change in Canada, Hydrol. Process., 16, 3707–3722, 2002.

Liston, G. E., and Sturm, M.: Winter precipitation patterns in arctic Alaska determined from a blowing-snow model and snow-depth observations, J. Hydrometeorol., 3, 646–659, 2002.

Magnuson, J. J., Robertson, D. M., Benson, B. J., Wynne, R. H., Livingstone, D. M., Arai, T., Assel, R. A., Barry, R. G., Virginia Card, V., Kuusisto, E., Granin, N. G., Prowse, T. D., Stewart, K. M., and Vuglinski, V. S.: Historical trends in lake and river ice cover in the Northern Hemisphere, Science, 289, 1743–1746, 2000.

Mellor, J.: Bathymetry of Alaskan Arctic lakes: A key to resource inventory with remote sensing methods, Ph. D. Thesis, Institute of Marine Science, University of Alaska, 1982.

Ménard, P., Duguay, C. R., Flato, G. M., and Rouse, W. R.: Simulation of ice phenology on Great Slave Lake, Northwest Territories, Canada, Hydrol. Process., 16, 3691–3706, 2002.

Mendez, J., Hinzman, L. D., and Kane, D. L.: Evapotranspiration from a wetland complex on the Arctic coastal plain of Alaska, Nord. Hydrol., 29, 303–330, 1998.

Morris, K., Jeffries, M. O., and Weeks, W. F.: Ice processes and growth history on Arctic and sub-Arctic lakes using ERS-1 SAR data, Polar Rec., 31, 115–128, 1995.

Morris, K., Jeffries, M. O., and Duguay, C.: Model simulation of the effects of climate variability and change on lake ice in central Alaska, USA, Ann. Glaciol., 40, 113–118, 2005.

National Climate Data Center (NCDC): Barrow, retrieved from http://www.ncdc.noaa.gov/cdo-web/datasets/ANNUAL/locations (last access: 28 April 2013), 2012.

Noguchi, K., Gel, Y. R., and Duguay, C. R.: Bootstrap-based test for trends in hydrological times series, with application to ice phenology data, J. Hydrol., 410, 150–161, 2011.

Ochilov, S., Svacina, N. A., Duguay, C. R., and Clausi, D. A.: Towards an automated lake ice monitoring system for SAR imagery, Abstract C51A-0474 presented at the 2010 Fall Meeting, AGU, San Francisco, Calif., 13–17 December, 2010.

Overland, J. E., Wang, M., Walsh, J. E., Christensen, J. H., Kattsov, V. M., and Chapman W. L.: Chapter 3: Climate model projections for the Arctic, Snow, Water, Ice and Permafrost in the Arctic (SWIPA), Oslo, Arctic Monitoring and Assessment Programme (AMAP), 2011.

Palecki, M. A. and Barry, R. G.: Freeze-up and break-up of lakes as an index of temperature changes during the transition seasons: A case study for Finland, J. Clim. Appl. Meteorol., 25, 893–902, 1986.

Prowse, T., Alfredsen, K., Beltaos, S., Bonsal, B., Duguay, C., Korhola, A., McNamara, J., Vincent, W. F., Vuglinsky, V., and Weyhenmeyer, G. A.: Arctic freshwater ice and its climatic role, Ambio, 40, 46–52, doi:10.1007/s13280-011-0214-9, 2011.

Robertson, D. M., Ragotzkie, R. A., and Magnuson, J. J.: Lake ice records used to detect historical and future climatic changes, Clim. Change, 21, 407–427, 1992.

Romanovsky, V. E., Smith, S. L., and Christiansen, H. H.: Permafrost thermal state in the polar Northern Hemisphere during the International Polar Year 2007–2009: A synthesis, Permafrost Periglac., 21, 106–116, 2010.

Rovansek, R. J., Hinzman, L. D., and Kane, D. L.: Hydrology of a tundra wetland complex in the Alaskan Arctic coastal plain, USA., Arctic Alpine Res., 28, 311–317, 1996.

Schindler, D. W. and Smol, J. P.: Cumulative effects of climate warming and other human activities on freshwaters of Arctic and Subarctic North America, Ambio, 35, 160–168, 2006.

Schindler, D. W., Beaty, K. G., Fee, E. J., Cruikshank, D. R., DeBruyn, E. R., Findlay, D. L., and Turner, M. A.: Effects of climatic warming on lakes of the central boreal forest, Science, 250, 967–970, 1990.

Sellmann, P. V., Weeks, W. F., and Campbell, W. J.: Use of side-looking airborne radar to determine lake depth on the Alaskan North Slope, CRREL Special Report (US Army Cold Regions Research and Engineering Laboratory), 230, 1975.

Sen, P. K.: Estimates of the regression coefficient based on Kandall's tau, J. Am. Stat. Assoc., 63, 1379–1389, 1968.

Serreze, M. C., Walsh, J. E., Chapin III, F. S., Osterkamp, T., Dyurgerov, M., Romanovsky, V., Oechel, W. C., Morison, J., Zhang, T., and Barry, R. G.: Observational evidence of recent change in the northern high-latitude environment, Clim. Change, 46, 159–207, 2000.

Serreze, M. C., Holland, M. M., and Stroeve, J.: Perspectives on the Arctic's shrinking sea-ice cover, Science, 315, 1533–1536, 2007.

Smith, L. C., Sheng Y., MacDonald, G. M., and Hinzman, L. D.: Disappearing Arctic lakes, Science, 308, p. 1429, doi:10.1126/science.1108142, 2005.

Smol, J. P. and Douglas, M. S. V.: Crossing the final ecological threshold in High Arctic ponds, P. Natl. Acad. Sci. USA, 104, 12395–12397, 2007.

Sobiech, J. and Dierking, W.: Monitoring lake-ice decay with imaging radar in the Siberian Arctic, International Symposium on Seasonal Snow and Ice, Lahti, Finland, 28 May 2012–1 June 2012, 2012.

Sturm, M. and Liston, G. E.: The snow cover on lakes of the Arctic Coastal Plain of Alaska, USA., J. Glaciol., 49, 370–380, 2003.

Trenberth, K. E., Jones, P. D., Ambenje, P., Bojariu, R., Easterling, D., Klein Tank, A., Parker, D., Rahimzadeh, F., Renwick, J. A., Rusticucci, M., Soden, B., and Zhai, P.: Observations: Surface and Atmospheric Climate Change, in: Climate Change 2007: The Physical Science Basis. Contribution of Working Group I to the Fourth Assessment Report of the Intergovernmental Panel on Climate Change, edited by: Solomon, S., Qin, D., Manning, M., Chen, Z., Marquis, M., Averyt, K. B., Tignor, M., and Miller, H. L., Cambridge University Press, Cambridge, United Kingdom and New York, NY, USA, 2007.

Vavrus, S. J., Wynne, R. H., and Foley, J. A.: Measuring the sensitivity of Southern Wisconsin lake ice to climate variations and lake depth using a numerical model, Limnol. Oceanogr., 41, 822–831, 1996.

Walsh, S. E., Vavrus, S. J., Foley, J. A., Fisher, V. A., Wynne, R. H., and Lenters, J. D.: Global patterns of lake ice phenology and climate: Model simulations and observations, J. Geophys. Res.-Atmos., 103, 28825–28837, 1998.

Walsh, J. E., Overland, J. E., Groisman, P. Y., and Rudolf, B.: Chapter 2: Arctic climate: Recent variations, Snow, Water, Ice and Permafrost in the Arctic (SWIPA), Oslo, Arctic Monitoring and Assessment Programme (AMAP), 2011.

Walter, K. M., Zimov, S. A., Chanton, J. P., Verbyla, D., and Chapin III, F. S.: Methane bubbling from Siberian thaw lakes as a positive feedback to climate warming, Nature, 443, 71–75 doi:10.1038/nature05040, 2006.

Walter, K. M., Engram, M., Duguay, C. R., Jeffries, M. O., and Chapin, F. S.: The potential use of synthetic aperture radar for estimating methane ebullition from Arctic lakes, J. Am. Water. Resour. As., 44, 305–315, 2008.

Weeks, W. F., Fountain, A. G., Bryan, M. L., and Elachi, C.: Differences in radar returns from ice-covered North Slope lakes, J. Geophys. Res., 83, 4069–4073, 1978.

Wendler, G., Chen, L., and Moore, B.: The First decade of the new century: a cooling trend for most of Alaska, The Open Atmospheric Science Journal, 6, 111–116, 2012.

White, D. M., Prokein, P., Chambers, M. K., Lilly, M. R., and Toniolo, H.: Use of synthetic aperture radar for selecting Alaskan lakes for winter water use, J. Am. Water. Resour. As., 44, 276–284, 2008.

Young, K. L. and Woo, M. K.: Hydrological response of a patchy high arctic wetland, Nord. Hydrol., 31, 317–338, 2000.

Zhang, T. and Jeffries, M. O.: Modeling interdecadal variations of lake ice thickness and sensitivity to climatic change in northernmost Alaska, Ann. Glaciol., 31, 339–347, 2000.

11

Geophysical mapping of palsa peatland permafrost

Y. Sjöberg[1], P. Marklund[2], R. Pettersson[2], and S. W. Lyon[1]

[1]Department of Physical Geography and the Bolin Centre for Climate Research, Stockholm University, Stockholm, Sweden
[2]Department of Earth Sciences, Uppsala University, Uppsala, Sweden

Correspondence to: Y. Sjöberg (ylva.sjoberg@natgeo.su.se)

Abstract. Permafrost peatlands are hydrological and bio-geochemical hotspots in the discontinuous permafrost zone. Non-intrusive geophysical methods offer a possibility to map current permafrost spatial distributions in these environments. In this study, we estimate the depths to the permafrost table and base across a peatland in northern Sweden, using ground penetrating radar and electrical resistivity tomography. Seasonal thaw frost tables (at ~ 0.5 m depth), taliks (2.1–6.7 m deep), and the permafrost base (at ~ 16 m depth) could be detected. Higher occurrences of taliks were discovered at locations with a lower relative height of permafrost landforms, which is indicative of lower ground ice content at these locations. These results highlight the added value of combining geophysical techniques for assessing spatial distributions of permafrost within the rapidly changing sporadic permafrost zone. For example, based on a back-of-the-envelope calculation for the site considered here, we estimated that the permafrost could thaw completely within the next 3 centuries. Thus there is a clear need to benchmark current permafrost distributions and characteristics, particularly in under studied regions of the pan-Arctic.

1 Introduction

Permafrost peatlands are widespread across the Arctic and cover approximately 12 % of the arctic permafrost zone (Hugelius et al., 2013; Hugelius et al., 2014). They often occur in sporadic permafrost areas, protected by the peat cover, which insulates the ground from heat during the summer (Woo, 2012). In the sporadic permafrost zone, the permafrost ground temperature is often close to 0 °C, and therefore even small increases in temperature can result in thawing of permafrost. In addition, permafrost distribution and thaw-ing in these landscapes are influenced by several factors other than climate, including hydrological, geological, morphological, and erosional processes that often combine in complex interactions (e.g., McKenzie and Voss, 2013; Painter et al., 2013; Zuidhoff, 2002). Due to these interactions, peatlands are often dynamic with regards to their thermal structures and extent, as the distribution of permafrost landforms (such as dome-shaped palsas and flat-topped peat plateaus) and talik landforms (such as hollows, fens, and lakes) vary with climatic and local conditions (e.g., Sannel and Kuhry, 2011; Seppälä, 2011; Wramner, 1968). This dynamic nature and variable spatial extent has potential implications across the pan-Arctic as these permafrost peatlands store large amounts of soil organic carbon (Hugelius et al., 2014; Tarnocai et al., 2009). The combination of large carbon storage and high potential for thawing make permafrost peatlands biogeochemical hotspots in the warming Arctic. In light of this, predictions of future changes in these environments require knowledge of current permafrost distributions and characteristics, which is sparse in today's scientific literature.

While most observations of permafrost to date consist of temperature measurements from boreholes, advances in geophysical methods provide a good complement for mapping permafrost distributions in space. Such techniques can provide information about permafrost thickness and the extent and distribution of taliks, which can usually not be obtained from borehole data alone. As the spatial distribution and extent of permafrost directly influences the flow of water through the terrestrial landscape (Sjöberg et al., 2013), adding knowledge about the extent and coverage of permafrost could substantially benefit development of coupled hydrological and carbon transport models in northern latitudes (e.g., Jantze et al., 2013; Lyon et al., 2010). This may be particularly important for regions where palsa peatlands make up a large portion of the landscape mosaic and regional-scale differences exist in carbon fluxes (Giesler et al., 2014).

Geophysical methods offer non-intrusive techniques for measuring physical properties of geological materials; however, useful interpretation of geophysical data requires other types of complementary data, such as sediment cores. Ground penetrating radar (GPR) has been used extensively in permafrost studies for identifying the boundaries of permafrost (e.g., Arcone et al., 1998; Doolittle et al., 1992; Hinkel et al., 2001; Moorman et al., 2003), characterizing ground ice structures (De Pascale et al., 2008; Hinkel et al., 2001; Moorman et al., 2003), and estimating seasonal thaw depth and moisture content of the active layer (Gacitua et al., 2012; Westermann et al., 2010). Electrical resistivity tomography (ERT) has also been widely applied in permafrost studies (Hauck et al., 2003; Ishikawa et al., 2001; Kneisel et al., 2000), the majority of which focus on mountain permafrost. By combining two or more geophysical methods complementary information can often be acquired raising the confidence in interpretations of permafrost characteristics (De Pascale et al., 2008; Hauck et al., 2004; Schwamborn et al., 2002). For example, De Pascale et al. (2008) used GPR and capacitive-coupled resistivity to map ground ice in continuous permafrost and demonstrated the added value of combining radar and electrical resistivity measurements for the quality of interpretation of the data. While some non-intrusive geophysical investigations have been done in palsa peatland regions (Dobinski, 2010; Doolittle et al., 1992; Kneisel et al., 2007, 2014; Lewkowicz et al., 2011), the use of multiple geophysical techniques to characterize the extent of permafrost in palsa peatland environments has not been employed.

In this study we use GPR and ERT in concert to map the distribution of permafrost along three transects (160 to 320 m long) in the Tavvavuoma palsa peatland in northern Sweden. Our aim is to understand how depths of the permafrost table and base vary in the landscape and, based on resulting estimates of permafrost thickness, to make a first-order assessment of the potential time needed to completely thaw this permafrost due to climate warming. Furthermore, we hope to demonstrate the added value of employing complementary geophysical techniques in such landscapes. This novel investigation thus helps contribute to our understanding of the current permafrost distribution and characteristics across palsa peatlands, creating a baseline for future studies of possible coupled changes in hydrology and permafrost distribution in such areas.

2 Study area

Tavvavuoma is a large palsa peatland complex in northern Sweden at 68°28′ N, 20°54′ E, 550 m a.s.l. (Fig. 1) and consists of a patchwork of palsas, peat plateaus, thermokarst lakes, hummocks, and fens. Ground temperatures and

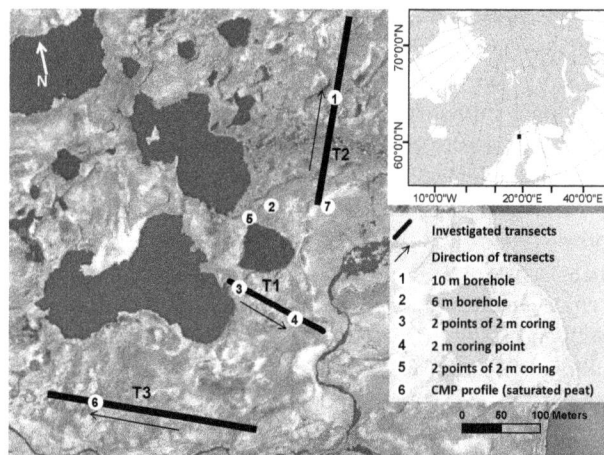

Figure 1. Location of the study site (inset), investigated transects, existing boreholes (Ivanova et al., 2011, points 1 and 2), coring points, and point of CMP measurement (described in Sect. 3.1; aerial photograph from Lantmäteriet, the Swedish land survey, 2012).

weather parameters have been monitored at the site since 2005 (Christiansen et al., 2010). Sannel and Kuhry (2011) have analyzed lake changes in the area and detailed local studies of palsa morphology have been conducted by Wramner (1968, 1973).

Tavvavuoma is located on a flat valley bottom, in piedmont terrain with relative elevations of surrounding mountains about 50 to 150 m above the valley bottom. Unconsolidated sediments, observed from two borehole cores (points 1 and 2 in Fig. 1), are of mainly glaciofluvial and lacustrine origin and composed of mostly sands, loams, and coarser-grained rounded gravel and pebbles (Ivanova et al., 2011). The mean annual air temperature is −3.5 °C (Sannel and Kuhry, 2011), and the average winter snow cover in Karesuando, a meteorological station approximately 60 km east of Tavvavuoma, is approximately 50 cm, although wind drift generally gives a thinner snow cover in Tavvavuoma (Swedish Meteorological and Hydrological Institute, http://www.smhi.se/klimatdata/meteorologi).

Permafrost occurs primarily under palsas and peat plateaus in Tavvavuoma, where the average thickness of the active layer is typically 0.5 m (Christiansen et al., 2010; Sannel and Kuhry, 2011). The mean annual temperature in permafrost boreholes is 2 °C (Christiansen et al., 2010). However, no observations of the depth to the permafrost base have been presented for the area. Warming of the air temperature of about 2 °C has been observed in direct measurements from the region over the past 200 years (Klingbjer and Moberg, 2003). In light of this warming, winter precipitation (mainly snow) in northern Sweden shows increasing trends over the past 150 years (Alexandersson, 2002). Furthermore, permafrost is degrading across the region and northern Sweden (Sjöberg et al., 2013). For example, peatland active layer thickness in Abisko (located about 60 km southwest of Tavvavuoma) is increasing according to direct observation over the past 30 years (Åkerman and Johansson, 2008) and inference from hydrologic shifts over the past century (Lyon et al., 2009). This regional permafrost degradation has led to changes in palsas as well. Regionally, reductions in both areas covered by palsas and palsa height have been observed (Sollid and Sorbel, 1998; Zuidhoff, 2002; Zuidhoff and Kolstrup, 2000). In Tavvavuoma, both growth and degradation of palsas have been observed in detailed morphological studies during the 1960s and 1970s (Wramner, 1968, 1973), and expansion and infilling of thermokarstic lakes have been observed through remote sensing analyses (Sannel and Kuhry, 2011). Palsa degradation and infilling of lakes with fen vegetation have been the dominating processes during recent years (Sannel and Kuhry, 2011; Wramner et al., 2012).

3 Theory and methods

Measurements of permafrost extent and structure were made with both GPR and ERT between 20 and 26 August 2012 along three transects covering the main permafrost landforms in the Tavvavuoma area (Fig. 1). The ERT transects were somewhat extended (i.e., slightly longer) compared to the GPR transects to increase the penetration depth along the overlapping parts of the transects.

Transect T1 was 160 m long and crossed a peat plateau that was raised approximately 1.5 m above the surrounding landscape (Fig. 1). It further crossed two thermokarst depressions (centered at 45 and 130 m) within the peat plateau. Transect T2 was 320 m long, but the southern part covering about 180 m could not be measured with GPR due to dense vegetation cover (mainly *Salix* sp.). Transect T2 started on a peat plateau surface at the edge of a drained lake and continued north over a fen (110–180 m) and a small stream (140 m). The northern part, measured with both ERT and GPR, crossed a palsa (200 m) that was raised about 4 m above the surrounding landscape. This palsa has been described via a borehole profile (Ivanova et al., 2011; point 1 in Fig. 1). Transect T2 then continued across two fens (250 and 290 m) separated by a lower palsa (270 m). Transect T3 was 275 m long. It started on a relatively low palsa and stretched over a flat area covered by hummocks and thermokarst depressions.

In addition to the geophysical investigations (details of which are described in the following sections), the depth to the permafrost table (the active layer) was probed every 2 m along all transects using a 1 m steel rod. Sediment cores were retrieved at four points along T1 and two points along T3 down to 2 m. These cores were used to locate the depth to the peat–mineral substrate interface and the depth to the permafrost table (at points 3, 4, 5, and 6 in Fig. 1). The topography was measured along the transects using a differential GPS with supplemental inclinometer observations along profiles where only ERT was used. The position of the transects

was measured using a tape measure and marked at regular intervals to ensure that locations of GPR and ERT transects coincided.

3.1 Ground penetrating radar

Ground penetrating radar (GPR) can be used to map near surface geology and stratigraphy because of differences in dielectric properties between different subsurface layers or structures. An electromagnetic pulse is transmitted through the ground and the return time of the reflected pulse is recorded. The resolution and penetration depth of the radar signal depends on the characteristics of the transmitted pulse and the choice of antennas, which usually range between 10 and 1000 MHz. Higher frequencies will yield a higher resolution but a smaller penetration depth; however, the penetration depth will also depend on dielectric and conductive properties of the ground material. Mapping of permafrost using GPR becomes possible due to the difference in permittivity between unfrozen and frozen water.

In this study, measurements were made with a Malå GeoScience ProEx GPR system using 200 MHz unshielded antennas along T1 and T2. The transmitting and receiving antennas were held at a constant distance of 0.6 m (common offset) and the sampling time window was set to 621 ns, with recorded traces stacked 16 times. Measurements were made at every 10 cm along the length of these two transects. Along T3, measurements were made using 100 MHz unshielded antennas with a 1 m antenna separation and measurements made every 0.2 s while moving the antennas along the transect. The sampling time window for T3 was 797 ns and traces were stacked 16 times. The GPR data were processed for a time-zero correction and with a dewow filter, a vertical gain, and a normal-moveout correction for antenna geometry using the software ReflexW (version 6.1, Sandmeier Geophysical Research, 2012, http://www.sandmeier-geo.de).

The depths to the permafrost table and the interface between peat and mineral substrates were calculated by converting the two-way travel time to known substrate transitions using estimated velocities for the speed through three different substrate materials: dry peat, saturated peat, and saturated mineral substrate (see Fig. 2 for conceptual sketch of these substrate layers and velocity profiles). To account for uncertainty due to small-scale heterogeneity of these ground materials, in addition to the optimal "representative" velocity identified, the likely maximum and minimum velocities for each substrate were considered in the GPR depth conversions (Table 1). The end product here is a range of plausible substrate velocities accounting for potential uncertainties such that any resultant interpretation about subsurface conditions and interface locations can be considered robust. The velocity in dry peat (found in the active layer of palsas, hummocks, and peat plateaus) was calibrated using the active layer thickness measurements made with a steel rod. The minimum and maximum velocities were obtained by subtracting and

Table 1. Velocities used for converting two-way travel times to depth in GPR data.

Material	Velocity (m ns^{-1})	Method/source
Dry peat – representative	0.049	Calibration against every second field measurements* of active layer depths
Dry peat – min	0.046	Representative estimate minus 1 standard deviation of field measurements*
Dry peat – max	0.052	Representative estimate plus 1 standard deviation of field measurements*
Saturated peat – representative	0.036	Calibration against coring (points 3 and 5, in Fig. 1)
Saturated peat – min	0.033	Velocity in pure water (Davis and Annan, 1989)
Saturated peat – max	0.049	Representative estimate for dry peat
Saturated mineral – representative	0.060	Velocity in sand and clay from Davis and Annan (1989)
Saturated mineral – min	0.053	Calculated from Joseph et al. (2010) for saturated loams and sands
Saturated mineral – max	0.073	Highest estimated velocity from CMP analysis

* Field measurement using a 1 m steel rod.

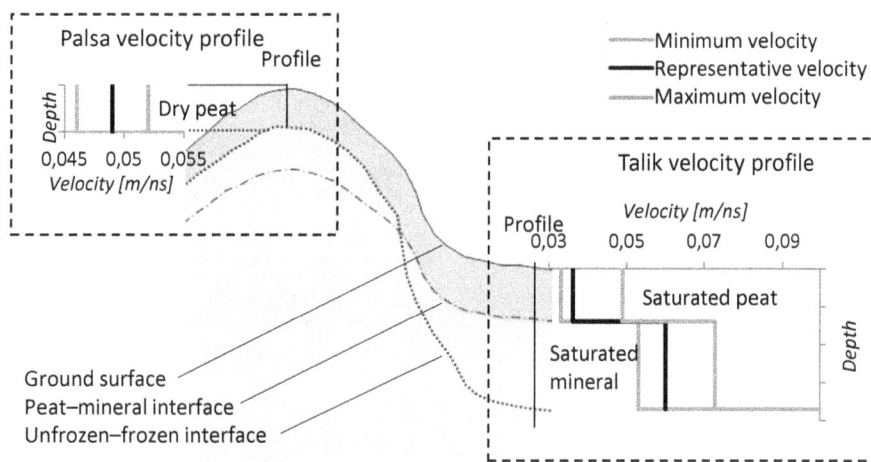

Figure 2. Conceptual sketch of typical distribution of ground substrates and associated estimated velocities for a palsa and talik ground profile.

adding 1 standard deviation of the measured depths, respectively. For velocities in saturated peat that was found in taliks such as fens, the thickness of the saturated peat layer identified by coring with a 2 m steel pipe (points 3 and 5, Fig. 1) was used. The velocity in pure water was used as the minimum velocity and the representative velocity for dry peat was used as the maximum velocity for saturated peat.

To obtain velocities for unfrozen saturated mineral substrate, a common midpoint (CMP) GPR profile was measured on a drained lake surface (point 7 in Fig. 1). Coring down to 2 m with a steel pipe at this location revealed the existence of an unfrozen saturated peat layer down to 1.75 m depth and unfrozen mineral soils consisting of mainly sand and silt below that depth. CMP analysis is a widely used method to estimate local GPR signal velocities through ground materials. By moving GPR transmitting and receiving antennas apart incrementally between measurements, the same point in space is imaged with different antenna offsets, making it possible to back out material velocity estimates from the hyperbolic shape of the recorded reflectors.

The measured reflectors must be relatively flat so that the signal moves through the same materials at the same depths independent of antenna offset. For the CMP measurement, 100 MHz unshielded antennas were moved apart in 10 cm increments along a 15 m transect with a time window of 797 ns and 16 stacks of each trace. The data were processed in ReflexW software (version 6.1, Sandmeier Geophysical Research, 2012, http://www.sandmeier-geo.de) for a time-zero correction, a dewow filter, and a vertical gain. Semblance analysis (Neidell and Tanner, 1971) was used to identify appropriate reflectors from which velocities could be estimated. Figure 3 shows the estimated velocity profile, recorded CMP radargram, and semblance plot for the CMP transect. Although a relatively flat reflector was identified for the CMP measurement, the results from the semblance analysis does not show one clear reflector and associated velocity at the identified depth of the peat–mineral interface. Instead, a wide range of possible velocities are shown in the semblance plot for the top ∼ 200 ns, likely due to high heterogeneity in ground substrates and/or water content. Due to the difficulty

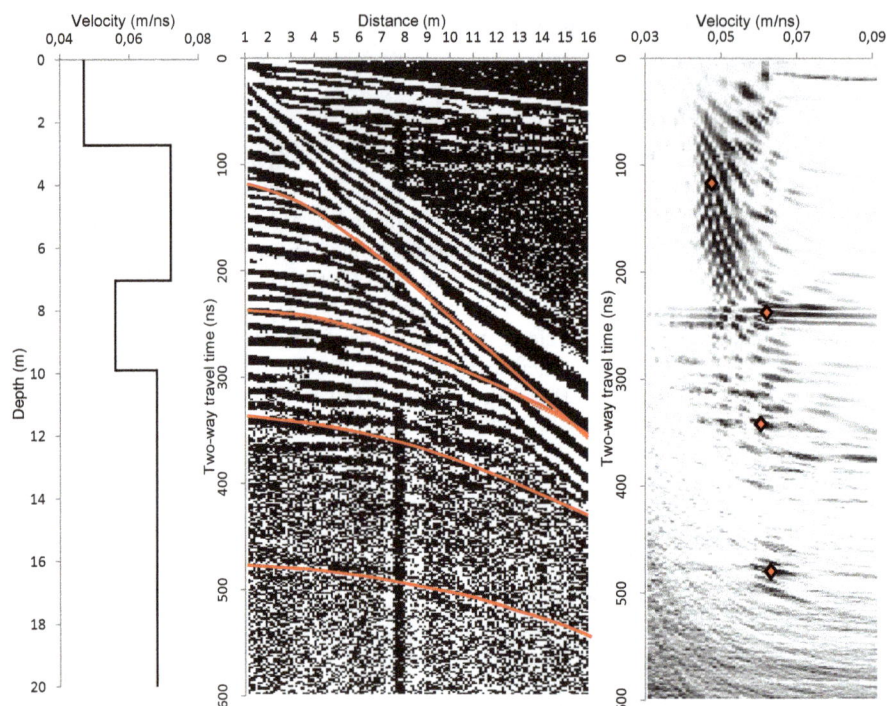

Figure 3. Estimated velocity profile, recorded CMP radargram, and semblance plot for the CMP transect measured on the drained lake surface. The semblance plot shows more likely velocities in darker shades of grey with the velocities from the reflectors (red lines in radargram) used for generating the velocity profile indicated by black and red diamonds.

in constraining the material velocities for the deeper layers using this method, these results were only used for estimating a probable maximum velocity in unfrozen mineral sediments (as this was higher than most literature values). This maximum velocity estimate was complemented with literature values for the representative and minimum velocities.

3.2   Electrical resistivity tomography

Direct-current electrical resistivity measurements are based on a measured potential difference between two electrodes (ΔV) inserted with galvanic coupling to the ground and, similarly, two electrodes where current is injected into the ground (I) with a known geometric factor (k) depending on the arrangement of the electrodes. This gives a value of the apparent resistivity (ρ_a) of the ground subsurface as

$$\rho_a = k\Delta V/l. \tag{1}$$

During a tomographic resistivity survey, many of these measurements are made in lateral and vertical directions (by increasing the electrode spacing). The acquired data are subsequently modeled to generate an image of the resistivity distribution under the site. Values of resistivity vary substantially with grain size, porosity, water content, ice content, salinity, and temperature (e.g., Reynolds, 2011); thus, the resistivity of permafrost also varies to a large degree. This makes

ERT techniques useful in detecting the sharp contrast between frozen and unfrozen water content within sediments.

At the Tavvavuoma site, measurements of electrical resistivity were made with the Terrameter LS from ABEM and an electrode spacing of 2 m for the T1 transect and 4 m for the T2 and T3 transects. The Wenner array configuration for the electrodes was used due to its high signal-to-noise ratio and for its accuracy in detecting vertical changes over other common array types (Loke, 2010). For the inverse modeling, the smoothness-constrained least-square method was applied (Loke and Barker, 1996). The inversion progress was set to stop when the change in root mean squared error from the previous iteration was less than 5 % (implying convergence of the inversion). The software Res2dinv (v.3.59.64, Geotomo Software, Loke, 2010) was used for the inverse modeling during this study.

To assess the quality and reliability of the resistivity modeling for the Tavvavuoma site, the depth of investigation (DOI) method (Oldenburg and Li, 1999) was used. This appraisal technique uses the difference between two inverted models where the reference resistivity parameter is varied to calculate a normalized DOI-index map. From these values a depth at which the surface data are no longer sensitive to the physical properties of the ground can be interpreted. The method has previously been applied in permafrost studies (e.g., Fortier et al., 2008; Marescot et al., 2003). To calculate the DOI-index we used a symmetrical two-sided difference scheme where 0.1 and 10 times the average apparent resistivity of the resistivity model was considered (respectively) for the initial reference resistivity parameter. Normalized DOI values higher than 0.1 indicate that the model is likely not constrained by the data and should be given little significance in subsequent model interpretation.

To further validate the ERT interpretations, one shorter transect with 0.5 m electrode spacing was conducted over a palsa. This was used to acquire a local resistivity value for the interface between unfrozen and frozen sediments at the bottom of the active layer. This value (1700 Ωm) allowed us to map permafrost boundaries in the ERT images, with all resistivity values > 1700 Ωm interpreted as permafrost. However, as the resistivity of the ground varies with other sediment physical properties and the sediment distribution is complex at the site, the resistivity boundary value for permafrost will naturally vary along transects and with depth. For instance, sands generally have maximum values for the unfrozen state close to 1200 Ωm and for some gravels this can reach up to 3000 Ωm (Hoekstra et al., 1974). Finer sediments, such as clays and silts, have lower values ranging from ca. 80 to 300 Ωm (Hoekstra et al., 1974). At our site sands dominate, but there is also evidence of loams. Lewkowicz et al. (2011) report a resistivity of 1000 Ωm at the base of permafrost under a palsa in similar, but somewhat finer, sediment conditions in southern Yukon. This value from Lewkowicz et al. (2011) was thus used as a possible minimum resistivity value for the permafrost boundary in the interpretations, while the local resistivity estimate (1700 Ωm) was used as a maximum and representative value. All resistivity values < 1000 Ωm were thus interpreted as unfrozen ground and the values between 1000 and 1700 Ωm represent a range of uncertainty for the location of the interface between frozen and unfrozen sediments. Again, the motivation here was to account for potential uncertainty allowing for robust interpretation.

3.3 Calculations of active layer thickness and future thaw rates

To help put the geophysical measurements and their potential implications for this peatland palsa region in context, the thickness of the active layer as well as first-order estimate of long-term thaw rates were estimated using a simple equation for 1-D heat flow by conduction, the Stefan equation (as described by Riseborough et al., 2008):

$$Z = \sqrt{\frac{2\lambda I}{Ln}}, \qquad (2)$$

where Z is the thaw depth, λ is thermal conductivity, I is the thawing degree day index (as described by Nelson and Outcalt, 1987), L is the volumetric latent heat of fusion, and n is the saturated porosity of the ground substrate. As a talik is by definition unfrozen ground occurring in a permafrost area, Eq. (2) was used to confirm that ground identified as

talik in Tavvavuoma through the GPR and ERT images did not correspond to locations of deeper active layer relative to surrounding positions (i.e., provide a confirmation that these sites would not freeze during winter).

Calculations of active layer depths in fens were made using a sinusoidal annual air temperature curve generated from the average temperature of the warmest and the coldest months of the year as input. The effect of the snow cover, which would give higher ground-surface temperatures in the winter, was not explicitly taken into consideration in this simple calculation as we did not have any direct estimates of snow cover available for the transects. As such, these calculations are simply a first-order approximation. Representative properties for saturated peat (the most common material in the uppermost part of the ground in suspected taliks) were chosen, including a thermal conductivity of $0.5\,\mathrm{W\,m^{-1}\,K^{-1}}$ and a saturated fraction of 0.80 (Woo, 2012).

In addition, a first-order approximation of long-term thaw rates was carried out. An instantaneous increase in air temperature of 2 °C was assumed, which represents a warming within current climate projections for the 21st century, although at the low end of projections for Arctic warming (IPCC, 2013). Material properties for this calculation were based on information on deeper sediment layers from the 10 m borehole (Ivanova et al. 2011, point 1 in Fig. 1). A saturated fraction of 0.5, representative of sand slightly oversaturated with ice, was used. To account for some of the uncertainty in this rough estimate, a range of likely minimum and maximum values for thermal conductivity (2 and $3\,\mathrm{W\,m^{-1}\,K^{-1}}$, respectively) for this material were used to estimate a range of thaw rates. The annual freezing degree days were subtracted from the annual thawing degree days, I in Eq. (2), and the number of days necessary to thaw the estimated local thickness of permafrost was estimated. This is a simple estimate since, clearly, the Stefan equation is neither designed to calculate long-term thaw rates nor does such an estimate consider any density-dependent feedbacks and/or subsequent hydroclimatic shifts. Regardless, combined with estimates of permafrost thickness made in our geophysical investigation, the aim of this back-of-the-envelope calculation was to provide an order-of-magnitude estimate for the time it could potentially take permafrost to completely thaw at this site to help place it in a pan-arctic context.

4 Results

4.1 GPR data

In the GPR images the permafrost table was clearly detectable under the palsa and peat plateau surfaces along all transects (Fig. 4). The interface between peat and mineral substrates was only detectable in unfrozen sediments. Deeper reflections, interpreted as the permafrost table under suprapermafrost taliks, were found under the fens and surface

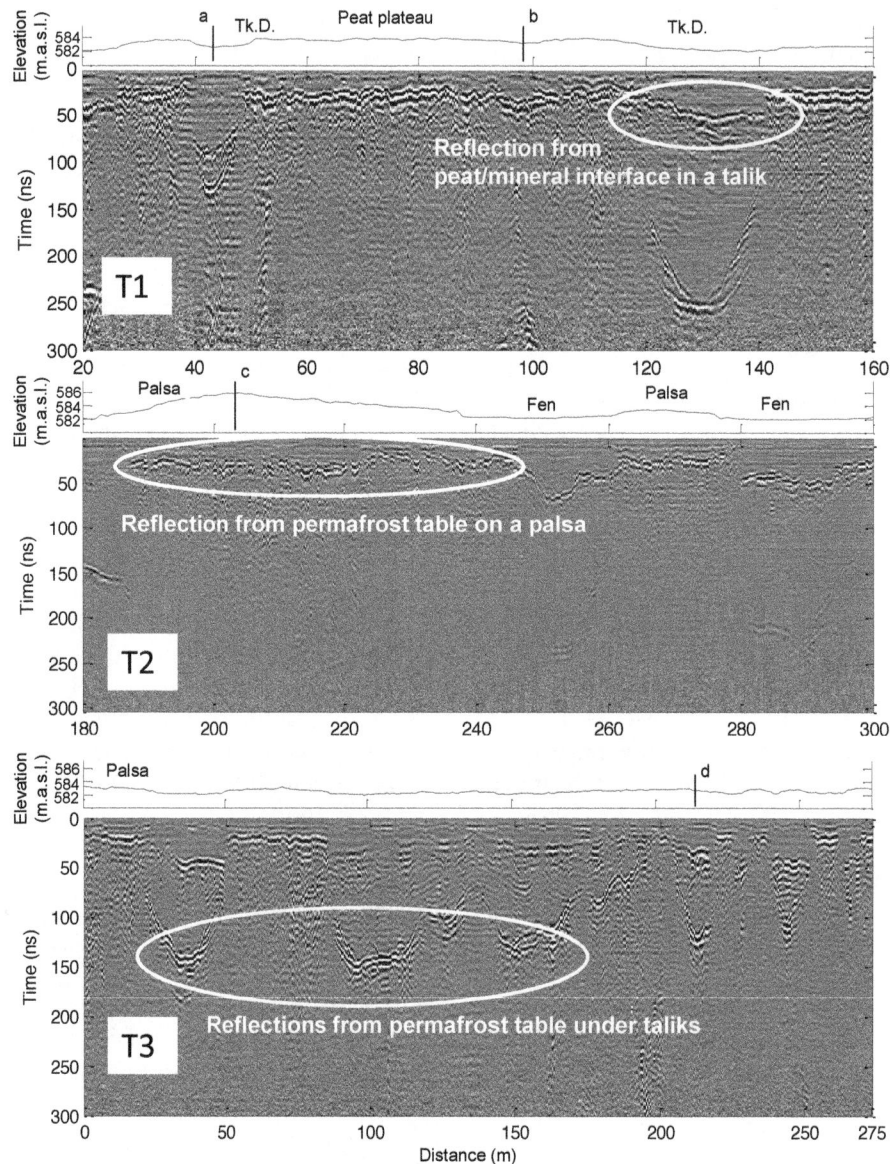

Figure 4. Elevation profiles and GPR images for T1, T2, and T3 with selected reflections marked as examples of interfaces that were identified for this study. Landforms are indicated on top of elevation profiles along T1 and T2 (Tk.D is thermokarst depression) together with coring points in T1 (a is point 3 in Fig. 1, and b is point 4 in Fig. 1) and T3 (d is point 6 in Fig. 1) as well as the 10 m borehole in T2 (c is point 1 in Fig. 1). No landforms are indicated along T3 after the first palsa (0–25 m) due to the complex micro topography of hummocks and thermokarst depressions along this transect.

depression in all transects. At the beginning of both transects T1 and T2, deep reflections that end abruptly were present in the images at about 250 and 150 ns, respectively. In T1, this corresponds to a wet fen bordering a lake; for T2 it corresponds to a fen bordering a stream. The proximity to these water bodies suggests that these are likely not reflections from the permafrost table. The base of the permafrost could not be detected at any point in the GPR images likely because of loss of signal strength at depth.

4.2   ERT data

The inverted resistivity sections showed areas of high resistivity (1000–100 000 Ωm) where permafrost could be expected due to the sharp contrast to surrounding surfaces. This suggests permafrost boundaries are detectable for both the extent of the horizontal distribution and the vertical extent to the base of permafrost (Fig. 5). The highest resistivity values were found under the peat plateau in T1 and under the palsas in T2 and T3. Low resistivity values were found under the fens in all transects. DOI values increase with depth for

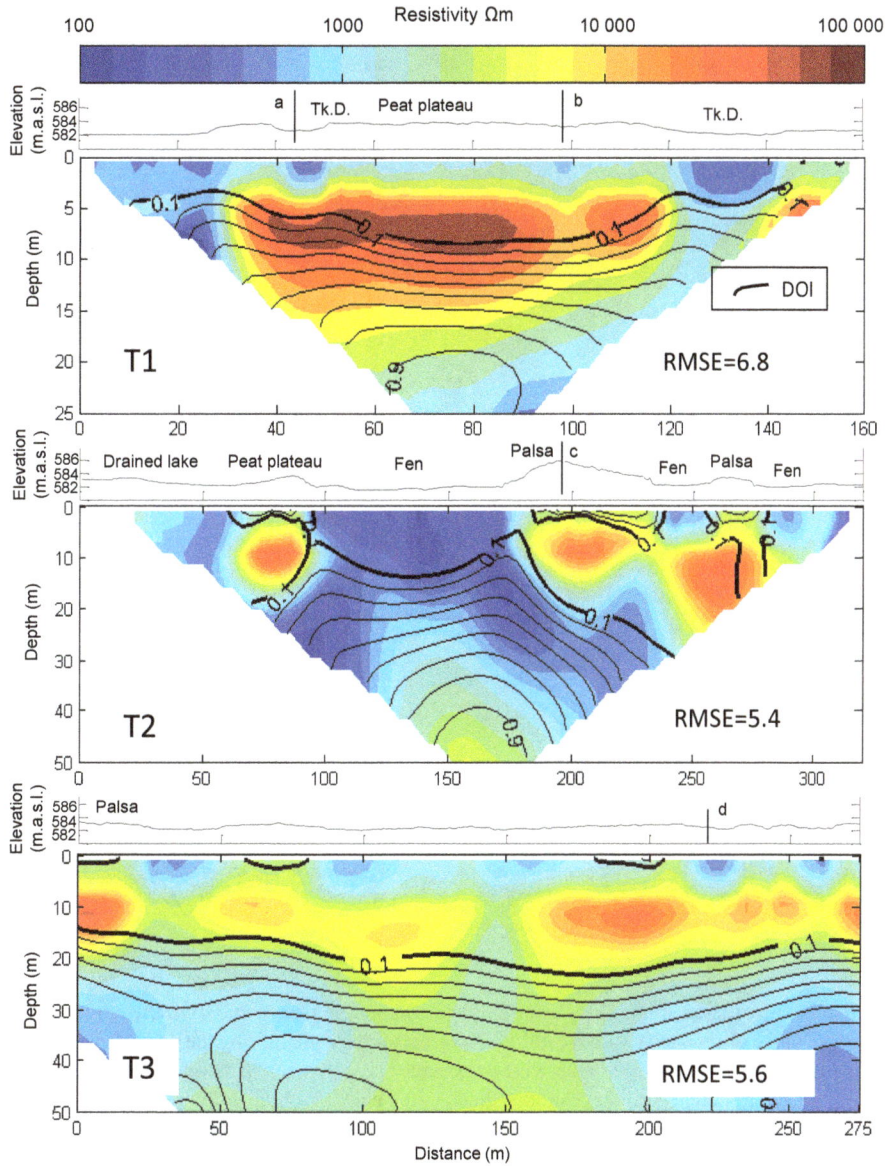

Figure 5. Elevation profiles and ERT results for T1, T2, and T3. DOI < 0.1 (black lines) indicates that the model is well constrained by the data. Landforms are indicated on top of elevation profiles along T1 and T2 (Tk.D = thermokarst depression) together with coring points in T1 (a is point 3 in Fig. 1, and b is point 4 in Fig. 1) and T3 (d is point 6 in Fig. 1) as well as the 10 m borehole in T2 (c is point 1 in Fig. 1). No landforms are indicated along T3 after the first palsa (0–25 m) due to the complex micro topography of hummocks and thermokarst depressions along this transect.

all transects, allowing the permafrost base to be interpreted only along parts of T2. In contrast, under T1 and T3 the DOI rapidly increases under the peat plateau and hummocks. Due to the wide electrode spacing adopted (2 and 4 m), the permafrost table under the active layer is too shallow to be visible in the ERT data.

4.3 Geophysical interpretations

Permafrost occurs under the palsa and peat plateau surfaces along T1 and T2 as well as under the hummocks along T3

(Fig. 6). The active layer depths estimated from the GPR data closely matched the depths measured in the field (Table 2). This is expected since measured active layer depths were used to derive the velocity of the radar signal in the dry peat in the active layer. The depth to the base of the permafrost could only be estimated with good confidence along parts of T2 and is on average 15.8 m from the ground surface and at least 25 m at its deepest point. Along transects T1 and T3 the deepest permafrost was found at 8.4 and 23.4 m, respectively; however, the permafrost base could not be identified with confidence below this depth.

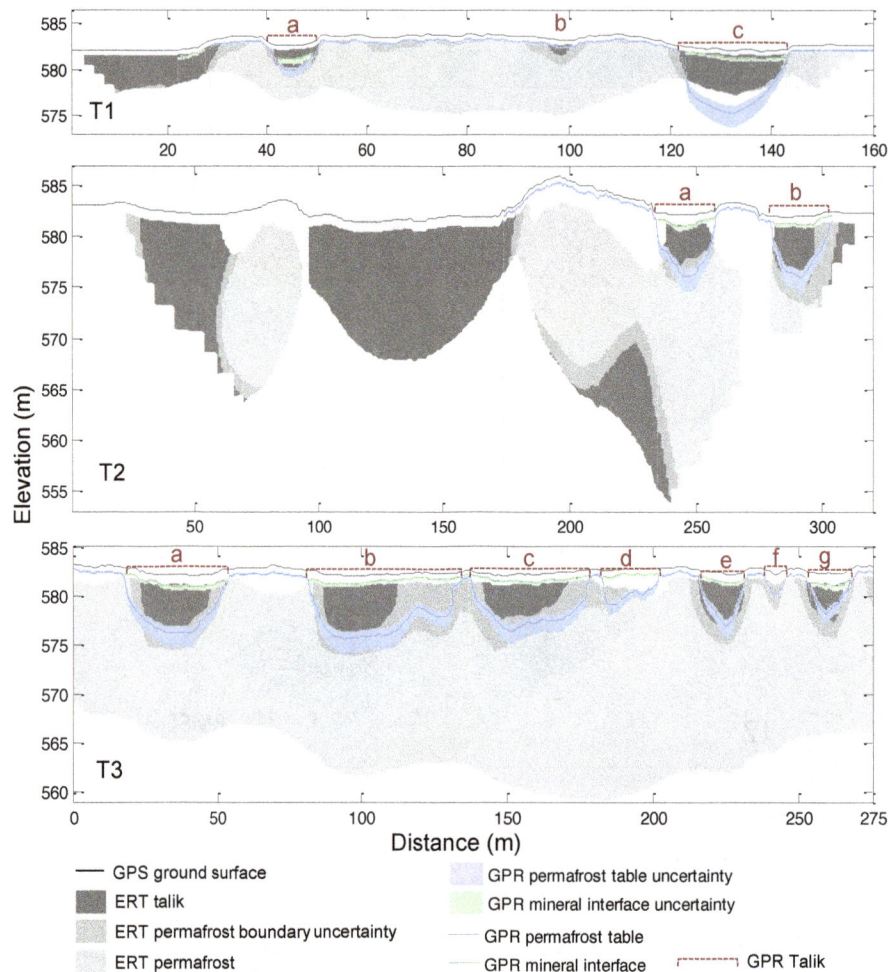

Figure 6. Interpreted permafrost distribution along T1, T2, and T3. Uncertainty intervals come from the range of estimated signal velocities for GPR (Table 1) and from the range of resistivity values (1000–1700 Ωm) used for identifying the permafrost boundary for ERT. In sections marked GPR Talik (red dotted line), GPR depth conversions have been made using saturated peat velocities down to the peat–mineral interface (green line) and then using saturated mineral substrate velocities down to the permafrost table (blue line). In the remaining parts of transects, the dry peat velocities have been used down to the permafrost table. No interpretations of ERT data with DOI > 0.1 have been made and therefore the permafrost base is only visible along parts of T2. Note the differences in scale in the x direction between figures and the vertical exaggeration.

Potential taliks (Table 3 and Fig. 6) are numerous and occur in both wet fens, such as all taliks along T2, and relatively dry depressions in the terrain, such as all taliks along T1. The sediment cores used for estimating the GPR representative signal velocity in saturated peat were taken in both a relatively dry location and a wet fen, but the calculated velocities were nearly identical, indicating that the soil moisture at depth was similar at both locations. Most of T3 was underlain by taliks and these were found under both wet fens and drier surface depressions. The taliks range in depth from 2.1 m (T3f, numbering from Table 3 and Fig. 6) to 6.7 m (T1c) based on the GPR data and are slightly deeper, although within the range of uncertainty based on the ERT results. From the ERT data, T1c is in fact interpreted as a potential through-going talik. Talik T1b was only detected

from the ERT data, and taliks T3b–T3d appear as one large talik in the ERT data.

4.4 Calculations of active layer thickness and future thaw rates

The active layer depths calculated using the Stefan equation support the interpretation that identified taliks do not freeze during winter. The seasonal frost penetration depth was estimated to be 0.72 m, which is about the same as the average peat depth along the transects and much less than the estimated minimum depth of the taliks (2.1 m). While a shallower peat depth would give a deeper frost penetration, it is unlikely that the seasonal frost penetration is > 2.1 m in the Tavvavuoma area. This ancillary estimate confirms

Table 2. Range of interpreted depths (m) of active layer, peat–mineral interface, and permafrost base averaged along transects at Tavvavuoma.

	T1			T2			T3		
	Min[a]	Representative[b]	Max[c]	Min[a]	Representative[b]	Max[c]	Min[a]	Representative[b]	Max[c]
				Active layer					
Observed[d]		0.51			0.52			0.56	
GPR	0.50	0.53	0.57	0.48	0.51	0.54	0.52	0.56	0.59
				Peat–mineral interface					
GPR	0.77	0.84	1.14	0.68	0.74	1.01	0.63	0.69	0.93
				Permafrost base					
ERT	–	–		15.8	17.3		–	–	

[a] GPR: using the estimated minimum velocity (Table 1). ERT: using 1000 Ωm resistivity boundary (talik). [b] GPR: using representative estimate velocity (Table 1). ERT: using 1700 Ωm resistivity value. [c] GPR: using the estimated maximum velocity (Table 1). ERT using 1000 Ωm resistivity boundary (permafrost base). [d] Depth from manual field measurement using a steel probe.

Table 3. Estimated depths (m) of taliks at deepest point. Numbering is the same as in Fig. 6.

Talik	GPR min[a]	GPR representative[b]	GPR max[c]	ERT min[a]	ERT representative[b]
T1a	2.4	2.7	3.4	2.5	3.1
T1b	–	–	–	1.6	2.8
T1c	6.0	6.7	8.3	> 4.7	> 4.7
T2a	5.4	6.1	7.6	5.4	6.9
T2b	5.3	6.0	7.4	6.9	8.8
T3a	5.3	5.9	7.4	5.8	7.8
T3b	5.7	6.4	8.0	6.3	8.2
T3c	5.1	5.7	7.0	4.8	7.9
T3d	3.1	3.5	4.4	–	4.0
T3e	4.6	5.2	6.4	5.4	7.2
T3f	2.0	2.1	2.2	–	3.8
T3g	3.7	4.1	5.2	5.0	6.8

[a] GPR: using the estimated minimum velocity (Table 1). ERT: using 1000 Ωm resistivity boundary (talik). [b] GPR: using representative estimate velocity (Table 1). ERT: using 1700 Ωm resistivity value. [c] GPR: using the estimated maximum velocity (Table 1). ERT using 1000 Ωm resistivity boundary (permafrost base).

the aforementioned geophysical interpretation. Furthermore, assuming a 2 °C instantaneous temperature increase at the site, a first-order approximation of the long-term thaw rate was calculated to be 6–8.5 cm yr^{-1}. At this rate, the time to completely thaw permafrost, assuming the estimated average thickness along T2 (15.3 m), was calculated to be 175–260 years.

5 Discussion

5.1 Permafrost and talik distribution at Tavvavuoma

The spatial pattern of permafrost and taliks in Tavvavuoma is closely linked to the distribution of palsas, peat plateaus, fens, and water bodies. This suggests that local factors, such as soil moisture, groundwater flow, ground ice content, sediment distributions, and geomorphology, strongly influence the local ground thermal regime (see e.g., Delisle and Allard, 2003; McKenzie and Voss, 2013; Woo, 2012; Zuidhoff, 2002). The relative elevation of permafrost landforms, as well as permafrost resistivity values and sediment distributions, suggests that there is a large variation in ground ice content in the area. Surface elevations of palsas and peat plateaus are highest along T2 and lowest along T3, indicating a higher ice content of the underlying ground along T2, which is likely related to differences in ground substrates between the transects. Coring (< 2 m) across the site and existing borehole descriptions (Ivanova et al., 2011) confirm that the ground contains a larger fraction of coarse glaciofluvial sand and gravel, which are not susceptible to frost heave, closer to T3 as compared to T2.

Lewkowicz et al. (2011) used the height of palsas and permafrost thickness, estimated by ERT, to calculate excess ice fractions (EIF, defined as the ratio of palsa height to permafrost thickness) in permafrost mounds in southern Yukon. In Tavvavuoma, the highest palsa at T2 is approximately 4 m high and underlain by 16 m thick permafrost at the highest point. This corresponds to an EIF of 0.25, which is comparable to the EIFs reported by Lewkowicz et al. (2011) that generally ranged between 0.2 and 0.4. In contrast, along T3 the relative heights of permafrost landforms are lower and the permafrost is thicker for most of the transect. Similarly calculated EIFs along T3 were on average < 0.03 and at maximum < 0.09 but are likely lower in reality as the base of the permafrost is at a greater depth than what could be detected in our study. The relatively low resistivity of the permafrost along T3 further supports interpretations for lower ice content in this permafrost. Permafrost with low ice content is more susceptible to thaw, as less energy is needed for latent heat exchange. This provides a possible explanation for why taliks are more widespread along T3, as permafrost with a low ice content would have reacted more rapidly to warming in the area.

The calculated thaw rate of 6–8.5 cm yr^{-1} is considerably higher than the ca. 1 cm yr^{-1} deepening of the active layer observed in the region (Åkerman and Johansson, 2008) and inferred from hydrological records (Lyon et al., 2009). One possible reason for this is that these observations were made in the relatively ice-rich top layer of peat, while for the calculations in this study a medium with higher thermal conductivity and lower ice content was used to represent the lower mineral sediment layer. The 2 °C instantaneous step change in temperature could have further contributed to the higher thaw rates compared to the ones observed. As thawing is driven by gradients in heat it can be argued that permafrost thaw rates should increase with warmer air temperatures. Considering this, the calculated time of complete permafrost thaw of about 175–260 years can be considered reasonable in at least 1 order of magnitude. However, much more rapid palsa degradation has been observed in the region (Zuidhoff, 2002) due to block and wind erosion processes and thermal influence on palsas from expanding water bodies, and very rapid decay of palsa surface areas has been observed in both southern Norway and the Canadian Arctic (Payette et al., 2004; Sollid and Sorbel, 1998). The coupled erosion, hydrological, and thermal processes are not represented in the Stefan equation but can be of great importance for permafrost thaw rates (McKenzie and Voss, 2013; Painter et al., 2013; Zuidhoff, 2002). There is clearly a need for quantification of the relative importance of these processes for permafrost thaw to better understand expected future changes in these environments.

5.2 On the complementary nature of the geophysical techniques

Several previous studies have shown the benefits of combining more than one geophysical technique for mapping permafrost (e.g., De Pascale et al., 2008; Hauck et al., 2004; Schwamborn et al., 2002); in this study the GPR and ERT data also provided complementary information that allowed for interpretations that would not have been possible by using only one of the two data sets. Of course, combining multiple techniques for inference compounds our estimate uncertainties. To attain more precise estimates of depths to the different interfaces, deeper coring data would have been necessary for both more accurate signal velocity estimates for the GPR and for local resistivity values of the ground materials. The fact that ERT depth estimates are consistently higher than the GPR estimates suggest that either the resistivity boundary value for permafrost is in fact lower than our local estimate, or that GPR signal velocities are higher than the values used in this study. Since our local permafrost resistivity estimate was made in peat at the permafrost table, which can have a

very high ice content compared to deeper sediment layers, it is a more likely explanation for this discrepancy.

GPR and ERT yielded somewhat overlapping data but the two data sets have different strengths and therefore complement each other well. The GPR data worked well for identifying the permafrost table with high confidence, especially in the top 2 m where sediment cores could be easily obtained for validation and signal velocity estimates. This suitability of GPR for identifying permafrost interfaces in the top 1–2 m has been shown in several studies (e.g., Doolittle et al., 1992; Hinkel et al., 2001; Moorman et al., 2003). The ERT data, using the setup in this study, do not yield data in the uppermost part of the ground and also have higher uncertainty where resistivity contrasts are high (Fig. 5), which makes them less suited for the active layer and shallow taliks. With the ERT data it is, however, possible to image relatively deep in the ground where the GPR cannot penetrate. By combining both GPR and ERT the active layer, the base of permafrost, and potential taliks could be identified along at least parts of the transects, which could not have been achieved with good confidence by either of the two methods alone.

6 Concluding remarks

Peat plateau complexes offer an interesting challenge to the Cryosphere community as they are clear mosaics combining local-scale differences manifested as permafrost variations. As such variation occurs both horizontally and vertically in the landscape, geophysical techniques offer a good possibility to record current permafrost conditions across scales. Furthermore, by combining methods, such as GPR and ERT as demonstrated here, complementary and independent views of the permafrost extents can be acquired. The results of this study show a heterogeneous pattern of permafrost extent reflecting both local and climatic processes of permafrost formation and degradation. To improve our understanding of landscape–permafrost interactions and dynamics will require a community effort to benchmark variability across the scales and environments within the pan-Arctic. This is particularly important in lesser-studied regions and across the sporadic permafrost zone where changes are occurring rapidly.

Author contributions. Ylva Sjöberg designed the study, carried out the GPR measurements and analysis, and did the main writing of the manuscript. Per Marklund carried out the ERT measurements and analysis, did the main writing for the sections on ERT methods and ERT results, and commented on the whole manuscript. Rickard Pettersson provided input on the geophysical techniques and analyses and commented on the whole manuscript. Steve Lyon provided input on the project design and commented on the whole manuscript, including language and style.

Acknowledgements. This study was kindly supported by Sveriges Geologiska Undersökning (SGU), the Bolin Centre for Climate Research, Lagrelius fond, Göran Gustafssons Stiftelse för natur och miljö i Lappland, and Svenska Sällskapet för Antropologi och Geografi. The authors are grateful to Peter Jansson and Britta Sannel for lending us the equipment necessary for this study and to Romain Pannetier, Kilian Krüger, Matthias Siewert, Britta Sannel, and Lars Labba for fieldwork support. We are also grateful to Andrew Parsekian for technical advice on the GPR survey design and María A. García Juanatey for consultation when processing the ERT data.

Edited by: J. Boike

References

Åkerman, H. J. and Johansson, M.: Thawing permafrost and thicker active layers in sub-arctic Sweden, Permafrost Periglac., 19, 279–292, doi:10.1002/ppp.626, 2008.

Alexandersson, H.: Temperature and precipitation in Sweden 1860–2001, SMHI, Norrköping, Sweden, 2002.

Arcone, S. A., Lawson, D. E., Delaney, A. J., Strasser, J. C., and Strasser, J. D.: Ground-penetrating radar reflection profiling of groundwater and bedrock in an area of discontinuous permafrost, Geophysics, 63, 1573–1584, doi:10.1190/1.1444454, 1998.

Christiansen, H. H., Etzelmuller, B., Isaksen, K., Juliussen, H., Farbrot, H., Humlum, O., Johansson, M., Ingeman-Nielsen, T., Kristensen, L., Hjort, J., Holmlund, P., Sannel, A. B. K., Sigsgaard, C., Akerman, H. J., Foged, N., Blikra, L. H., Pernosky, M. A., and Odegard, R. S.: The Thermal State of Permafrost in the Nordic Area during the International Polar Year 2007–2009, Permafrost Periglac., 21, 156–181, doi:10.1002/ppp.687, 2010.

Davis, J. L. and Annan, A. P.: Ground-penetrating radar for high-resolution mapping of soil and rock, Geophys. Prospect., 37, 531–551, doi:10.1111/j.1365-2478.1989.tb02221.x, 1989.

Delisle, G. and Allard, M.: Numerical simulation of the temperature field of a palsa reveals strong influence of convective heat transport by groundwater, in: Proceeding of the 8th International Permafrost Conference, 21–25 July 2003, Zurich, Switzerland, 181–186, 2003.

De Pascale, G. P., Pollard, W. H., and Williams, K. K.: Geophysical mapping of ground ice using a combination of capacitive coupled resistivity and ground-penetrating radar, Northwest Territories, Canada, J. Geophys. Res.-Earth., 113, F02S90, doi:10.1029/2006jf000585, 2008.

Dobinski, W.: Geophysical characteristics of permafrost in the Abisko area, northern Sweden, Pol. Polar Res., 31, 141–158, doi:10.4202/ppres.2010.08, 2010.

Doolittle, J. A., Hardisky, M. A., and Black, S.: A ground-penetrating radar study of Goodstream palsas, Newfoundland, Canada, Arct. Alp. Res., 24, 173–178, doi:10.2307/1551537, 1992.

Fortier, R., LeBlanc, A. M., Allard, M., Buteau, S., and Calmels, F.: Internal structure and conditions of permafrost mounds at Umiujaq in Nunavik, Canada, inferred from field investigation and electrical resistivity tomography, Can. J. Earth Sci., 45, 367–387, doi:10.1139/e08-004, 2008.

Gacitua, G., Tamstorf, M. P., Kristiansen, S. M., and Uribe, J. A.: Estimations of moisture content in the active layer in an Arctic ecosystem by using ground-penetrating radar profiling, J. Appl. Geophys., 79, 100–106, doi:10.1016/j.jappgeo.2011.12.003, 2012.

Giesler, R., Lyon, S. W., Mörth, C.-M., Karlsson, J., Karlsson, E. M., Jantze, E. J., Destouni, G., and Humborg, C.: Catchment-scale dissolved carbon concentrations and export estimates across six subarctic streams in northern Sweden, Biogeosciences, 11, 525–537, doi:10.5194/bg-11-525-2014, 2014.

Hauck, C., Vonder Muhll, D., and Maurer, H.: Using DC resistivity tomography to detect and characterize mountain permafrost, Geophys. Prospect., 51, 273–284, doi:10.1046/j.1365-2478.2003.00375.x, 2003.

Hauck, C., Isaksen, K., Muhll, D. V., and Sollid, J. L.: Geophysical surveys designed to delineate the altitudinal limit of mountain permafrost: An example from Jotunheimen, Norway, Permafrost Periglac., 15, 191–205, doi:10.1002/ppp.493, 2004.

Hinkel, K. M., Doolittle, J. A., Bockheim, J. G., Nelson, F. E., Paetzold, R., Kimble, J. M., and Travis, R.: Detection of subsurface permafrost features with ground-penetrating radar, Barrow, Alaska, Permafrost Periglac., 12, 179–190, doi:10.1002/ppp.369, 2001.

Hoekstra, P., Selimann, P., and Delaney, A.: Airborne resistivity mapping of permafrost near Fairbanks, Alaska, US Army CRREL, Hanover, New Hampshire, 51, 1974.

Hugelius, G., Bockheim, J. G., Camill, P., Elberling, B., Grosse, G., Harden, J. W., Johnson, K., Jorgenson, T., Koven, C. D., Kuhry, P., Michaelson, G., Mishra, U., Palmtag, J., Ping, C.-L., O'Donnell, J., Schirrmeister, L., Schuur, E. A. G., Sheng, Y., Smith, L. C., Strauss, J., and Yu, Z.: A new data set for estimating organic carbon storage to 3 m depth in soils of the northern circumpolar permafrost region, Earth Syst. Sci. Data, 5, 393–402, doi:10.5194/essd-5-393-2013, 2013.

Hugelius, G., Strauss, J., Zubrzycki, S., Harden, J. W., Schuur, E. A. G., Ping, C.-L., Schirrmeister, L., Grosse, G., Michaelson, G. J., Koven, C. D., O'Donnell, J. A., Elberling, B., Mishra, U., Camill, P., Yu, Z., Palmtag, J., and Kuhry, P.: Estimated stocks of circumpolar permafrost carbon with quantified uncertainty ranges and identified data gaps, Biogeosciences, 11, 6573–6593, doi:10.5194/bg-11-6573-2014, 2014.

IPCC: Climate Change 2013: The Physical Science Basis, in: Contribution of Working Group I to the Fifth Assessment Report of the Intergovernmental Panel on Climate Change, edited by: Stocker, T. F., Qin, D., Plattner, G.-K., Tignor, M., Allen, S. K., Boschung, J., Nauels, A., Xia, Y., Bex, V., and Midgley, P. M., Cambridge University Press, Cambridge, UK and New York, NY, USA, 1535 pp., 2013.

Ishikawa, M., Watanabe, T., and Nakamura, N.: Genetic differences of rock glaciers and the discontinuous mountain permafrost zone in Kanchanjunga Himal, eastern Nepal, Permafrost Periglac., 12, 243–253, doi:10.1002/ppp.394, 2001.

Ivanova, N. V., Kuznetsova, I. L., Parmuzin, I. S., Rivkin, F. M., and Sorokovikov, V. A.: Geocryological Conditions in Swedish Lapland, in: Proceedings of the 4th Russian Conference on Geocryology, 7–9 June 2011, Moscow State University, Moscow, Russia, 77–82, 2011.

Jantze, E. J., Lyon, S. W., and Destouni, G.: Subsurface release and transport of dissolved carbon in a discontinuous permafrost region, Hydrol. Earth Syst. Sci., 17, 3827–3839, doi:10.5194/hess-17-3827-2013, 2013.

Joseph, S., Giménez, D., and Hoffman, J. L.: Dielectric permittivity as a function of water content for selected New Jersey soils, New Jersey Geological and Water Survey, Trenton, NJ, available at: http://www.state.nj.us/dep/njgs/geodata/dgs10-1.htm (last access: 17 December 2013), 2010.

Klingbjer, P. and Moberg, A.: A composite monthly temperature record from Tornedalen in northern Sweden, 1802–2002, Int. J. Climatol., 23, 1465–1494, doi:10.1002/joc.946, 2003.

Kneisel, C., Hauck, C., and Vonder Muhll, D.: Permafrost below the timberline confirmed and characterized by geoelectrical resistivity measurements, Bever Valley, eastern Swiss Alps, Permafrost Periglac., 11, 295–304, doi:10.1002/1099-1530(200012)11:4<295::aid-ppp353>3.0.co;2-l, 2000.

Kneisel, C., Saemundsson, D., and Beylich, A. A.: Reconnaissance surveys of contemporary permafrost environments in central Iceland using geoelectrical methods: Implications for permafrost degradation and sediment fluxes, Geogr. Ann. A, 89, 41–50, doi:10.1111/j.1468-0459.2007.00306.x, 2007.

Kneisel, C., Emmert, A., and Kästl, J.: Application of 3D electrical resistivity imaging for mapping frozen ground conditions exemplified by three case studies, Geomorphology, 210, 71–82, doi:10.1016/j.geomorph.2013.12.022, 2014.

Lewkowicz, A. G., Etzelmuller, B., and Smith, S. L.: Characteristics of Discontinuous Permafrost based on Ground Temperature Measurements and Electrical Resistivity Tomography, Southern Yukon, Canada, Permafrost Periglac., 22, 320–342, doi:10.1002/ppp.703, 2011.

Loke, M. H.: Rapid 2-D Resistivity & IP inversion using the least-squares method (RES2DINV ver. 3.59 for Windows XP/Vista/7, manual), Geotomo Software, Malaysia, 2010.

Loke, M. H. and Barker, R. D.: Rapid least-squares inversion of apparent resistivity pseudosections by a quasi-Newton method, Geophys. Prospect., 44, 131–152, doi:10.1111/j.1365-2478.1996.tb00142.x, 1996.

Lyon, S. W., Destouni, G., Giesler, R., Humborg, C., Mörth, M., Seibert, J., Karlsson, J., and Troch, P. A.: Estimation of permafrost thawing rates in a sub-arctic catchment using recession flow analysis, Hydrol. Earth Syst. Sci., 13, 595–604, doi:10.5194/hess-13-595-2009, 2009.

Lyon, S. W., Mörth, M., Humborg, C., Giesler, R., and Destouni, G.: The relationship between subsurface hydrology and dissolved carbon fluxes for a sub-arctic catchment, Hydrol. Earth Syst. Sci., 14, 941–950, doi:10.5194/hess-14-941-2010, 2010.

Marescot, L., Loke, M. H., Chapellier, D., Delaloye, R., Lambiel, C., and Reynard, E.: Assessing reliability of 2D resistivity imaging in mountain permafrost studies using the depth of investigation index method, Near Surf. Geophys., 1, 57–67, 2003.

McKenzie, J. M. and Voss, C. I.: Permafrost thaw in a nested groundwater-flow system, Hydrogeol. J., 21, 299–316, doi:10.1007/s10040-012-0942-3, 2013.

Moorman, B. J., Robinson, S. D., and Burgess, M. M.: Imaging periglacial conditions with ground-penetrating radar, Permafrost Periglac., 14, 319–329, doi:10.1002/ppp.463, 2003.

Neidell, N. S. and Taner, T. M.: Semblance and other coherency measures for multichannel data, Geophysics, 36, 482–497, 1971.

Nelson, F. E. and Outcalt, S. I.: A computational method for prediction and regionalization of permafrost, Arct. Alp. Res., 19, 279–288, doi:10.2307/1551363, 1987.

Oldenburg, D. W. and Li, Y. G.: Estimating depth of investigation in dc resistivity and IP surveys, Geophysics, 64, 403–416, doi:10.1190/1.1444545, 1999.

Painter, S. L., Moulton, J. D., and Wilson, C. J.: Modeling challenges for predicting hydrologic response to degrading permafrost, Hydrogeol. J., 21, 221–224, doi:10.1007/s10040-012-0917-4, 2013.

Payette, S., Delwaide, A., Caccianiga, M., and Beauchemin, M.: Accelerated thawing of subarctic peatland permafrost over the last 50 years, Geophys. Res. Lett., 31, L18208, doi:10.1029/2004gl020358, 2004.

Reynolds, J. M.: An Introduction to Applied and Environmental Geophysics, 2nd Edn., John Wiley & Sons, Hoboken, 2011.

Riseborough, D., Shiklomanov, N., Etzelmuller, B., Gruber, S., and Marchenko, S.: Recent advances in permafrost modelling, Permafrost Periglac., 19, 137–156, doi:10.1002/ppp.615, 2008.

Sandmeier Geophysical Research: ReflexW software version 6.1, available at: http://www.sandmeier-geo.de (last access: March 2015), 2012.

Sannel, A. B. K. and Kuhry, P.: Warming-induced destabilization of peat plateau/thermokarst lake complexes, J.Geophys. Res.-Biogeo., 116, 156–181, doi:10.1029/2010jg001635, 2011.

Schwamborn, G. J., Dix, J. K., Bull, J. M., and Rachold, V.: High-resolution seismic and ground penetrating radar-geophysical profiling of a thermokarst lake in the western Lena Delta, northern Siberia, Permafrost Periglac., 13, 259–269, doi:10.1002/ppp.430, 2002.

Seppälä, M.: Synthesis of studies of palsa formation underlining the importance of local environmental and physical characteristics, Quatern. Res., 75, 366–370, doi:10.1016/j.yqres.2010.09.007, 2011.

Sjöberg, Y., Frampton, A., and Lyon, S. W.: Using streamflow characteristics to explore permafrost thawing in northern Swedish catchments, Hydrogeol. J., 21, 121–131, doi:10.1007/s10040-012-0932-5, 2013.

Sollid, J. L. and Sorbel, L.: Palsa bogs as a climate indicator – Examples from Dovrefjell, southern Norway, Ambio, 27, 287–291, 1998.

Tarnocai, C., Canadell, J., Mazhitova, G., Schuur, E. A. G., Kuhry, P., and Zimov, S.: Soil organic carbon stocks in the northern circumpolar permafrost region, Global Biogeochem. Cy., 23, GB2023, doi:10.1029/2008GB003327, 2009.

Westermann, S., Wollschläger, U., and Boike, J.: Monitoring of active layer dynamics at a permafrost site on Svalbard using multi-channel ground-penetrating radar, The Cryosphere, 4, 475–487, doi:10.5194/tc-4-475-2010, 2010.

Woo, M.-K.: Permafrost Hydrology, Springer, Heidelberg, 563 pp., 2012.

Wramner, P.: Studier av palsmyrar i Tavvavuoma och Laivadalen, Lappland – Studies of palsa mires in Tavvavuoma and Laivadalen, Lappland, Licenciate Thesis, Göteborg University, Gothenburg, Sweden, 1968.

Wramner, P.: Palsmyrar i Tavvavuoma, Lappland (Palsa mires in Tavvavuoma, Lapland), GUNI report 3, Göteborg University, Gothenburg, Sweden, 1973.

Wramner, P., Backe, S., Wester, K., Hedvall, T., Gunnarsson, U., Alsam, S., and Eide, W.: Förslag till övervakningsprogram för Sveriges palsmyrar – Proposed monitoring program for Sweden's palsa mires, Länsstyrelsen i Norrbottens Län, Luleå, Sweden, 2012.

Zuidhoff, F. S.: Recent decay of a single palsa in relation to weather conditions between 1996 and 2000 in Laivadalen, northern Sweden, Geogr. Ann. A, 84, 103–111, doi:10.1111/1468-0459.00164, 2002.

Zuidhoff, F. S. and Kolstrup, E.: Changes in palsa distribution in relation to climate change in Laivadalen, northern Sweden, especially 1960–1997, Permafrost Periglac., 11, 55–69, doi:10.1002/(sici)1099-1530(200001/03)11:1<55::aid-ppp338>3.0.co;2-t, 2000.

12

Assessment of sea ice simulations in the CMIP5 models

Q. Shu[1,2]**, Z. Song**[1,2]**, and F. Qiao**[1,2]

[1]First Institute of Oceanography, State Oceanic Administration, Qingdao, 266061, China
[2]Key Lab of Marine Science and Numerical Modeling, SOA, Qingdao, 266061, China

Correspondence to: F. Qiao (qiaofl@fio.org.cn)

Abstract. The historical simulations of sea ice during 1979 to 2005 by the Coupled Model Intercomparison Project Phase 5 (CMIP5) are compared with satellite observations, Global Ice-Ocean Modeling and Assimilation System (GIOMAS) output data and Pan-Arctic Ice Ocean Modeling and Assimilation System (PIOMAS) output data in this study. Forty-nine models, almost all of the CMIP5 climate models and earth system models with historical simulation, are used. For the Antarctic, multi-model ensemble mean (MME) results can give good climatology of sea ice extent (SIE), but the linear trend is incorrect. The linear trend of satellite-observed Antarctic SIE is 1.29 (± 0.57) $\times 10^5$ km^2 decade^{-1}; only about 1/7 CMIP5 models show increasing trends, and the linear trend of CMIP5 MME is negative with the value of -3.36 (± 0.15) $\times 10^5$ km^2 decade^{-1}. For the Arctic, both climatology and linear trend are better reproduced. Sea ice volume (SIV) is also evaluated in this study, and this is a first attempt to evaluate the SIV in all CMIP5 models. Compared with the GIOMAS and PIOMAS data, the SIV values in both the Antarctic and the Arctic are too small, especially for the Antarctic in spring and winter. The GIOMAS Antarctic SIV in September is 19.1×10^3 km^3, while the corresponding Antarctic SIV of CMIP5 MME is 13.0×10^3 km^3 (almost 32 % less). The Arctic SIV of CMIP5 in April is 27.1×10^3 km^3, which is also less than that from PIOMAS SIV (29.5×10^3 km^3). This means that the sea ice thickness simulated in CMIP5 is too thin, although the SIE is fairly well simulated.

1 Introduction

The Coupled Model Intercomparison Project Phase 5 (CMIP5) provides a very useful platform for studying climate change. Simulations and projections by more than 60 state-of-the-art climate models and earth system models are archived under CMIP5. Assessment of the performance of CMIP5 outputs is necessary for scientists to decide which model outputs to use in their research and for model-developers to improve their models. Here, we focus on the assessment of sea ice simulations under the CMIP5 historical experiment. The CMIP5 data portal contains sea ice outputs from 49 coupled models. Many of these CMIP5 sea ice simulations have been evaluated and several valuable studies have been published.

For the Antarctic, the main problem of the CMIP5 models is their inability to reproduce the observed slight increase of sea ice extent (SIE). Turner et al. (2013) first assessed CMIP5 Antarctic SIE simulations using 18 models, and summarized that the majority of these models have too little SIE at the minimum sea ice period of February, and the mean of these 18 models' SIE shows a decreasing trend over 1979–2005, opposite to the satellite observation that exhibits a slight increasing trend. Polvani et al. (2013) used four CMIP5 models to study the cause of observed Antarctic SIE increasing trend under the conditions of increasing greenhouse gases and stratospheric ozone depletion. They concluded that it is difficult to attribute the observed trend in total Antarctic sea ice to anthropogenic forcing. Zunz et al. (2013) suggested that the model Antarctic sea ice internal variability is an important metric to evaluate the observed positive SIE trend. Using simulations from 25 CMIP5 models, Mahlstein et al. (2013) pointed out that internal sea ice variability is large in the Antarctic region and that both the observed and simulated trends may represent natural variation along with external forcing.

For the Arctic, CMIP5 models offer much better simulations. Stroeve et al. (2012) evaluated CMIP5 Arctic SIE trends using 20 CMIP5 models. They found that the seasonal cycle of SIE was well represented, and that the simulated SIE decreasing trend was more consistent with the observations over the satellite era than that of CMIP3 models but still smaller than the observed trend. They also noted that the spread in projected SIE through the 21st century from CMIP5 models is similar to that from CMIP3 models. Massonnet et al. (2012) examined 29 CMIP5 models and provided several important metrics to constrain the projections of summer Arctic sea ice projection. Liu et al. (2013) also pointed out that CMIP5 projections have large inter-model spread, but they also found that they could reproduce consistent Arctic ice-free time by reducing the large spread using two different approaches with 30 CMIP5 models.

Most evaluations of CMIP5 sea ice simulation in these studies are based only on some of CMIP5 models' outputs with some metrics, because other CMIP5 model outputs were not yet submitted. By now, all the CMIP5 participants have finished their model runs and submitted their model outputs. So, here we will evaluate all CMIP5 sea ice simulations with more metrics in both the Antarctic and the Arctic in an attempt to provide the community a useful reference. Generally speaking, our study shows the following: that the performance of Arctic sea ice simulation is better than that of Antarctic sea ice simulation, that sea ice extent simulation is better than sea ice volume simulation, and that mean state simulation is better than long-term trend simulation. If we want to get a similar result with all CMIP5 sea ice simulations, the number of models during analysis should be more than 22.

The rest of the paper is structured as follows. Section 2 presents sea ice data and analysis methodology used in this study. Model assessment is given in Sect. 3. Conclusions and discussion are provided in Sect. 4.

2 Data and methodology

Sea ice simulations of CMIP5 historical runs from 49 CMIP5 coupled models are now available. Monthly sea ice concentration (SIC) and sea ice thickness from these models are used in this study. These outputs are published by the Earth System Grid Federation (ESGF) (http://pcmdi9.llnl.gov/esgf-web-fe/) by each institute that is responsible for its model. Although there are several ensemble realizations of each CMIP5 model, the standard deviation between different ensemble realizations of each model is small (Turner et al., 2013; Table 1). We also plot the spatial patterns of SIC in February (Fig. S1 in the Supplement) and September (Fig. S2) from different ensemble realizations from GISS-E2-R which has 15 ensemble realizations and

more ensemble realizations than most CMIP5 models. We can see that the standard deviation between different ensemble realizations from the same model is comparable. So, here we only choose the first realization of each model for the analysis. CMIP5 historical runs cover the period from 1850 to 2005, but the continuous sea ice satellite record only started in 1979; so the period of 1979–2005 is chosen for the following analysis. Monthly satellite-observed SIC is used in this study, which is based on the National Aeronautics and Space Administration (NASA) team algorithm (Cavalieri et al., 1996) provided by the National Snow and Ice Data Centre (NSIDC) (http://nsidc.org/data/seaice/). Satellite-observed sea ice extent used here is also from NSIDC (ftp://sidads.colorado.edu/DATASETS/NOAA/G02135/). Sea ice volume (SIV) is an important index for assessment of sea ice simulation, although direct observations of SIV are very limited. SIV in the Antarctic used here is from the Global Ice-Ocean Modeling and Assimilation System (GIOMAS) (http://psc.apl.washington.edu/zhang/Global_seaice/index.html). SIV in the Arctic is from Pan-Arctic Ice Ocean Modeling and Assimilation System (PIOMAS) (http://psc.apl.washington.edu/wordpress/research/projects/arctic-sea-ice-volume-anomaly/). Note that SIV data from GIOMAS and PIOMAS are not observations but model simulations with data assimilation. Stroeve et al. (2014) compared observed sea ice thickness data in the Arctic with that of PIOMAS, and concluded that PIOMAS provides useful estimates of Arctic sea ice thickness and SIV, and that it can be used to assess the CMIP5 models' performances. But there are not enough observations to validate GIOMAS sea ice thickness in the Antarctic. The climatology and linear trends of CMIP5-simulated SIE, SIC and SIV are compared with satellite observations and GIOMAS and PIOMAS data. CMIP5 simulated SIE is computed as the total area of all grid cells where SIC exceeds 15 %. SIV is computed as the sum of the product of SIC, the area of grid cell and sea ice thickness of each grid cell. All gridded SIC and sea ice thickness are re-gridded onto $1.0°$ longitude by $1.0°$ latitude grids before the analysis is performed. In this study, spring is from March to May for the Arctic, and from September to November for the Antarctic. Summer, autumn and winter are defined accordingly.

3 Results

We select several metrics to assess the sea ice simulations in CMIP5 models. Mean state, seasonal cycle, the model internal variability, linear trends and simulation errors are used. For the Arctic sea ice, model mean state and seasonal cycle are important to Arctic sea ice projection (Massonnet et al., 2012). For the Antarctic sea ice, the model internal variability is an important metric to evaluate the observed positive SIE trend (Zunz et al., 2013). Annual mean SIE, SIE amplitude, standard deviation of detrended SIE anomaly (SIE

Table 1. Antarctic sea ice metrics in CMIP5 models, satellite observations and GIOMAS data set. Column (a) is mean annual SIE in million km^2. Column (b) is monthly SIE amplitude in million km^2. Column (c) is standard deviation of detrended monthly SIE anomaly in million km^2. Column (d) is linear trend in monthly SIE in 10^5 km^2 decade^{-1}, and the value in parentheses is 95 % confidence level. Column (e) is monthly SIE root mean square error in million km^2. Column (f) is mean annual SIV in 10^3 km^3. Column (g) is monthly SIV amplitude in 10^3 km^3. Column (h) is standard deviation of detrended monthly SIV anomaly in 10^3 km^3. Column (i) is linear trend in monthly SIV in 10^3 km^3 decade^{-1}, and the value in parentheses is 95 % confidence level. Column (j) is monthly SIV root mean square error in 10^3 km^3. The models with bold metric number have special performances, and details can be found in the text.

Data sources or CMIP5 models	(a)	(b)	(c)	(d)	(e)	(f)	(g)	(h)	(i)	(j)
Observations or GIOMAS	11.94	15.70	0.40	1.29 (0.57)	–	11.02	17.17	0.63	0.45 (0.09)	–
Multi-model ensemble mean (MME)	11.50	15.46	0.11	−3.36 (0.15)	0.71	7.73	10.31	0.10	−0.36 (0.01)	4.20
ACCESS1.0	12.10	19.12	0.59	−1.72 (0.83)	1.57	6.30	11.35	0.43	−0.15 (0.06)	5.20
ACCESS1.3	14.24	15.77	0.54	−0.97 (0.77)	2.31	10.71	9.78	0.67	−0.03 (0.09)	2.75
BCC-CSM1.1	13.42	19.32	**1.27**	**2.71 (1.78)**	2.11	7.13	11.51	0.92	**0.09 (0.13)**	4.41
BCC-CSM1-1-M	12.26	18.86	**1.06**	−20.03 (1.49)	1.52	5.65	9.98	0.71	−1.20 (0.10)	5.92
BNU-ESM	**20.60**	**23.46**	0.82	−9.60 (1.15)	9.19	18.49	22.48	0.87	−2.03 (0.12)	7.89
CanCM4	14.65	20.58	0.74	−2.79 (1.03)	3.40	3.09	4.81	0.28	−0.06 (0.04)	9.21
CanESM2	14.69	20.64	**0.96**	−7.74 (1.35)	3.42	3.09	4.82	0.40	−0.15 (0.06)	9.22
CCSM4	18.37	13.70	0.58	−7.34 (0.82)	6.64	19.34	18.63	1.12	−1.56 (0.16)	8.34
CESM1-BGC	17.67	14.05	0.49	−6.68 (0.69)	5.93	18.28	18.31	0.91	−1.19 (0.13)	7.28
CESM1-CAM5	14.06	14.78	0.47	−5.52 (0.66)	2.58	11.22	16.05	0.58	−0.97 (0.08)	1.13
CESM1-CAM5-1-FV2	13.01	14.11	0.58	−3.16 (0.82)	1.77	9.96	14.12	0.74	−0.22 (0.10)	1.89
CESM1-FASTCHEM	17.86	13.42	0.60	−8.78 (0.84)	6.14	18.41	18.15	1.18	−1.70 (0.17)	7.42
CESM1-WACCM	14.33	12.57	0.39	−6.45 (0.54)	2.95	11.55	13.15	0.66	−0.91 (0.09)	1.80
CMCC-CESM	11.84	19.43	**0.99**	**2.91 (1.39)**	2.01	6.70	11.18	0.71	**0.26 (0.10)**	4.91
CMCC-CM	11.81	16.84	0.67	−2.49 (0.94)	0.90	6.82	10.14	0.48	−0.05 (0.07)	4.97
CMCC-CMS	11.74	19.33	0.87	−1.52 (1.23)	1.83	6.31	10.70	0.59	−0.12 (0.08)	5.34
CNRM-CM5	7.78	16.98	0.77	−2.59 (1.09)	4.53	3.01	7.81	0.42	−0.10 (0.06)	8.79
CNRM-CM5-2	9.28	14.08	**1.08**	**4.29 (1.51)**	3.16	4.93	9.78	1.02	**0.38 (0.14)**	6.77
CSIRO-Mk3.6	15.92	12.11	0.67	−1.64 (0.95)	4.89	12.13	13.28	0.65	−0.29 (0.09)	2.62
EC-EARTH	10.66	17.18	0.66	−7.94 (0.92)	1.72	6.09	9.44	0.58	−0.66 (0.08)	5.75
FGOALS-g2	17.10	17.29	0.48	−1.47 (0.67)	5.28	15.65	13.89	0.74	−0.14 (0.10)	4.88
FIO-ESM	17.19	12.21	0.49	−8.53 (0.68)	5.61	21.23	13.98	1.16	−1.57 (0.16)	10.31
GFDL-CM2p1	8.00	15.38	0.81	−6.33 (1.14)	4.01	2.45	5.55	0.30	−0.19 (0.04)	9.57
GFDL-CM3	6.25	12.06	0.73	−6.82 (1.02)	5.82	1.92	4.16	0.37	−0.30 (0.05)	10.29
GFDL-ESM2G	8.11	14.34	0.63	−4.45 (0.88)	3.90	2.71	5.81	0.41	−0.24 (0.06)	9.31
GFDL-ESM2M	6.39	12.23	0.41	−1.61 (0.58)	5.65	1.81	4.20	0.16	−0.09 (0.02)	10.36
GISS-E2-H	6.21	10.62	0.38	−1.89 (0.53)	6.03	3.24	7.19	0.27	−0.24 (0.04)	8.65
GISS-E2-H-CC	12.18	19.07	0.75	−5.75 (1.05)	1.52	6.70	14.16	0.51	−0.54 (0.07)	4.57
GISS-E2-R	7.74	14.31	**1.01**	−3.39 (1.42)	4.31	3.06	6.17	0.47	−0.16 (0.07)	8.92
GISS-E2-R-CC	8.12	14.55	0.66	**0.82 (0.92)**	3.93	3.12	6.24	0.35	0.00 (0.05)	8.86
HadCM3	14.26	19.95	0.78	−2.74 (1.10)	3.28	14.70	21.87	0.83	−0.49 (0.12)	4.13
HadGEM2-AO	9.11	14.29	0.59	−5.31 (0.83)	3.20	5.58	9.70	0.49	−0.42 (0.07)	6.26
HadGEM2-CC	9.12	14.29	0.72	−0.85 (1.02)	3.25	5.50	9.68	0.61	−0.05 (0.09)	6.34
HadGEM2-ES	9.82	15.02	0.70	−3.25 (0.98)	2.60	6.16	10.33	0.61	−0.41 (0.09)	5.66
INMCM4	6.25	10.91	0.48	−4.00 (0.68)	6.04	2.81	6.12	0.38	−0.28 (0.05)	9.21
IPSL-CM5A-LR	9.66	19.06	0.84	−5.03 (1.17)	3.43	4.13	8.66	0.53	−0.26 (0.07)	7.70
IPSL-CM5A-MR	8.08	17.30	0.74	**1.69 (1.04)**	4.56	2.80	6.50	0.35	**0.01 (0.05)**	9.21
IPSL-CM5B-LR	3.34	8.09	0.42	**0.59 (0.59)**	9.09	1.22	3.32	0.20	**0.04 (0.03)**	11.10
MIROC4h	10.90	17.53	0.61	−7.96 (0.86)	1.33	5.35	9.74	0.41	−0.51 (0.06)	6.28
MIROC5	**3.23**	**6.62**	0.29	−1.03 (0.41)	9.29	1.40	3.15	0.16	−0.07 (0.02)	10.93
MIROC-ESM	12.65	19.12	0.64	−5.83 (0.91)	1.47	7.23	10.72	0.47	−0.48 (0.07)	4.46
MIROC-ESM-CHEM	13.38	19.80	0.53	−2.15 (0.74)	2.07	8.08	11.59	0.49	−0.21 (0.07)	3.61
MPI-ESM-LR	7.70	15.08	0.73	−2.95 (1.03)	4.50	3.41	6.35	0.38	−0.19 (0.05)	8.64
MPI-ESM-MR	7.90	15.62	0.84	**4.41 (1.17)**	4.28	3.54	7.06	0.48	**0.24 (0.07)**	8.39
MPI-ESM-P	7.91	15.69	0.75	−0.25 (1.06)	4.34	3.48	6.48	0.45	**0.05 (0.06)**	8.56
MRI-CGCM3	13.43	15.99	0.66	**1.52 (0.93)**	1.67	10.72	13.05	0.63	**0.22 (0.09)**	2.04
MRI-ESM1	13.24	16.32	0.75	−0.62 (1.05)	1.53	10.14	13.00	0.58	−0.03 (0.08)	2.25
NorESM1-M	13.08	14.19	0.57	−0.71 (0.80)	1.24	13.88	12.41	1.17	−0.07 (0.16)	3.66
NorESM1-ME	16.98	14.19	0.60	−3.77 (0.84)	5.24	17.57	16.82	1.40	−0.74 (0.20)	6.59

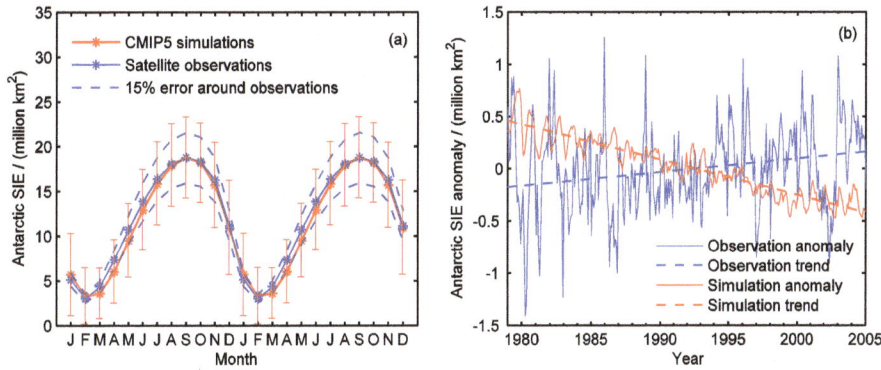

Figure 1. Climatology (**a**), anomaly and linear trend (**b**) of satellite-observed and CMIP5-simulated Antarctic sea ice extent during 1979–2005. Two annual cycles are plotted in (**a**). The error bar is the range of 1 standard deviation.

variability), SIE linear trend and CMIP5-simulated SIE root mean square (RMS) error are shown in Tables 1 and 2. The same metrics for SIV are also shown in Tables 1 and 2. Each CMIP5 model-simulated SIC and sea ice thickness are given in the Supplement. Detailed analyses for Antarctic and Arctic are as follows.

3.1 Assessment of Antarctic sea ice simulations

CMIP5 multi-model ensemble mean (MME) Antarctic climatological SIE compares well with the satellite-observed SIE, but the inter-model spread is large (Fig. 1a and Table 1). Satellite observations show that the Antarctic SIE has the minimum value of 3.0 million km^2 in February and the maximum value of 18.7 million km^2 in September; the annual mean SIE is 11.94 million km^2. CMIP5 MME SIE has the minimum and maximum values of 3.3 and 18.7 million km^2, and annual mean SIE of 11.50 million km^2, respectively. The seasonal cycle of observed SIE is well represented by the MME SIE of the 49 CMIP5 coupled models. Satellite-observed monthly SIE amplitude is 15.70 million km^2, and CMIP5 MME value is 15.46 million km^2. The simulated SIE errors are very small for each month. The simulated SIE errors are smaller than 15 % of the observations, except for March and April SIE values, which are a little less than 85 % of the observations. One standard deviation of CMIP5 simulations, which is greater than 15 % of the observations (Fig. 1a), shows that CMIP5 coupled models have a large spread each month in terms of Antarctic SIE. Table 1 also shows that CMIP5 models have a large spread. BNU-ESM has the largest annual mean and amplitude of SIE with the values of 20.60 and 23.46 million km^2, and MIROC5 has the smallest annual mean and amplitude of SIE with the values of 3.23 and 6.62 million km^2 (highlighted in Table 1 with bold font), respectively. BNU-ESM-simulated February SIE is even larger than MIROC5-simulated September SIE. Large SIE spread and small MME SIE errors indicate that we should use as many models as we can when using CMIP5 outputs.

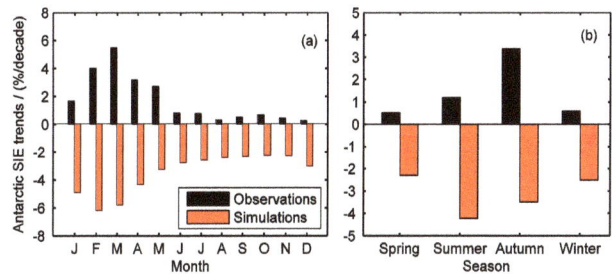

Figure 2. Monthly (**a**) and seasonal (**b**) linear trends of satellite-observed and CMIP5-simulated Antarctic sea ice extent during 1979–2005.

CMIP5 model-simulated and satellite-observed SICs in February and September during 1979–2005 are shown in Figs. S3 and S4. In February most models have an overly small SIC compared with satellite observations, especially in the Bellingshausen Sea and the Amundsen Sea. More than half of CMIP5 models have no sea ice in the Bellingshausen Sea or in the Amundsen Sea. CNRM-CM5, GFDL-CM2p1, GFDL-CM3, GFDL-ESM2G, GFDL-ESM2M, IPSL-CM5B-LR and MIROC5 almost have no sea ice in February in the Antarctic. But ACCESS1.3, BNU-ESM, CCSM4, CESM1-BGC, CESM1-FASTCHEM, CSIRO-Mk3.6, FGOALS-g2, FIO-ESM and NorESM1-ME have more sea ice than satellite observations. Although CMIP5 simulated MME SIE fits the observations well, the MME spatial pattern of SIC does not fit the observations so well. MME SICs in the Weddell Sea, the Bellingshausen Sea and the Amundsen Sea are smaller than the observations. In September, most CMIP5 models have better performance than that in February, and MME SIC also has a better spatial pattern.

Figures 1b and 2 show that linear trends of CMIP5 MME Antarctic SIE do not agree with the satellite observations. Many studies showed that Antarctic SIE has an increasing trend since the end of 1970s (Cavalieri et

Table 2. Arctic sea ice metrics in CMIP5 models, satellite observations and PIOMAS data set. Column (a) is mean annual SIE in million km^2. Column (b) is monthly SIE amplitude in million km^2. Column (c) is standard deviation of detrended monthly SIE anomaly in million km^2. Column (d) is linear trend in monthly SIE in $10^5 km^2 decade^{-1}$, and the value in parentheses is 95 % confidence level. Column (e) is monthly SIE root mean square error in million km^2. Column (f) is mean annual SIV in $10^3 km^3$. Column (g) is monthly SIV amplitude in $10^3 km^3$. Column (h) is standard deviation of detrended monthly SIV anomaly in $10^3 km^3$. Column (i) is linear trend in monthly SIV in $10^3 km^3 decade^{-1}$, and the value in parentheses is 95 % confidence level. Column (j) is monthly SIV root mean square error in $10^3 km^3$. The models with bold metric number have special performances, and details can be found in the text.

Data sources or CMIP5 models	(a)	(b)	(c)	(d)	(e)	(f)	(g)	(h)	(i)	(j)
Observations or PIOMAS	12.02	8.80	0.29	−4.35 (0.41)	–	21.85	16.17	1.02	−2.14 (0.14)	–
Multi-model ensemble mean (MME)	12.81	10.40	0.13	−3.71 (0.19)	1.07	18.45	17.50	0.35	−1.45 (0.05)	3.57
ACCESS1.0	12.13	10.33	0.41	−5.51 (0.57)	0.94	15.41	18.74	1.05	−1.58 (0.15)	6.60
ACCESS1.3	11.79	9.47	0.43	−0.78 (0.60)	0.73	18.81	17.02	1.02	−1.05 (0.14)	3.23
BCC-CSM1.1	14.86	15.39	0.69	−8.79 (0.97)	3.70	14.29	22.70	1.00	−2.01 (0.14)	8.02
BCC-CSM1-1-M	13.19	15.96	0.65	−5.19 (0.92)	2.87	11.04	20.69	0.87	−0.74 (0.12)	11.02
BNU-ESM	14.72	12.61	0.50	−4.41 (0.70)	3.19	23.03	19.79	1.23	−4.37 (0.17)	1.83
CanCM4	12.79	14.77	0.52	−4.97 (0.73)	2.49	11.41	15.35	0.97	−0.38 (0.14)	10.47
CanESM2	12.01	13.76	0.49	−6.80 (0.69)	1.91	**9.97**	14.21	0.63	−1.18 (0.09)	11.92
CCSM4	12.33	8.56	0.44	−1.34 (0.62)	0.42	20.27	16.16	1.51	−1.54 (0.21)	1.82
CESM1-BGC	12.10	7.96	0.41	−2.85 (0.58)	0.35	20.30	15.52	1.51	−2.63 (0.21)	1.86
CESM1-CAM5	12.33	8.35	0.38	−1.87 (0.53)	0.52	22.73	16.01	1.96	−1.22 (0.28)	1.35
CESM1-CAM5-1-FV2	12.52	8.68	0.42	−5.07 (0.59)	0.64	23.17	16.01	1.87	−3.63 (0.26)	1.49
CESM1-FASTCHEM	12.02	8.86	0.39	−3.70 (0.55)	0.25	18.27	15.86	1.37	−1.98 (0.19)	3.69
CESM1-WACCM	13.44	8.10	0.36	−2.88 (0.51)	1.51	27.32	9.47	2.07	0.09 (0.29)	6.27
CMCC-CESM	13.97	9.33	0.36	−2.63 (0.51)	2.12	28.75	11.93	1.38	−1.44 (0.19)	7.11
CMCC-CM	13.99	7.35	0.30	−5.09 (0.43)	2.06	**33.01**	9.87	1.73	−2.40 (0.24)	11.52
CMCC-CMS	12.64	7.92	0.34	−2.87 (0.48)	0.82	28.29	9.73	1.29	−1.18 (0.18)	6.89
CNRM-CM5	12.41	11.41	0.46	−7.58 (0.65)	1.11	14.44	20.22	0.99	−1.76 (0.14)	7.60
CNRM-CM5-2	14.20	10.65	0.45	−2.32 (0.63)	2.40	20.11	21.83	1.29	−0.96 (0.18)	2.76
CSIRO-Mk3.6	**16.13**	7.57	0.30	−5.33 (0.42)	4.20	25.94	12.16	0.81	−2.32 (0.11)	4.30
EC-EARTH	12.45	8.04	0.35	−3.84 (0.49)	0.57	24.01	12.44	1.90	−0.59 (0.27)	2.86
FGOALS-g2	11.68	**3.35**	0.13	−1.44 (0.18)	1.86	–	–	–	–	–
FIO-ESM	12.46	10.27	0.40	−2.23 (0.57)	1.00	18.94	18.96	1.86	−1.69 (0.26)	3.15
GFDL-CM2p1	12.58	12.85	0.54	−3.76 (0.75)	1.68	11.11	18.13	0.87	−1.01 (0.12)	10.80
GFDL-CM3	12.22	8.71	0.33	−2.89 (0.46)	0.41	15.25	15.47	1.31	−1.18 (0.18)	6.61
GFDL-ESM2G	**15.72**	13.72	0.48	−7.05 (0.68)	4.24	16.91	19.33	1.24	−1.77 (0.17)	5.17
GFDL-ESM2M	12.46	11.06	0.53	−0.31 (0.74)	0.98	12.13	16.11	1.02	−0.56 (0.14)	9.75
GISS-E2-H	12.96	14.87	0.54	−5.07 (0.75)	2.47	13.61	25.67	0.76	−0.91 (0.11)	9.10
GISS-E2-H-CC	13.94	14.24	0.60	−5.91 (0.84)	2.80	14.94	27.49	0.80	−1.29 (0.11)	8.23
GISS-E2-R	13.65	15.17	0.49	−6.31 (0.69)	2.89	15.50	29.32	0.75	−1.28 (0.11)	8.17
GISS-E2-R-CC	**15.13**	16.73	0.48	−5.65 (0.67)	4.28	17.16	31.86	0.76	−1.08 (0.11)	7.64
HadCM3	13.94	13.59	0.56	−4.74 (0.78)	2.78	21.07	26.96	0.87	−2.25 (0.12)	4.46
HadGEM2-AO	11.38	10.75	0.40	−3.81 (0.56)	1.15	16.58	20.16	0.84	−0.98 (0.12)	5.53
HadGEM2-CC	13.20	10.68	0.45	−3.10 (0.63)	1.45	21.56	21.55	0.96	−2.47 (0.13)	2.22
HadGEM2-ES	12.34	11.21	0.43	−6.03 (0.60)	1.14	18.85	21.13	1.00	−1.69 (0.14)	3.64
INMCM4	12.92	12.02	0.42	−0.21 (0.59)	1.61	15.20	22.08	0.96	−0.21 (0.13)	7.07
IPSL-CM5A-LR	12.72	10.07	0.44	−3.03 (0.62)	1.14	21.87	16.41	1.48	−0.96 (0.21)	1.66
IPSL-CM5A-MR	11.06	9.55	0.35	−2.85 (0.49)	1.25	14.83	16.32	0.92	−1.69 (0.13)	7.17
IPSL-CM5B-LR	14.06	8.28	0.40	−0.77 (0.56)	2.08	27.28	13.11	2.91	−1.37 (0.41)	6.25
MIROC4h	**10.66**	9.65	0.40	−3.11 (0.56)	1.47	10.86	16.48	0.82	−1.00 (0.12)	11.02
MIROC5	12.12	6.63	0.29	−6.78 (0.40)	0.65	25.31	14.88	1.09	−3.68 (0.15)	3.81
MIROC-ESM	**10.40**	8.05	0.34	−1.91 (0.47)	1.69	11.09	14.36	0.62	−1.04 (0.09)	10.79
MIROC-ESM-CHEM	**10.83**	7.89	0.46	−4.24 (0.65)	1.30	12.59	14.73	1.39	−1.69 (0.20)	9.29
MPI-ESM-LR	11.10	7.95	0.40	−2.48 (0.56)	1.01	15.07	16.87	0.85	−1.23 (0.12)	6.85
MPI-ESM-MR	11.07	8.00	0.40	−4.94 (0.56)	1.02	15.20	17.30	0.90	−1.75 (0.13)	6.74
MPI-ESM-P	**10.94**	8.27	0.34	−1.83 (0.48)	1.13	13.45	17.05	1.13	−0.80 (0.16)	8.46
MRI-CGCM3	**15.01**	15.27	0.47	−1.44 (0.66)	3.97	15.70	19.40	1.48	−0.55 (0.21)	6.33
MRI-ESM1	14.65	14.67	0.61	−4.07 (0.86)	3.52	15.21	18.89	1.74	−1.56 (0.24)	6.76
NorESM1-M	12.01	5.96	0.25	−1.98 (0.36)	0.90	23.77	11.23	1.57	−0.68 (0.22)	3.11
NorESM1-ME	12.47	5.99	0.31	−0.21 (0.43)	0.97	23.97	9.71	2.14	−0.46 (0.30)	3.69

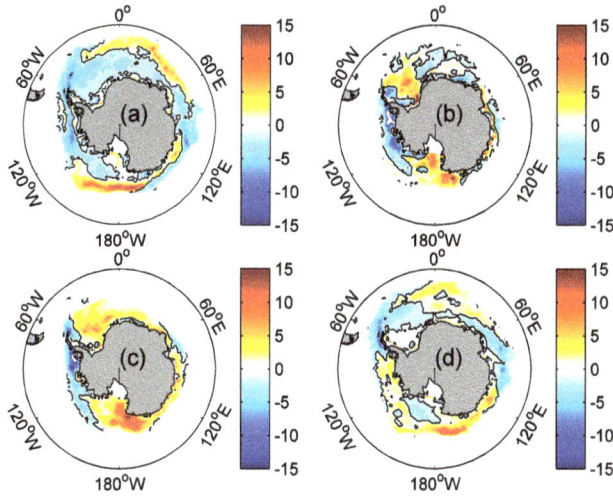

Figure 3. Linear trends (unit: % per decade) of satellite-observed Antarctic sea ice concentration during 1979 to 2005. (**a**) Spring, (**b**) summer, (**c**) autumn, and (**d**) winter.

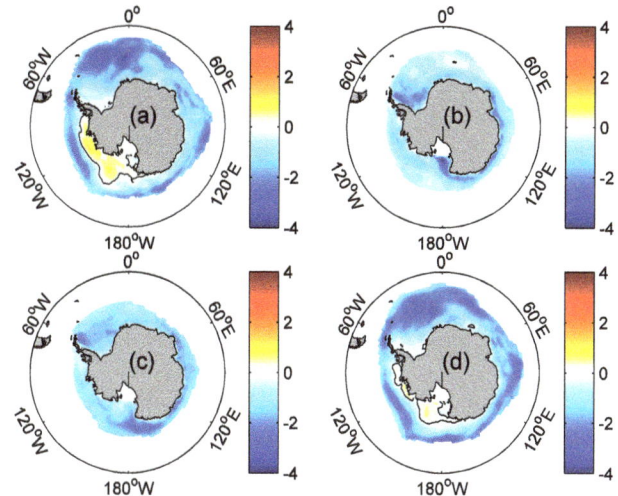

Figure 4. Linear trends (units: % per decade) of CMIP5-simulated Antarctic sea ice concentration during 1979–2005. (**a**) Spring, (**b**) summer, (**c**) autumn, and (**d**) winter.

al., 1997, 2003; Zwally et al., 2002; Turner et al., 2009). Satellite-observed Antarctic SIE has a small increasing linear trend with the rate of 1.29 (± 0.57) $\times 10^5$ km^2 decade^{-1} during 1979–2005, while CMIP5-simulated linear trend is -3.36 (± 0.15) $\times 10^5$ km^2 decade^{-1} (Fig. 1b). Only 8 out of 49 CMIP5 models have increasing linear trends as the observations (highlighted in Table 1 with bold font). They are BCC-CSM1.1, CMCC-CESM, CNRM-CM5-2, GISS-E2-R-CC, IPSL-CM5A-MR, IPSL-CM5B-LR, MPI-ESM-MR and MRI-CGCM3. This supports the conclusion by Polvani et al. (2013) that it is difficult to attribute the observed Antarctic SIE trends to anthropogenic forcing. From Table 1 we can see that several models (highlighted in Table 1 with bold font) such as BCC-CSM1.1, BCC-CSM1-1-M, CanESM2, CMCC-CESM, CNRM-CM5-2 and GISS-E2-R have large internal variabilities, and these models always have large linear trends. This mean that the satellite-observed positive SIE trend may represent natural variation along with external forcing (Mahlstein et al., 2013). Figure 2 shows that the monthly and seasonal trends of CMIP5-simulated Antarctic SIE also do not agree with the observations. Observed Antarctic SIE shows increasing trends in each month and each season, and the largest trend is in March and the autumn season. CMIP5 MME SIE, however, has decreasing trends in each month and each season, and the largest trend is in February and the summer season.

The trends of observed Antarctic SIC have large spatial differences (Fig. 3), but the simulated Antarctic SIC trends are almost decreasing everywhere (Fig. 4). Figure 3 shows that decreasing SIC is mainly in the Antarctic Peninsula, which is one of the three high-latitude areas showing rapid regional warming over the last 50 years (Vaughan et al., 2003). SIC also decreases in the Bellingshausen Sea and the Amundsen Sea in summer and autumn. The increasing SIC

is mainly in the Ross Sea all year round and in the Weddell Sea in summer and autumn. Figure 4 clearly shows that CMIP5 MME SIC has decreasing trend everywhere except in the coast of the Amundsen Sea and in part of the Ross Sea in spring and winter.

SIV depends on both sea ice coverage and sea ice thickness. SIV is more directly tied to climate forcing than SIE. So, SIV is an important climate indicator in climate study. The observed sea ice thickness records are mainly from submarine, aircraft and satellite. But the observations are not continuous spatially or temporally over a long period (Stroeve et al., 2014). For the Antarctic, the observed sea ice thickness data are quite limited. A climatological 2.5° × 5.0° gridded Antarctic sea ice thickness map was provided until 2008 (Worby et al., 2008). Recently, there have been several studies using satellite observations of sea ice thickness (Kurtz and Markus, 2012; Xie et al., 2013). These observations provide modelers with useful validation of their models. However, these data are not easily used to long-term simulation validations by now, because they are not long enough. Here, we use GIOMAS data, which are from a global ice-ocean model (Zhang and Rothrock, 2003) with data assimilation capability. What we should keep in mind is that GIOMAS sea ice thickness is not from observations and may also have large degrees of uncertainty. CMIP5-simulated and GIOMAS Antarctic sea ice thicknesses during 1979–2005 are shown in Fig. S5. GIOMAS outputs show that thick sea ice is mainly in the coasts of the Weddell Sea, the Bellingshausen Sea and the Amundsen Sea. CMIP5 MME sea ice thickness can reproduce similar spatial patterns, but most of CMIP5 MME sea ice thickness is thinner than GIOMAS sea ice thickness. The spatial pattern for each CMIP5 model has a large difference. BCC-CSM1.1, CESM1-CAM5-1-FV2,

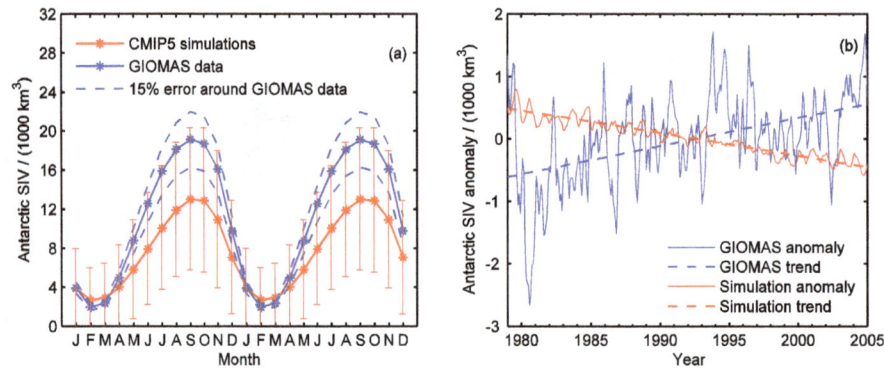

Figure 5. Climatology (**a**), anomaly and linear trend (**b**) of GIOMAS and CMIP5-simulated Antarctic sea ice volume during 1979–2005. Two annual cycles are plotted in (**a**). The error bar is the range of 1 standard deviation.

CMCC-CM, and CMCC-CMS fit GIOMAS sea ice thickness well. Several CMIP5 models such as CCSM4, CESM1-BGC, CESM1-FASTCHEM, FGOALS-g2 and FIO-ESM have overly thick sea ice near the coasts of Antarctica.

CMIP5 SIV simulations have more problems than the SIE simulations. The main problems of CMIP5 Antarctic SIV simulations include overly big SIV in summer, overly small SIV in winter, overly large model spread, and an incorrect linear trend compared with the GIOMAS data (Fig. 5). The annual mean SIV from GIOMAS is 11.02×10^3 km^3, but CMIP5 MME SIV is only 7.73×10^3 km^3 (Table 1). In February, Antarctic SIV from GIOMAS is 1.9×10^3 km^3, while the CMIP5 MME is 2.7×10^3 km^3. In September, GIOMAS SIV is 19.1×10^3 km^3, while CMIP5 MME is only 13.0×10^3 km^3, almost 32 % less than the GIOMAS. We can also see from Figure 5a that the model spread of Antarctic SIV in CMIP5 is very large. One standard deviation is greater than 15 % of the GIOMAS data in every month. We checked the correlation between SIE RMS error and SIV RMS error, and we can find that the models with small SIE RMS errors always have small SIV RMS errors (Table 1). It means that for the Antarctic models with a more realistic SIE mean state may result in a convergence of estimates of SIV. Figure 5b shows that GIOMAS SIV has an increasing trend of 0.45 (± 0.09) $\times 10^3$ km^3 decade^{-1}, while CMIP5 MME SIV has a decreasing trend of -0.36 (± 0.01) $\times 10^3$ km^3 decade^{-1}. If we check each CMIP5 model separately, we will also find that only 8 out of the 49 CMIP5 models have increasing SIV trend that is consistent with the GIOMAS. They are BCC-CSM1.1, CMCC-CESM, CNRM-CM5-2, IPSL-CM5A-MR, IPSL-CM5B-LR, MPI-ESM-MR, MPI-ESM-P and MRI-CGCM3 (highlighted in Table 1 with bold font).

3.2   Assessment of Arctic sea ice simulations

CMIP5 shows a quite good annual cycle of Arctic SIE, but the model error in winter is larger than that in summer and model spread is large (Fig. 6a). Arctic SIE reaches the maximum value of 15.7 million km^2 in March, and it reaches the minimum value of 6.9 million km^2 in September; the annual mean value is 12.02 million km^2. The MME climatological SIE compares well with the satellite-observed SIE. CMIP5 MME SIE reaches the maximum value of 17.2 million km^2, and reaches the minimum value of 6.8 million km^2, and the annual mean value is 12.81 million km^2. The modeled error is less than 15 % of the observations in every month. CMIP5 MME SIE is bigger than the satellite observation in spring, and the modeled error is quite small at other times. The model spread is large, with 1 standard deviation greater than 15 % of the observed SIE in every month (Fig. 6a). CSIRO-MK3.6, GFDL-ESM2G, GISS-E2-R-CC and MRI-CGCM3 have large annual mean SIE with the values larger than 15 million km^2 (highlighted in Table 2 with bold font). CSIRO-MK3.6 has more sea ice in the Barents Sea in summer (Fig. S6). GFDL-ESM2G, GISS-E2-R-CC and MRI-CGCM3 have more sea ice in winter (Fig. S7). MIROC4h, MIROC-ESM, MIROC-ESM-CHEM and MPI-ESM-P have small annual mean SIE with the values less than 11 million square kilometers (highlighted in Table 1 with bold font). Arctic SIE amplitudes from CMIP5 models also have a large spread. GISS-E2-R-CC has the largest amplitude with the value of 16.73 million km^2, and FGOAL-g2 has the smallest amplitude with the value of only 3.35 million km^2 (highlighted in Table 2 with bold font). Compared with the Antarctic variability, CMIP5-simulated Arctic SIE variability has a small spread (column c in Table 2).

CMIP5 MME SIE shows a decreasing trend that is consistent with the satellite observation, though the decreasing rate is a little smaller than that of the observation (Figs. 6b and 7). The satellite-observed SIE linear trend over the period of 1979–2005 is -4.35 (± 0.41) $\times 10^5$ km^2 decade^{-1}, while CMIP5 MME SIE linear trend is only -3.71 (± 0.19) $\times 10^5$ km^2 decade^{-1}. BCC-CSM1.1 has the largest trend of -8.79 (± 0.97) $\times 10^5$ km^2 decade^{-1}. A total of 31 out of the 49 CMIP5 models have smaller decreasing rate than the observation, and NorESM1-ME has the smallest trend of -0.21 (± 0.43) $\times 10^5$ km^2 decade^{-1}. Both observed

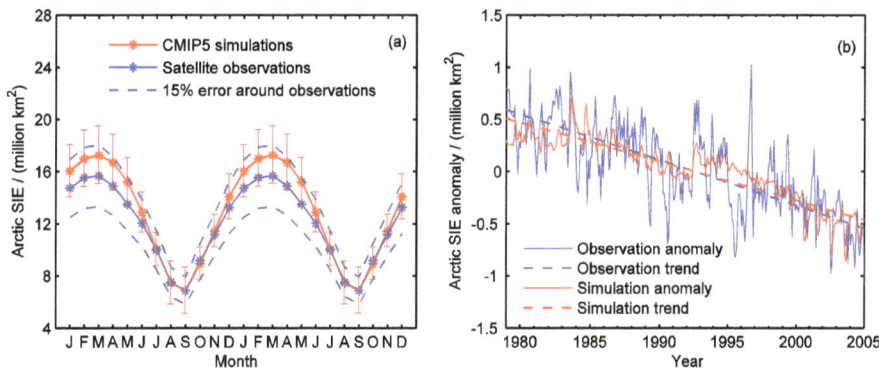

Figure 6. Climatology (**a**), anomaly and linear trend (**b**) of satellite-observed and CMIP5-simulated Arctic sea ice extent during 1979–2005. Two annual cycles are plotted in (**a**). The error bar is the range of 1 standard deviation.

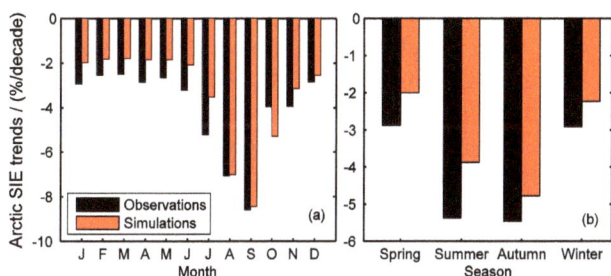

Figure 7. Monthly (**a**) and seasonal (**b**) linear trends of satellite-observed and CMIP5-simulated Arctic sea ice extent during 1979–2005.

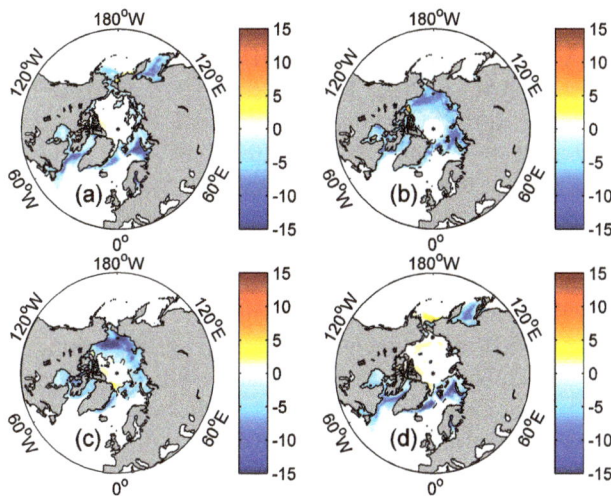

Figure 8. Linear trends (units: % per decade) of satellite-observed Arctic sea ice concentration during 1979–2005. (**a**) Spring, (**b**) summer, (**c**) autumn, and (**d**) winter.

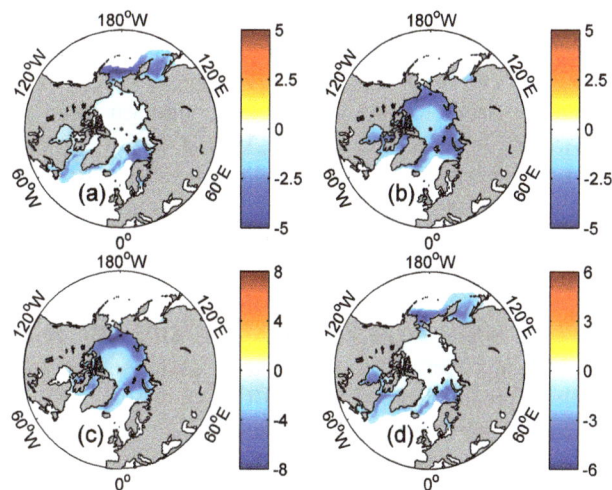

Figure 9. Linear trends (units: % per decade) of CMIP5-simulated Arctic sea ice concentration during 1979–2005. (**a**) Spring, (**b**) summer, (**c**) autumn, and (**d**) winter.

simulated SIE has a small reduction in summer, especially in July (Fig. 7). The satellite-observed SIE decreasing rate is 5.22 % per decade in July, while the CMIP5-simulated decreasing rate is 3.54 % per decade. The largest decreasing rate is in September; the observed trend is −8.61 % per decade, and the simulated trend is −8.46 % per decade.

Figures 8 and 9 show that the spatial patterns of CMIP5-simulated SIC reduction rate are consistent with the observations from 1979 to 2005, but the decreasing rates are smaller than the observations. In spring and winter, the observed decreasing SIC is mainly in the Okhotsk Sea, Baffin Bay, Greenland Sea and Barents Sea; CMIP5-simulated decreasing SIC is also in these regions. In summer and autumn, the main decreasing SIC is in the Chukchi Sea, the Barents Sea, and the Kara Sea (Figs. 8 and 9), and CMIP5 MME SIC has similar characteristics. However, CMIP5 simulations have larger trends in the central Arctic Ocean.

and CMIP5-simulated SIE in autumn have the largest decreasing trend. The CMIP5-simulated difference between the summer and autumn SIE-decreasing trend is, however, larger than that of the observations. The main reason is that CMIP5-

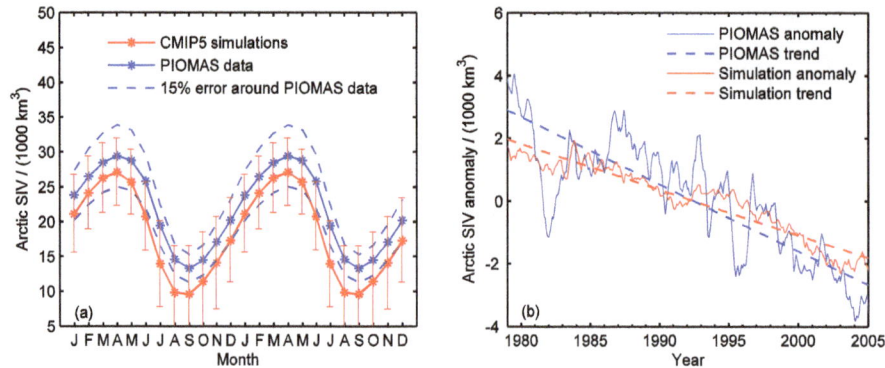

Figure 10. Climatology (**a**), anomaly and linear trend (**b**) of PIOMAS and CMIP5-simulated Arctic sea ice volume during 1979–2005. Two annual cycles are plotted in (**a**). The error bar is the range of 1 standard deviation.

Compared with PIOMAS sea ice thickness, the main problems of CMIP5 simulations are smaller Arctic SIV all year round and an overly large model spread (Fig. 10). In spring, the Arctic has the largest SIV. Long-term mean PIOMAS SIV is at its maximum in April at $29.5 \times 10^3 \, km^3$, and the corresponding CMIP5 MME is $27.1 \times 10^3 \, km^3$. Long-term mean PIOMAS SIV reaches its minimum in September at $13.3 \times 10^3 \, km^3$, and the corresponding CMIP5 MME is $9.6 \times 10^3 \, km^3$. Amplitude of SIV from PIOMAS is $16.17 \times 10^3 \, km^3$, and CMIP5 MME can give good amplitude of SIV with $17.50 \times 10^3 \, km^3$. CMIP5 SIV model spread is also very large: 1 standard deviation for each month is greater than 15 % of GIOMAS SIV. CanESM2 has the smallest SIV of $9.97 \times 10^3 \, km^3$, and CMCC-CM has the largest SIV of $33.01 \times 10^3 \, km^3$. Figure S8 shows that sea ice thickness in BCC-CSM1-1-M, CanCM4, CanESM2, GFDL-CM2p1, GISS-E2-H, GISS-E2-H-CC, GISS-E2-R, GISS-E2-R-CC, MIROC4h, MIROC-ESM, and MIROC-ESM-CHEM is significantly undervalued. Sea ice thickness in CESM1-WACCM, CMCC-CESM, CMCC-CM, FGOALS-g2, IPSL-CM5B-LR, NorESM1-M, NorESM1-ME is significantly overvalued. Based on PIOMAS, the linear trend of Arctic SIV during 1979–2005 is $-2.14 \, (\pm 0.14) \times 10^3 \, km^3 \, decade^{-1}$. CMIP5 MME trend has the same sign but with smaller value, at $-1.45 \, (\pm 0.05) \times 10^3 \, km^3 \, decade^{-1}$. Unlike most of CMIP5 models, CESM1-WACCM SIV has a slight positive trend during 1979–2005. The reason may be CESM1-WACCM SIV has large variability ($2.07 \times 10^3 \, km^3$), and its internal variability is not in phase with the natural observed variability.

4   Conclusions and discussion

The first ensemble realizations of the 49 CMIP5 historical simulations are evaluated in terms of the performance of sea ice. Our results show that the Arctic sea ice simulations are better than the Antarctic sea ice simulations, and SIE simulations are better than SIV simulations. CMIP5 MME SIV

is too little in winter and spring, because the sea ice thickness in CMIP5 models is too thin in winter and spring compared with the GIOMAS and PIOMAS data. In the Antarctic, MME can reproduce good mean state and monthly amplitude for SIE, but for SIV MME mean state and amplitude are smaller. In the Arctic, MME can reproduce good mean state and monthly amplitude for both SIE and SIV. CMIP5 simulations have very different variability (indicated by standard deviation of detrended monthly SIE and SIV) for different models. From Tables 1 and 2 we can conclude that the performance of each model is different. For the Antarctic, ACCESS1.0, BCC-CSM1.1, CESM1-CAM5-1-FV2, CMCC-CM, EC-EARTH, GISS-E2-H-CC, MIROC-ESM, MIROC-ESM-CHEM, MRI-CGCM3, MRI-ESM1 and NorESM1-M can give better SIE and SIV mean state. For the Arctic, ACCESS1.3, CCSM4, CESM1-BGC, CESM1-CAM5, CESM1-CAM5-1-FV2, CESM1-FASTCHEM, EC-EARTH, MIROC5, NorESM1-M and NorESM1-ME can give better mean state of SIE and SIV. The Arctic SIE linear trends of BNU-ESM, CanCM4, CESM1-FASTCHEM, EC-EARTH, GFDL-CM2p1, HadCM3, HadGEM2-AO, MIROC-ESM-CHEM, MPI-ESM-MR and MRI-ESM1 are closed to the observations.

Both satellite-observed Antarctic SIE and GIOMAS Antarctic SIV show increasing trends over the period of 1979–2005, but CMIP5 MME Antarctic SIE and SIV have decreasing trends. Only eight models' SIE and eight models' SIV show increasing trends. Can these few CMIP5 models reproduce the correct Antarctic sea ice trend? If we use these eight CMIP5 models to plot Antarctic SIC trends (not shown) as in Fig. 4, we will find that these eight CMIP5 model mean SIC trends have different spatial patterns with the observations (Fig. 3), although their model mean SIE and SIV have increasing trends. Satellite-observed Antarctic SIE has increased trends, but when we use the satellite-observed sea ice record, we should also keep in mind that it may also have a large degree of uncertainty. Eisenman et al. (2014) point

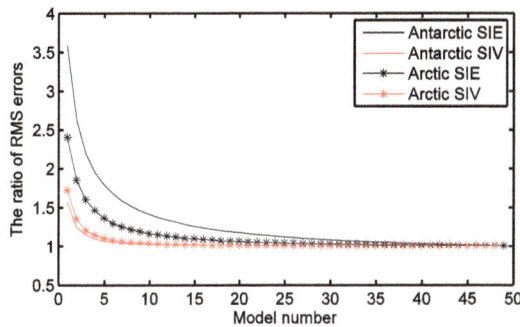

Figure 11. The ratio of SIE and SIV RMS errors between the errors calculated using different number of CMIP5 models and the error calculated using all 49 CMIP5 models.

out that sensor transition may cause a substantial change in the long-term trend.

We can see that the CMIP5 MME does a good job in terms of climatological mean, but their inter-model spread is large. The number of models used in published studies is usually less than the total CMIP5 models. How many models can give as similarly good simulations as all the available CMIP5 models do? We first choose the CMIP5 models randomly. The model number changes from 1 to 49. We then calculate the SIE and SIV RMS errors between MME and observations or GIOMAS and PIOMAS data sets. For each fixed model number, we choose these models randomly at many different times, and then calculate the mean of the RMS errors. Figure 11 shows the ratio of SIE and SIV RMS errors between the errors calculated using different number of CMIP5 models and the errors calculated using all 49 CMIP5 models. We can see that the model errors decrease quickly as the model number increases; and the more models we use, the smaller error we have. For a fixed model number, the ratios of SIE are larger than the ratios of SIV, and Antarctic SIE has the largest ratio. When the model number is greater than 30, the model errors do not change much anymore. If we choose a criterion of RMS error larger than 15 % of all the model RMS error, the model number of 22 is the critical number for Arctic SIE. It means that more than 22 CMIP5 models should give similar MME as all 49 CMIP5 models.

In this study, satellite observations, PIOMAS and GIOMAS data during the period of 1979–2005 are used to assess the sea ice simulations from CMIP5 models. We always expect the models can capture the observed trends during this period. But we should note that simulations without data assimilation are always out of phase with the natural variability seen in the observations. So the differences between simulations and observations can either be due to model biases or natural climate variability (Stroeve et al., 2014).

Acknowledgements. Satellite-observed sea ice concentration data are provided by http://nsidc.org/data/seaice/, sea ice extent are from ftp://sidads.colorado.edu/DATASETS/NOAA/G02135/, GIOMAS sea ice date are downloaded from http://psc.apl.washington.edu/zhang/Global_seaice/index.html, and PIOMAS sea ice date are from http://psc.apl.washington.edu/wordpress/research/projects/arctic-sea-ice-volume-anomaly/. CMIP5 sea ice simulations are downloaded from http://pcmdi9.llnl.gov/esgf-web-fe/. The authors thank the above-listed data providers. This work is supported by the National Basic Research Program of China (973 Program) under Grant 2010CB950500, National Natural Science Foundation of China (Grant Numbers 41406027 and 41306206), the Project of Comprehensive Evaluation of Polar Areas on Global and Regional Climate Changes (CHINARE2014-04-04, CHINARE2014-04-01, and CHINARE2014-01-01), and Polar Strategic Research Foundation of China (20120103).

Edited by: J. Stroeve

References

Cavalieri, D. J., Parkinson, C. L., Gloersen, P., and Zwally, H.: Sea Ice Concentrations from Nimbus-7 SMMR and DMSP SSM/I-SSMIS Passive Microwave Data, NASA DAAC at the National Snow and Ice Data Center, Boulder, Colorado, USA, 1996.

Cavalieri, D. J., Gloersen, P., Parkinson, C. L., Comiso, J. C., and Zwally, H. J.: Observed hemispheric asymmetry in global sea ice changes, Science, 278, 1104–1106, 1997.

Cavalieri, D. J., Parkinson, C. L., and Vinnikov, K. Y: 30-Year satellite record reveals contrasting Arctic and Antarctic decadal sea ice variability, Geophys. Res. Lett., 30, 1970, doi:10.1029/2003GL018031, 2003.

Eisenman, I., Meier, W. N., and Norris, J. R.: A spurious jump in the satellite record: has Antarctic sea ice expansion been overestimated?, The Cryosphere, 8, 1289–1296, doi:10.5194/tc-8-1289-2014, 2014.

Kurtz, N. and Markus, T.: Satellite observations of Antarctic sea ice thickness and volume, J. Geophys. Res., 117, C08025, doi:10.1029/2012JC008141, 2012.

Liu, J., Song, M., Horton, R. M., and Hu, Y.: Reducing spread in climate model projections of a September ice-free Arctic, P. Natl. Acad. Sci., 110, 12571–12576, 2013.

Mahlstein, I., Gent, P. R., and Solomon, S.: Historical Antarctic mean sea ice area, sea ice trends, and winds in CMIP5 simulations, J. Geophys. Res.-Atmos., 118, 5105–5110, 2013.

Massonnet, F., Fichefet, T., Goosse, H., Bitz, C. M., Philippon-Berthier, G., Holland, M. M., and Barriat, P.-Y.: Constraining projections of summer Arctic sea ice, The Cryosphere, 6, 1383–1394, doi:10.5194/tc-6-1383-2012, 2012.

Polvani, L. M. and Smith, K. L.: Can natural variability explain observed Antarctic sea ice trends? New modeling evidence from CMIP5, Geophys. Res. Lett., 40, 3195–3199, 2013.

Stroeve, J., Barrett, A., Serreze, M., and Schweiger, A.: Using records from submarine, aircraft and satellites to evaluate climate model simulations of Arctic sea ice thickness, The Cryosphere, 8, 1839–1854, doi:10.5194/tc-8-1839-2014, 2014.

Stroeve, J. C., Kattsov, V., Barrett, A., Serreze, M., Pavlova, T., Holland, M., and Meier, W. N.: Trends in Arctic sea ice extent from CMIP5, CMIP3 and observations, Geophys. Res. Lett., 39, L16502, doi:10.1029/2012GL052676, 2012.

Turner, J., Comiso, C., Marshall, G. J., Lachlan-Cope, T. A., Bracegirdle, T., Maksym, T., Meredith, M. P., Wang, Z., and Orr, A.: Non-annular atmospheric circulation change induced by stratospheric ozone depletion and its role in the recent increase of Antarctic sea ice extent, Geophys. Res. Lett., 36, L08502, doi:10.1029/2009GL037524, 2009.

Turner, J., Bracegirdle, T. J., Phillips, T., Marshall, G. J., and Hosking, J. S.: An Initial Assessment of Antarctic Sea Ice Extent in the CMIP5 Models, J. Climate, 26, 1473–1484, doi:10.1175/JCLI-D-12-00068.1, 2013.

Vaughan, D. G., Marshall, G. J., Connolley, W. M., Parkinson, C., Mulvaney, R., Hodgson, D. A., King, J. C., Pudsey, C. J., and Turner, J.: Recent rapid regional climate warming on the Antarctic Peninsula, Climatic Change, 60, 243–274, 2003.

Worby, A. P., Geiger, C. A., Paget, M. J., Van Woert, M. L., Ackley, S. F., and DeLiberty, T. L.: Thickness distribution of Antarctic sea ice, J. Geophys. Res.-Oceans, 113, C05S92, doi:10.1029/2007JC004254, 2008.

Xie, H., Tekeli, A. E., Ackley, S. F., Yi, D., and Zwally, H. J.: Sea ice thickness estimations from ICESat altimetry over the Bellingshausen and Amundsen Seas, 2003–2009, J. Geophys. Res.-Oceans, 118, 2438–2453, 2013.

Zhang, J. and Rothrock, D.: Modeling global sea ice with a thickness and enthalpy distribution model in generalized curvilinear coordinates, Mon. Weather Rev., 131, 845–861, 2003.

Zunz, V., Goosse, H., and Massonnet, F.: How does internal variability influence the ability of CMIP5 models to reproduce the recent trend in Southern Ocean sea ice extent?, The Cryosphere, 7, 451–468, doi:10.5194/tc-7-451-2013, 2013.

Zwally, H. J., Comiso, J. C., Parkinson, C. L., Cavalieri, D. J., and Gloersen, P.: Variability of Antarctic sea ice 1979–1998, J. Geophys. Res.-Oceans, 107, 9-1–9-19, 2002.

Permissions

The contributors of this book come from diverse backgrounds, making this book a truly international effort. This book will bring forth new frontiers with its revolutionizing research information and detailed analysis of the nascent developments around the world.

We would like to thank all the contributing authors for lending their expertise to make the book truly unique. They have played a crucial role in the development of this book. Without their invaluable contributions this book wouldn't have been possible. They have made vital efforts to compile up to date information on the varied aspects of this subject to make this book a valuable addition to the collection of many professionals and students.

This book was conceptualized with the vision of imparting up-to-date information and advanced data in this field. To ensure the same, a matchless editorial board was set up. Every individual on the board went through rigorous rounds of assessment to prove their worth. After which they invested a large part of their time researching and compiling the most relevant data for our readers.

The editorial board has been involved in producing this book since its inception. They have spent rigorous hours researching and exploring the diverse topics which have resulted in the successful publishing of this book. They have passed on their knowledge of decades through this book. To expedite this challenging task, the publisher supported the team at every step. A small team of assistant editors was also appointed to further simplify the editing procedure and attain best results for the readers.

Apart from the editorial board, the designing team has also invested a significant amount of their time in understanding the subject and creating the most relevant covers. They scrutinized every image to scout for the most suitable representation of the subject and create an appropriate cover for the book.

The publishing team has been an ardent support to the editorial, designing and production team. Their endless efforts to recruit the best for this project, has resulted in the accomplishment of this book. They are a veteran in the field of academics and their pool of knowledge is as vast as their experience in printing. Their expertise and guidance has proved useful at every step. Their uncompromising quality standards have made this book an exceptional effort. Their encouragement from time to time has been an inspiration for everyone.

The publisher and the editorial board hope that this book will prove to be a valuable piece of knowledge for researchers, students, practitioners and scholars across the globe.

List of Contributors

F. Günther
Alfred Wegener Institute Helmholtz Centre for Polar and Marine Research, Potsdam, Germany

P. P. Overduin
Alfred Wegener Institute Helmholtz Centre for Polar and Marine Research, Potsdam, Germany

I.A. Yakshina
Ust-Lensky State Nature Reserve, Tiksi, Yakutia, Russia

T. Opel
Alfred Wegener Institute Helmholtz Centre for Polar and Marine Research, Potsdam, Germany

A.V. Baranskaya
Lab. Geoecology of the North, Faculty of Geography, Lomonosov Moscow State University, Moscow, Russia

M. N. Grigoriev
Melnikov Permafrost Institute, Russian Academy of Sciences, Siberian Branch, Yakutsk, Russia

M. J. Siegert
Bristol Glaciology Centre, School of Geographical Sciences, University of Bristol, Bristol, BS8 1SS, UK

N. Ross
School of Geography, Politics and Sociology, Newcastle University, Newcastle upon Tyne, NE1 7RU, UK

H. Corr
British Antarctic Survey, Cambridge CB3 0ET, UK

B. Smith
Applied Physics Lab, Polar Science Center, University of Washington, Seattle, WA 98105, USA

T. Jordan
British Antarctic Survey, Cambridge CB3 0ET, UK

R. G. Bingham
School of GeoSciences, University of Edinburgh, Edinburgh EH8 9XP, UK

F. Ferraccioli
British Antarctic Survey, Cambridge CB3 0ET, UK

D. M. Rippin
Environment Department, University of York, York YO10 5DD, UK

A. Le Brocq
Geography, College of Life and Environmental Sciences, University of Exeter, Exeter EX4 4RJ, UK

M. Fuchs
Department of Physical Geography, Stockholm University, 106 91 Stockholm, Sweden
Alfred Wegener Institute Helmholtz Centre for Polar and Marine Research, Telegrafenberg A43, 14473 Potsdam, Germany

P. Kuhry
Department of Physical Geography, Stockholm University, 106 91 Stockholm, Sweden
Alfred Wegener Institute Helmholtz Centre for Polar and Marine Research, Telegrafenberg A43, 14473 Potsdam, Germany

G. Hugelius
Department of Physical Geography, Stockholm University, 106 91 Stockholm, Sweden
Alfred Wegener Institute Helmholtz Centre for Polar and Marine Research, Telegrafenberg A43, 14473 Potsdam, Germany

S. Kern
Center for Climate System Analysis and Prediction CliSAP, University of Hamburg, Hamburg, Germany

K. Khvorostovsky
Nansen Environmental and Remote Sensing Center NERSC, Bergen, Norway

H. Skourup
Danish Technical University-Space, Copenhagen, Denmark

E. Rinne
Finnish Meteorological Institute FMI, Helsinki, Finland

Z. S. Parsakhoo
Center for Climate System Analysis and Prediction CliSAP, University of Hamburg, Hamburg, Germany

V. Djepa
University of Cambridge, Cambridge, UK
now at: Institute for Meteorology and Geophysics, University of Cologne, Cologne, Germany

P. Wadhams
University of Cambridge, Cambridge, UK
now at: Institute for Meteorology and Geophysics, University of Cologne, Cologne, Germany

S. Sandven
Nansen Environmental and Remote Sensing Center NERSC, Bergen, Norway

J. Fiddes
Department of Geography, University of Zurich, Zurich, Switzerland

S. Endrizzi
Department of Geography, University of Zurich, Zurich, Switzerland

S. Gruber
Department of Geography and Environmental Studies, Carleton University, Ottawa, Canada

C. Mu
College of Earth and Environmental Sciences, Lanzhou University, Lanzhou Gansu 730000, China

T. Zhang
College of Earth and Environmental Sciences, Lanzhou University, Lanzhou Gansu 730000, China

Q. Wu
State Key Laboratory of Frozen Soil Engineering, Cold and Arid Regions Environmental and Engineering Research Institute, CAS, Lanzhou Gansu 730000, China

X. Peng
College of Earth and Environmental Sciences, Lanzhou University, Lanzhou Gansu 730000, China

B. Cao
College of Earth and Environmental Sciences, Lanzhou University, Lanzhou Gansu 730000, China

X. Zhang
College of Earth and Environmental Sciences, Lanzhou University, Lanzhou Gansu 730000, China

B. Cao
College of Earth and Environmental Sciences, Lanzhou University, Lanzhou Gansu 730000, China

G. Cheng
State Key Laboratory of Frozen Soil Engineering, Cold and Arid Regions Environmental and Engineering Research Institute, CAS, Lanzhou Gansu 730000, China

P. J. Griewank
Max Planck Institute for Meteorology, Bundesstr. 53, 20146 Hamburg, Germany

D. Notz
Max Planck Institute for Meteorology, Bundesstr. 53, 20146 Hamburg, Germany

V. Zunz
Université catholique de Louvain, Earth and Life Institute, Georges Lemaître Centre for Earth and Climate Research, Louvain-la-Neuve, Belgium

H. Goosse
Université catholique de Louvain, Earth and Life Institute, Georges Lemaître Centre for Earth and Climate Research, Louvain-la-Neuve, Belgium

J. Lehtiranta
Finnish Meteorological Institute, Marine Research Programme, Helsinki, PB 503, 00101 Finland

S. Siiriä
Finnish Meteorological Institute, Marine Research Programme, Helsinki, PB 503, 00101 Finland

J. Karvonen
Finnish Meteorological Institute, Marine Research Programme, Helsinki, PB 503, 00101 Finland

C. M. Surdu
Department of Geography & Environmental Management and Interdisciplinary Centre on Climate Change (IC3), University of Waterloo, Waterloo, Canada

C. R. Duguay
Department of Geography & Environmental Management and Interdisciplinary Centre on Climate Change (IC3), University of Waterloo, Waterloo, Canada

L. C. Brown
Climate Research Division, Environment Canada, Toronto, Canada

D. Fernández Prieto
EO Science, Applications and Future Technologies Department, European Space Agency (ESA), ESA-ESRIN, Frascati, Italy

Y. Sjöberg
Department of Physical Geography and the Bolin Centre for Climate Research, Stockholm University, Stockholm, Sweden

P. Marklund
Department of Earth Sciences, Uppsala University, Uppsala, Sweden

R. Pettersson
Department of Earth Sciences, Uppsala University, Uppsala, Sweden

S. W. Lyon
Department of Physical Geography and the Bolin Centre for Climate Research, Stockholm University, Stockholm, Sweden

Q. Shu
First Institute of Oceanography, State Oceanic Administration, Qingdao, 266061, China
Key Lab of Marine Science and Numerical Modeling, SOA, Qingdao, 266061, China

Z. Song
First Institute of Oceanography, State Oceanic Administration, Qingdao, 266061, China
Key Lab of Marine Science and Numerical Modeling, SOA, Qingdao, 266061, China

F. Qiao
First Institute of Oceanography, State Oceanic Administration, Qingdao, 266061, China
Key Lab of Marine Science and Numerical Modeling, SOA, Qingdao, 266061, China

www.ingramcontent.com/pod-product-compliance
Lightning Source LLC
Chambersburg PA
CBHW050458200326
41458CB00014B/5223